THE HIGHEST
STRANGENESS

Richard Freeman

Edited by Guin Palmer
Typeset by Jonathan Downes,
Cover and Layout by SPiderKaT for CFZ Communications
Using Microsoft Word 2000, Microsoft Publisher 2000, Adobe Photoshop CS.

First published in Great Britain by CFZ Press

CFZ Press
Myrtle Cottage
Woolsery
Bideford
North Devon
EX39 5QR

ISBN: 978-1-909488-72-4

For my old mate Jon 'Orrin' Hare who's aim is to live the strangest life he can!

FOREWORD

I'm still surprised these days when I hear that certain people on have a defined interest in one particular aspect of the Fortean realm be it hauntings, ufology or the hunt for elusive beasts around the world. I've always enjoyed a variety of topics from all kinds of angles in the world of the strange though I'm hard pressed to think of which particular story or witness account made me think that things were not just as cut and dried as we can sometimes perceive.

Maybe it was the wonderful world of Gef, the infamous talking Mongoose from the Isle of Man, a story I still find myself revisiting time and again, trying to see if I'd missed something the may have even evaded the legendary Harry Price and Nandor Fodor. Alas, this has so far eluded me, but I live in hope.

Could it have been legends of the infamous man who could not die, the Count St. Germain, as he sauntered his way around the 18th and 19th centuries, ingratiating himself into numerous royal courts, spheres of influence and even the musical world. Maybe the strange case at Hopkinsville or the Dover Demon?

Rains of strange objects, spontaneous human combustion, lights in the sky, odd cryptids that defy explanation even in that particular field of strange creatures. My appetite for the weirder aspects of the strange has grown throughout the years as I become aware of more and more cases that even my open mind struggles to comprehend.

Sometimes, even the most dedicated investigator or researcher can often ignore or pass over high strangeness or unusual events because they do not fit into the frame work of their particular field or interest. I remember speaking with Tony Healy for one particular episode of my show and as we discussed the number of Bigfoot investigators that invaded Bluff Creek in the weeks and months after October 1967, he mentioned how they'd ignored locals wanting to discuss black panther sightings, a local river monster and even ufos.

Simply put, sometimes even those searching for the strange ignore the high strangeness that lies beyond the framework of their interest. Not so my good friend Richard Freeman. In this book you will find accounts and stories that defy rational explanation. Encounters that stretch the minds of even the most diligent of researchers. Sometimes even strange things can strike people as being too strange.

Richard has trawled through archived reports, long lost newspaper articles, periodicals and more to uncover some of the weirdest, oddest and simply incredible accounts that simply leave us confused, amused, perplexed and even a little bit uneasy. I can only image how many sleepless nights collating this book must have given him.

Whilst I've come across some strange stories and encounters, there are so many in this collection I'd never heard of before. From strange monsters to unusual ghosts, peculiar UFO incidents and much more besides, I have to say the collections of odd talking animals and frightening, and aggressive puppets left me feeling amused and terrified at the same time.

I'll never view plush toys in the same way again and I think once you've finished this wonderful book, you'll probably join me in placing them all in a securely locked cupboard for a while.

A wonderous journey into the world of the high strangeness awaits you in the pages ahead, so brace yourself as you realise you think of as strange is nothing compared to some of the accounts you are about to read.

Paul Bestall
Host of Mysteries and Monsters, February 2023

INTRODUCTION

"From ancient worlds I come
To see what man has done
What's fact and what is fiction
To judge the contradiction."

Klaus Nomi, *Keys of Life*.

I think I must define 'high strangeness' in the context of Forteana, before we begin. Forteana (anomalous phenomena that take their name from the American author and researcher Charles Hoy Fort, 1874-1932) by their very nature, are strange. Sightings of sea serpents, rains of fish or frogs, hauntings, ball lightning and so forth, could under no circumstances be described as every day events. But high strangeness cases are the ones that stick out from even the weird crowd of what Fort called the 'parade of the damned'. In this book we will meet giant phantom crabs, hairy sex dwarfs, crawling ghost trees, space apes, latter day dragons, a talking mongoose, a girl haunted by a muppet, cyborg sea monsters, rabbit killing goblins and the ghost of Elvis Presley in Jarrow, to name but a few.

This is a departure from my general output. I'm a cryptozoologist, I write and research mysterious animals, creatures unknown to science, but as flesh and blood as you or I. It's a branch of zoology that is only just becoming accepted by mainstream science. I must admit that I did have some trepidation in the writing of this book, no matter how much fun it was. Tackling high strangeness cases would not do my zoological credibility much good. But just what is the job of science? Science is about gathering and examining data. We can't simply cherry pick and ignore the cases that just seem too strange.

We humans have a love of putting things in boxes and putting artificial barriers between things in order to sort them out. These boxes and barriers, as you will see, are false and are there just for the convenience of people. I'm just as guilty of this as the next man, as I have written this book in sections. This was simply a way to make it easier for me as a writer, and you as a reader. The informed reader may well say that to corral Fortean data like this, is tantamount to herding cats, and they would be right. The phenomena on a whole strikes me as being like the old story of the blind men and the elephant, wherein each man simply describes the one part of the elephant that he is feeling. Each chapter of this book is simply part of the greater whole and phenomena bleeding into each other. Some readers may even take issue with which category I place certain cases in. But remember, these categories are simply there for the management of data. I have broken the work up into chapters on monsters, ghosts, UFOs and miscellaneous entities, and finally I try to make some kind of sense of this weird cavalcade, if any conclusions *can* be gained.

The colourful character Tony 'Doc' Shiels once said *"Why are all you Fortean gobshites trying to work this all out? Don't you know this is just the way the universe works?"*

I'm inclined to agree with him. In this day and age scepticism has become mystique. A healthy scepticism has always been welcome in Forteana, but today there seems to be a glut of knee jerk sceptics, both online and in the written word. Many of these have done little or no research in the field, and simply gainsay anything not 100% proven by modern science. Science however, is continually dynamic. At one time meteorites, powered flight, and giant squids were deemed impossible. We are in danger of losing our sense of wonder. High strangeness is the medicine for this malady, as it is wonder in its purest form.

CHAPTER ONE: MONSTERS

There is a false notion that our planet is utterly mapped and explored down to the last inch. This is demonstrably untrue. Even in the age of satellite mapping and Google Earth, there are great swathes of the planet that are unexplored. Satellite photography cannot see through forest canopies or the surface of the sea. Many parts of the planet are hostile to mankind and few, if any, people live in them. Strange creatures seem to lurk in these places and sometimes much closer to home.

Animals like the giant squid, the okapi, the mountain gorilla and the Komodo dragon were dismissed as pure fantasy, but as it turns out, they were very real creatures. Others like the coelacanth were thought long extinct, but in fact were still thriving.

The field of cryptozoology concerns itself with the hunt for unknown animals. Some, such as the Mongolian deathworm, are utterly unknown to modern science. Others like the Tasmanian wolf are known species that are thought extinct, but seem to be still persisting in remote regions. Still others are individuals from known, living species that have grown far larger than the 'expert' deems possible, like the giant anaconda. It is a field criminally dismissed by the narrow minded, but gaining more respect and credibility as the years pass, and large new creatures like the Vu-Quang ox and the pygmy tapir are discovered.

The term cryptozoology itself was coined by Belgian zoologist Dr Bernard Heuvelmans (1916-2001) and Scottish zoologist and explorer Ivan T. Sanderson (1911-1973). These men are considered the fathers of the discipline. The target creatures are known as cryptids, though the public tend to call them monsters. This latter term may be quite fitting, as the original meaning of the Latin '*monstrum*' means a revelation or warning.

I myself have hunted these elusive beasts, tracking creatures like the yeti, the deathworm, the almasty, the giant anaconda, the Tasmanian wolf and the orang-pendek to name but a few. I have come away convinced that these creatures are real, flesh and blood animals.

There seems to be however, a secondary class of monsters that are much stranger. These appear not to be biological entities in the way we understand. It may seem an odd concept to the uninitiated. These creatures may kill and feed, leave droppings and tracks, but they also have some very weird traits that mark them out as something odd. I hope this will become apparent as the text unfolds.

WHAT LURKS BELOW?

What is it about water that fascinates and frightens man? Perhaps it is the fact that it is an alien and dangerous environment for us. A human can drown in an inch of water in the right conditions. Maybe it is some atavistic memory of the time when snatching a drink of water, meant taking your life in your hands for our remote ancestors. Predators lurk below the surface. Crocodiles still kill hundreds of people around the world each year in the modern age. The largest of them can approach thirty feet in length, over two tons in weight, and have a bite eight times stronger than a great white shark. Humans have always been on their menu. This fear may have been a survival mechanism passed down in our genes. Today many lakes are populated with monsters. It seems that any body of water where the bottom is invisible, has a story attached, be it a monster, ghost or piece of folklore.

The open oceans are unfathomably larger, covering 71% of the Earth's surface. In fact, our planet would be better named 'Water' than Earth. The sea has always been home to monsters and some of them have turned out to be real. The giant squid was dismissed as a legend for years, one of its early champions, the great French naturalist Pierre Denys de Montfort, was ostracised by his colleagues for his research into 'sea monsters and became a scientific pariah. He died of starvation, in a Parisian gutter in 1820. But he was vindicated when the giant squid was finally proven to exist in 1854. It can weigh a ton, reach sixty feet long, have eyes as large as footballs and a beak that can bite through inch thick steel cables. It revels in the scientific bi-nominal *Architeuthis dux*. Pierre Denys de Montfort gave the world a real monster and the scientific community owe him an apology.

This all begs the question 'what else is down there?'

There are certainly large, unknown animals still lurking in our seas. Various attempts have been made to classify them by different scientists over the years, but this is all speculation until a specimen falls into our hands.

In freshwater it is much harder to conceive of populations of huge, undiscovered predators. Some have postulated that there may be gigantic, sterile eels, who have no compulsion to head out to sea and spawn. They stay in fresh water getting older and older and bigger and bigger. Nobody knows just how big. Others have suggested massive sturgeon or catfish. Indeed, I myself investigated a case of a 'car sized', swan eating monster, in a tiny lake in Lancashire, England back in 2002. It turned out to be an eight foot long Wels catfish.

But other stories point to something of an altogether different nature. Something much stranger even than giant eels or century old sturgeons.

THE GREAT BEAST AND THE GREAT BEAST

In 1899, a flamboyant character arrived in Inverness. He purchased the brooding Boleskine House, on the shores of Loch Ness, for twice the amount the building was worth, becoming the 'Laird of Boleskine'. The man was Aleister Crowley, the self-styled Great Beast, and he had good reasons for paying over the odds for the remote foreboding house.

Crowley was born in Leamington, Warwickshire, in 1875. Rebelling against his ultra-strict Christian upbringing, he became the most renowned and colourful character in British occultism. His magickal and sexual experimentation shocked the prudish Victorian society. Crowley revelled in this, and wove an intricate web of myths about himself.

He chose Boleskine on account of its occult architecture. Previously he had scoured Britain for an abode to suit his needs and found none. Once in Boleskine, he intended to carry out the ritual of Abra-Melin - an ancient rite that took eighteen months to perform.

The ritual hearkened back to the 1400s. The ritual was translated by a Jewish scholar called Abraham the Jew, from a North African manuscript. Abraham was wandering the Middle-East looking for true magicians from whom to learn. He finally came upon a wizened mage called Abra-Melin, who passed the rite onto him. It dealt with the summoning of a guardian angel and the summoning and binding of the four great dukes of Hell, Lucifer, Satan, Leviathan and Belial. Leviathan, it should be noted is an ancient sea dragon, written of in Hebrew texts. The serpent is linked with the more ancient Lotan, in the legends of what is now Syria and the still more ancient Tiamat, the Babylonian dragon of chaos and the even earlier Sumerian water dragon Kur. These latter beast gods had power over the ancient waters and fertility.

The ritual demanded idiosyncratic architecture, and Crowley had previously tried to replicate this in his London flat. Though not having the desired effect, strange things happened there. In 'The Great Beast's' own words:

> *"During this time magical phenomena were of constant occurrence. I had two temples in my flat; one white, the walls being lined with six huge mirrors, each six feet by eight; the other black, a mere cupboard in which stood an altar supported by the figure of a Negro standing on his hands. The presiding genius of this place was a human skeleton, which I fed from time to time with blood, small birds and the like. The idea was to give it life, but I never got further than causing the bones to become covered in a*

viscous slime."

Exactly whose skeleton it was, and how Crowley came by it, is unclear, but it featured in one of his most notorious and amusing spells. Althoea Gyles - a local artist and lover of one of Crowley's poet rivals, W. B. Yeats - was sent to visit Crowley. Yeats had her scratch his foe's hand with a brooch and carried a drop of 'The Beast's' blood back to her spouse. Yeats allegedly used this as a spell component to invade Crowley's dreams. But more of the precious fluid was needed and Gyles was sent back. This time, 'The Beast' was ready and had already sprinkled Gyles apartment with a "magick potion." Upon arriving at Crowley's flat, she became overcome with lust for the skeleton and made love to the blood and slime festooned cadaver. After this "boneing" she was promptly rejected by Yeats.

Yet more happened at Crowley's flat, apparently on account of the occult décor:

"The demons connected with Abra-Melin do not wait to be evoked; they come unsought. One night Jones and I were out to dinner. I noticed while leaving the white temple that the latch of the Yale lock had not caught. Accordingly, I pulled the door to and tested it. As we went out, we noticed semi-solid shadows on the stairs; the whole atmosphere was vibrating with the forces we had been using. (We were trying to condense them into sensible images.) When we came back nothing had been disturbed in the flat; but the temple door was wide open, the furniture disarranged and some of the symbols flung about the room. We restored order and then observed that the semi-materialized beings were marching around the room in almost unending procession.

When I finally left the flat for Scotland, it was found that there was no way to take the mirrors out except by way of the black temple. This had, of course been completely dismantled before the workmen arrived. But the atmosphere remained and two of them were put out of action for several hours. It was almost a weekly experience, by the way, to hear of the casual callers fainting or being sized with dizziness, cramp of apoplexy on the staircase. It was a long time before those rooms were re-let. People fled instinctively at the presence of something uncanny. Similarly, later on, when I gave up my rooms on Victoria Street, a pushing charlatan thought to better himself by taking them. With this object he went to see them. A few seconds later he was leaping headlong down the five flights of stairs, screaming in terror. He had sufficient genuine sensitiveness to feel the forces, without possessing the knowledge, courage and will required to turn them to account, or even endure their impact."

Crowley's attempts to perform the ritual at Boleskine failed. He was called away by Samuel Liddell MacGregor Mathers, founder of the Hermetic Order of the Golden Dawn, half way through the ritual. The semi-formed shadows that he evoked in

London seemed to have been called again, however John Symonds - his biographer - recounts that the house's lodge and terrace became peopled by shadowy shapes. The place seemed to have a strange and violent effect on people. A workman employed to renovate the villa went berserk and attacked Crowley, who had to knock the man out and lock him in a coal-shed. His lodge-keeper - who had been a teetotaller - went on a three-day drinking-binge and tried to murder his own wife and children. Crowley finally left in 1918, but some believe he left something behind.

Subsequent owners of Boleskine have also reported disturbances. Musician and former member of the group Led Zeppelin, Jimmy Page, bought the house in the 1970s. His friend and custodian of Boleskine, Malcolm Dent, has experienced the house's dark side on several occasions:

> *"Most of the oddities occurred during upheavals in the house. I am not talking about wallpapering, but structural alterations. Any time there was anything major, it was almost as though the house didn't like it. If we didn't get on with the job and get it finished, something would let us know about it. We would be wakened up during the night with heavy doors banging all over the place and carpets and rugs being rolled up. It was though it was a reminder to get on quickly and get the job over. "*

Another time, Malcolm and some friends, saw a statue of The Devil rise up from a mantelpiece, float to the ceiling, then smash to the floor. The most horrifying event happened early one morning when the disturbances reached a crescendo:

> *"I was awakened in the wee small hours and just knew something was wrong. I was petrified. Something outside the bedroom door was snorting, snuffling and banging. It sounded like a huge beast. I had this clear picture in my mind of what it looked like, but there was no way I was going to open the door. I had a knife on my bedside table and I opened the blade and just sat there. The blade was so small it wouldn't have done any good, but I was so frightened that I just had to have something to hang on to. The noise went on for some time but even when it stopped, I still could not move. I sat in bed for hours and even when daylight came, it took a lot of courage to open that door. Whatever was there, I have no doubt it was pure evil."*

Could it be that the daemon-summoning ritual had worked in a way that Crowley had not foreseen?

Modern day wizard, surrealist, and performer, Anthony 'Doc' Shiels, thinks this may well be the case. Whilst engaged in the magickal "Monstermind" experiment described later, Doc made the acquaintance of a man named Patrick Kelly. Kelly claimed to have photographed a lake monster in Lough Leane, in 1981. This however was not the most fantastic of his claims.

He said that he was a direct descendent of Edward Kelly, the notorious scryer of Dr John Dee. Dee was the court magician to Queen Elizabeth 1st, and claimed to 'speak with the dead' via a young medium, whom he had trained.

The modern day Kelly also claimed his father, Laurence, had met Aleister Crowley in Paris in 1933, shortly after he had left the Abbey of Thelema. Crowley told Laurence that he was very interested in the Loch Ness Monster - whose first major flap of the 20th century was then occurring.

Finally, Patrick Kelly and his father both claimed to have seen the Loch Ness Monster on May 1st, 1969, close to Boleskine. Fantastic assertions indeed - but at least for this last one there may be *some* evidence. In June of the same year, three American students were exploring the 17th century cemetery below Boleskine House. They came upon a curious object. It was an old tapestry wrapped around a conch-shell. The tapestry was decorated with serpent-like symbols, embroidered in gold thread. It measured four feet by five feet, and seemed to be old and threadbare. There were reddish-stains at each corner, as if objects had been placed there.

All in all, it looked like an altar-cloth. The shell was about five inches long, white, and inscribed with two parallel grooves and a lotus blossom. When blew it produced a harsh braying sound. The objects were taken to the Victoria and Albert Museum to be studied by experts. The tapestry was latter identified as being Turkish in origin. The snake-like symbols were Turkish script for serpent. We should also note that today, Lake Van in the east of Turkey, is said to be inhabited by a dragon.

Lotus flowers - like the one on the conch-shell - along with roast swallows, were said to be the favourite food of dragons in China. They were often used as offerings to dragons in oriental lakes, to appease them and ensure rainfall. Could Patrick Kelly and his father have been performing some kind of ritual at Loch Ness? Perhaps they were disturbed and had to leave behind their artefacts, as they hurriedly retreated. If this were the case, then it seems that the Kellys were successful in their endeavour.

Tony 'Doc' Shiels was a figure that loomed large in the Forteana of the 70s and 80s. He appeared in a number of tabloid stories and wrote on a regular basis for *Fortean Times*. Doc instigated the largest monster raising experiment ever in 1977. Doc had been in contact with seven professional psychics from around the world for some time. The group called themselves Psychic Seven International or P-S-I. The seven decided to try and contact or call up aquatic dragons by using their powers. 'Doc' took part in a number of spectacular rituals involving sky-clad [naked] witches, beside various bodies of water in Ireland and Scotland. His colleagues concentrated on other lakes around the globe. The experiment commenced on the last day of January, or more importantly the Pagan feast of Imbolc.

Doc saw the results for himself on the 21st of May. He had travelled to Scotland with his wife Chris. And on the 20th he had performed a ritual invocation.

The following day, he was in the car-park of the Inchnacardoch Lodge with four friends, when they all saw three humps gliding through the water, about 900 feet away, in the direction of Fort Augustus. The humps slipped with hardly a ripple beneath the surface. None of the group had a camera on them at the time. Doc, though elated at the beast's presence, was understandably frustrated. Little did he know that he would later that day take what are widely regarded to be the finest pictures of the Loch Ness Monster ever obtained.

Doc and Chris had hitch-hiked to Drumnadrochit, and from there walked to Urquhart Castle, overlooking the Loch. At four p.m., he was ensconced in a ruined tower, looking over the water from a window, this time with his camera.

> *"Quite suddenly, a small dark head on the end of a long sinuous neck broke the surface of the water, about a hundred yards away. It was, undoubtedly, the Loch Ness Monster, proudly erectile, ready to be snapped. I instantly raised my camera and shot two pictures during the few seconds the creature was visible. Its neck was four or five feet long, greenish brown, with a yellowish underside.*
>
> *Its open mouthed head was tiny in relation to the muscular neck. The animal turned away from me, straightening its neck before sinking vertically.*
>
> *I stood there mesmerized by the brief dreamlike vision. My heart beating rapidly, hands shaking as I lowered the camera, whispering expletives, ecstatic."*

Doc's friend, David Clarke, arranged for the high speed Ektachrome film to be developed by Newquay Colour Services, who handled most of the colour transparency work for Cornish Life magazine. The two snaps were startling in their clarity, showing the muscles in the beast's neck and its open mouth. The gleam of an eye even seems to be visible.

Some however have dismissed the picture as a fake, possibly a close up photo of a painting on silk.

The results of Monstermind worldwide though, were spectacular - 1977 was a year filled with monsters. A gigantic aquatic dragon was sighted off San Francisco Bay. Champ, the monster of Lake Champlain appeared. A twelve metre (forty foot) monster was seen in Lake Kol-Kol in Kazakhstan (in what was then Soviet Central Asia). Miss M Lindsay took two pictures of Morag, the monster of Loch Morar, on 31st of January. On the same day, a monster was spotted in Loch Sheil by John Smith. The Loch Ness Monster was seen on at least three occasions by Mr and Mrs Alex McLeod, Pat Scott-Innes, and a Mr Flemming and his daughter Helen. Three colour photographs of

Morgawr - a Cornish sea serpent were taken by Gerald Bennet from Parson's Beach - and the monster was spotted again by Ray Hopley, off Trefusis Point.

A spectacular set of sightings, but they came at a cost. Doc Shiels had a theory of 'psychic backlash', a kind of intense bad luck, akin to a curse that some who investigate paranormal beasts find themselves under. He believed that the people involved in the Monstermind experiment, had become victims of this power. Dr David Hoy - one of the American participants in Monstermind - suffered a heart-attack. Another member - Major Leslie May M.B.E, of Edinburgh - also fell ill, and many other members of the team from the former USSR, Mexico, and India have seemingly vanished. Doc has heard nothing of them since 1977. Doc himself was attacked by a mob in Plymouth, accidentally set his beard on fire, had a son involved in a motorbike crash, had his daughter thrown by a usually docile horse, had another daughter stricken by abdominal pains, and lost two cats to some unidentifiable malady. Ted Holliday - another long-term monster hunter reported a similar "curse" that seemed to have banjaxed him some years before.

THE EXORCISM OF LOCH NESS

The idea that the Loch Ness Monster was a malevolent supernatural entity, reached its peak in the early 1970s. In 1973, one man believed things had gone on too long, and decided to exorcise Loch Ness. He was the Reverend Dr Donald Omand. Dr Omand was perhaps the 20th century's most renowned exorcist. During his long career he'd dealt not just with ghosts and demonic possession, but with latter-day vampires, phantom black dogs and areas of the sea where people were drawn by a strange siren-like urge to drown themselves. These cases, fantastic as they are, pale into children's games, when compared to the Doctor's strangest case.

Dr Omand's first encounter with a lake monster happened in 1967, whilst on a caravanning holiday on the shores of Long Loch, in Ross-shire. One morning, he set out to walk to the village of Ardelve. His route took him past Loch Duich. As he looked out over the loch, the calm water suddenly became violently disturbed, foaming and heaving. For one absurd moment, the Reverend thought a submarine was breaching in the loch. But the object revealed itself to be some immense, aquatic animal, with two huge humps, that reared out of the water. Then - just as swiftly as it had surfaced - the beast sank, leaving only concentric ripples as a clue to its manifestation.

It was not until the following year, that Dr Omand began to suspect that these monsters were not conventional flesh and blood creatures. In June 1968, he had a far more alarming encounter with a sea-serpent. The Reverend was holidaying with his friend Captain Jan Andersen, in Norway. Andersen had offered to show Dr Omand the "eeriest waterway in Norway" - the 'Fjord of Trolls' (note the link with hairy giants). The entrance to the fjord is hidden, and was used by British craft during The Second World War. The men travelled along the long, narrow waterway, screened on either side by gargantuan cliffs. But it was only on their return journey, that the exorcist began to feel something was badly wrong. A feeling of growing menace began to creep over the

area. As they approached the entrance to the fjord, the water began to seethe.
"What on Earth is it?" asked the Reverend.

"It can be only one thing", replied the Captain. *"It would be useless to try and avoid it."*

Two massive humps appeared, much like the ones the Doctor had seen in Scotland, but much closer. The massive animal bore down on their boat with terrifying speed. And the frightened cleric braced himself for an impact that would turn the vessel to matchwood.

"It will not hurt us, they never do", shouted The Captain. Sure enough the monster veered to the starboard at the last moment and submerged.

"Shall we follow it?" inquired the Reverend, eager to see more of the fantastic animal. The Captain's reply was cryptic:

"Sufficient unto the day is the evil thereof. The further we kept away from that thing the better. When I said evil it's what I really meant. This is the third time I have seen it. On the other occasions it was further north, closer to the North Cape. They are what our ancestors called the Sea Serpents. Today people regard them as existing only in legend. But when you have seen them you believe in them."

Dr Omand questioned his reasoning.

"But why are they evil? That one might have easily capsized our boat, but it did us no harm."

The Captain's answer was even stranger than before:

"They don't do physical harm. They want to convince any who see them that they are harmless. The evil they do is to men's characters. The serpent in the Garden of Eden was no ordinary snake, and what you have just witnessed is no ordinary creature."

Dr Omand enquired as to their true nature.

"The explanation for these extraordinary appearances, in my submission, lies not in the field of science, but in the realm of the supernatural. What has been seen, and is still visible to some on occasion, is not a concrete present day monster, but a projection into our day and age of something which had its habitat in Loch Ness and its surroundings, millions of years ago."

Commenting on modern witnesses he says:

"What they saw was not something that was taking place at that precise moment. The

gigantic creature they were so privileged to see was no longer in the land of the living. It was something seen out of time. The Loch Ness Monster is not physical but psychical, a spectre of something which existed in the waters and on the shores of the vast lake in the dim recesses of the past."

He had his ideas backed up in 1972, whilst attending a meeting of the Organisation of Enquiry into Psychical Disorder, in Sweden. An eminent Scandinavian neurologist delivered a report concerning the monster of Lake Storjsson. The report was about the malevolent effect that the monster seemed to have on those who hunted for it, or who had seen it regularly. It resulted in shocking moral degeneration. Similar patterns were found, or so the neurologist claimed, in Irish and Scottish cases.

At almost the same time, Dr Omand received a letter from F. W. (Ted) Holliday - the renowned monster hunter, who had recently come to similar conclusions to the Doctor. The letter congratulated him on his insight. Ted was a journalist who have served with the Royal Air Force during the war. In the 50s and early 60s he had written books on angling, but had then become interested in lake monsters. In his first book *The Great Orm of Loch Ness,* published in 1968, he postulated that the monster was some form of colossal invertebrate, related to slugs and snails. This is highly unlikely for physiological reasons. But by 1973 he had changed his mind.

Ted Holliday believed that lake monsters were dragons in the literal sense. Evil paranormal serpents, that were also linked to UFO sightings, that he also thought were supernatural in nature. He detailed these findings in two books *The Dragon and the Disc* and *The Goblin Universe.* Some encounters he had heard of seemed to generate unnatural levels of terror in the witnesses.

Mr George Spicer, a tailor from London, was on a motoring holiday with his wife. On August 4th 1933, they were driving along the newly constructed road along the loch side between Dores and Foyers when a burly, elephant grey creature with a writhing, elongate neck emerged from the bracken and crossed the road in front of their car, before plunging into the loch. Mr Spicer likened it to a giant snail without a shell.

"The only way I could describe it was prehistoric in its form, it looked malformed, ugly and quite appalling really. I don't care what people say about us imagining it or being tired and it not being what we thought. We know what we saw, we did see that thing and it wasn't anything small or that one would expect to see crossing the road in front of your motor car. It covered the width of the road, it was a frightening experience for us both, one we shall never forget or be allowed to forget."

Mr Richard Jenkyns and his wife saw the horror from their loch side house on the 30th September, 1974. The couple watched the monster through binoculars for half an hour. It seemed to be around eighteen metres (sixty feet) in length. The monster had a lasting effect on them. Later Richard commented:

"I felt the beast was obscene. This feeling of obscenity still persists and the whole thing put me in mind of a gigantic stomach with a long writhing gut attached."

Mrs Greta Finlay - an Inverness housewife - had similar feelings toward the thing she encountered at close range on August 20th, 1952. She had gone fishing with her young son, and was on the north-east shore of the loch, near Aldourie pier, off Tor Point.

"I was sitting outside the caravan when I heard a continual splashing in the water. After several moments passed and realizing this was not the usual wash from a boat I walked round. To my surprise I saw what I believe to be the Loch Ness Monster. My son and I stood looking at this creature in amazement. Although I was terrified, we stood and watched it until it submerged, which it did very quickly causing waves to break on the shore. We had an excellent view as it was so close to the shore. Its skin was dark in colour and looked very tough. The neck was long and held erect. The head was about the same width as the neck. There were two projections from it, each with a blob at the end. This was not a pleasant experience. I certainly never want to see the monster again."

Mrs Finlay was interviewed by the late Tim Dinsdale - perhaps the greatest 'Nessie'-hunter of them all. She confessed to him that she had been paralysed with fear, and that her son had been so utterly horrified, that he had given up fishing all together. She stated that she would never want to see the monster again, even if it were behind six inch steel bars.

This "terror effect" is not confined to the Loch Ness Monster. The most extreme case involved a relatively diminutive lake in Ireland.

In 1954, Miss Georgina Carberry - a librarian from Clifden - and three friends, were on a trout fishing trip to Lough Fadda. This small Connemara lough, lies in Derrygimlagh - a bog of some thirty-five square miles, dotted with small lakes connected via streams. The friends enjoyed a good day's fishing, but in the late afternoon an event occurred that changed their lives forever. She was later interviewed by Ted Holiday who reproduced the exchange in his books.

Holiday: *Could you tell us what you saw on Lough Fadda when you had the sighting of this object?*

Carberry: *Well, it was a very long object. We sighted it rising...coming out from an island towards us...swimming along slowly. So we kept watching it and eventually, through time, it got very near to us. Anyway, next thing, we began to get a bit worried. I was sitting nearest to it and it came in I suppose, oh, to within twenty yards of us. I was the first to move and jumped back and the other three who were sitting behind me did likewise.*

And as soon as we moved it just came right around...swung right round a rock which was near the shore and dived and we could see these awful big rings in the water as it was sinking. Before two minutes it had gone practically up to the island again when it surfaced. We could distinctly see two big humps showing behind its head out of the water. And the tail we noticed, when it swung around the rock, 't'was kind of a fork - a V-shaped tail. And the mouth which was open when it came in quite close to us at the shore and the eyes and that I can't really remember. But I distinctly remember that the whole body had movement in it.

Holiday: *What do you mean by "movement"?*

Carberry: *It seemed like...wormy. You know - creepy. The body seemed to have movement all over it all the time.*

Holiday: *What did the head look like? Was it like any thing you've seen before?*

Carberry: *No, it wasn't. The only thing the mouth resembled, open, was like shark-shaped.*

Holiday: *Did you see teeth? What did the mouth look like?*

Carberry: *Just...oh, a huge great mouth. I can remember white inside but as regards teeth and eyes, I can't remember what sort of eyes it had because we were so frightened to see such an unusual object.*

Holiday: *How high was it standing out of the water...the head and humps you mentioned?*

Carberry: *Oh, they came right up. As it swam towards us we could see the two humps behind the head very clearly.*

Holliday: *And how high out of the water did the head stand?*

Carberry: *Well...higher than the humps. It was a fair distance out of the water.*

Holiday: *Was there a neck to speak of?*

Carberry: *There was, yes. It seemed to come up just in one long...(curve?)*

Holiday: *And you jumped back?*

Carberry: *We certainly did! I was the first to move being the nearest to it.*
Holiday: *Could you see how wide it was or whether it was a narrow object of...?*

Carberry: *Well, now, it was fairly wide in the girth – a good span.*

Holiday: *And the head was up – clearly visible?*

Carberry: *Oh very clear... and the mouth.*

Holiday: *Was it that dark that it was silhouettish?*

Carberry: *Oh, no, it was a bright evening...a beautiful evening. I can always remember it was so fine – one of the fine summers we had some time ago.*

Holiday: *Were you effected by this at all?*

Carberry: *I certainly was! I don't think I went back to that lake for six or seven years after. And when we went back we would never go alone. Never alone.*

To scare someone away from a place for seven years takes a lot, but the fear exerted by this being didn't end at that. Whilst driving home, Carberry found herself watching for the monster, in fear that it had slithered out of the bog and was pursuing them. She suffered from nightmares about the thing for weeks afterwards. One of the other witnesses had a mental breakdown, and was hospitalised. The reader should also take note of the immense and seemingly unnatural level of fear these water serpents can generate in witnesses. The idea of pursuit by a monster, even if unseen, is a factor we see again and again in reports of different types of creature.

Back in Scotland, Loch Morar is another lake said to harbour a fear generating monster, this one has been named Morag. Whist hunting for the creature with my friend Dave Curtis back in 2004, we stopped at a lovely hunting lodge built in 1840, called Garramore House. It was used to train spies in WW2. Its current proprietor, Julia Moore, was a fascinating lady who knew my favourite author Mervyn Peake. She told us of a sighting that has never been recorded before and was unknown outside of the village. Apparently two youths from Yorkshire were on a fishing holiday about four years before. They were out on a boat, one keeping watch, the other operating the tiller. The lad on watch shouted out that there was a tree approaching the boat at an alarming rate. Both saw what looked like a tree trunk racing towards the boat. They feared a collision but at the last moment the trunk arched up and dove down into the depths. The creature scared the boys so much that they made the bank in record time, packed their tent and returned home on the same day.

Ronald Watson runs an excellent blog called Loch Ness Mystery, in which he showcases little known accounts of sightings of lake monsters. One such was sent to him by a reader called' Tricia'. She was on a boating and fishing holiday with her family. Their encounter occurred in the last week of August 1969, between 5.30 and 6.30 in the evening. Tricia was in a motorboat with her father and three younger brothers. The boat broke down in the middle of the loch.

"Ten minutes or so into the journey, at which point we were in the middle of the Loch, and to my recollection both shores were equidistant, the outboard motor packed in. We as kids were not particularly concerned with this because we were used to this type of event. My Dad regularly took us on fishing expeditions to many other lochs, sea and rivers where outboard motors would pack in and be fixed by him.

As we sat patiently while he worked on this motor I looked to my right and spotted two or three protrusions from the water, about two feet high. I thought these were rocks at first, but I remember having a feeling of unexplained fear. I turned to my sister who was sitting next to me and said "Look over there, what is that?" at which point my Dad said "Shut up, Patricia!" with gritted teeth.

This was upsetting to me as my Dad was very rarely angry at us kids and we were taught all our lives never to say "Shut up!" to anybody. This was a golden rule in our house. My Dad had, of course, spotted the same protrusions and kept this to himself, I and my sister stared at this sight in the water in confused wonderment.

My brothers at this point were oblivious down to their age and preoccupation with trying to untangle the fishing tackle which my dad had tasked them with. I said again "Dad, what is that in the water?" to which he replied again "Shut up!" in an angry voice. This prompted my brothers to look over and join in the debate.

The feeling of fear was now with us all. Sometime later, my Dad explained to us his mind was not only on trying to fix the motor but also trying to figure out how he could get us to shore safely away from this unexplained "thing" in the water. I am smiling as I write this but also empathising with his thoughts. He must have been frantic, he did however on the outside remain calm for our sake! The upshot is, he finally got the motor going. As we started to travel, the mysterious protrusions (which had stayed with us for the duration of time it had taken Dad to fix the motor in the same position), disappeared below the water leaving a slight swell, which we felt in the boat.

We trundled along again for which felt like hours till we reached the jetty where we had started out from. On landing, there was a flurry of activity and our Mum was crying and very agitated. She, of course, was witness to two fishermen relaying to a crowd of folks, including reporters, their

experience on the Loch that same day, some hours before.
From my recollection, they also had an up close and personal experience
with (I presume) the same "thing" we had witnessed. They however, said
this "thing" had hit their boat and they hit back with an oar and a gun
shot."

The fishermen in question were Duncan McDonnel and William Simpson, whose encounter was probably the best known of all accounts from Loch Morar. A thirty foot, brownish creature with rough skin, a body like a huge eel and a tortoise like face, collided with their boat. McDonnel hit it with an oar, breaking it in the process, and Simpson shot at it with a rifle. The bullets seemed to have little effect on the creature and it slowly sunk out of sight.

We will see this extreme fear again with other kinds of monster and with other Fortean phenomena.   The legendary Australian researchers Tony Healy and Paul Cropper, borrowing the term from the H. P Lovecraft story '*The Very Old Folk*', dubbed this intense, blind terror 'The Nameless Dread' and it may give us a clue to the nature of high strangeness cases.

With encouragements from Ted and others, the Doctor made up his mind to exorcise Loch Ness. The Reverend believed that the monster's manifestations were not in themselves evil, but rather that evil had attached itself to the phenomenon and to the area. He believed that he could purge this evil, and leave the monsters intact. This theory was not shared by Ted Holliday, who believed the creatures themselves to be overwhelmingly evil.

"I don't know for certain, but I think their character was described in the
first book of Genesis.. I am not sure even that it exists physically or not-
there are things which do not exist and yet may be visible to man."

The more Dr Omand thought on Andersen's words, the more convinced he became that the monsters were paranormal in nature. The Doctor decided to seek the advice of a fellow exorcist, Reverend Dom Robert Petipierre, a monk of the Anglican Order of St Benedict. Dom Robert took a large map of The Loch, and drew a cross upon it. The top of the crucifix was at the Inverness end of the loch, and the base near Fort Augustus. The intersection terminated on the left at Drumnadrochit and on the right at a point between Inverfahigaig and Dores. The men planned preliminary exorcisms at each of these points. The final rite was to be carried out in the centre of the cross - in the middle of the loch, in a boat. All the points of entry and exit along waterways, were 'bound against evil' to stop the contamination spreading during the ritual.

Between them, the exorcists drew up a rite from German, Spanish, Roman, Greek, and English sources. On June 2nd, 1973, the ritual took place. Accompanied by Holliday, Dr Omand exorcised all of the points, and eventually on a small boat he rowed to the centre of the dark peaty loch. There, floating on 800 feet of cold, black, water, he gave the final exorcism:

> *"I adjure thee, thou ancient serpent, by the judge of the quick and the dead, by Him who made thee and the world, that thou cloak thyself no more in manifestation of prehistoric demons, which henceforth shall bring no sorrow to the children of men."*

After the ceremony, Dr Omand felt drained and fell into a deep sleep. He believed his exorcism to have been a success and subsequently went on to exorcise Lake Storjsson in Sweden.

Quite a story, all in all. What are we to make of it? Are these the ramblings of a fundamentalist Christian madman, with his world view set in The Dark Ages, or was the Doctor really grappling with some supernatural force in the form of a monster? Exorcism is not confined to Christianity - many other faiths have practising exorcists. Muslim imams cast out "djinn", pagan wizards and witches drive out malignant spirits with spells.

Despite his optimism Dr Omand seems to have failed, for monsters are still seen in Loch Ness and Lake Storjsson.

Boleskine House recently caught fire and three quarters of it were burnt to the ground. The whole thing is quite a story and we will be meeting some of the players again further on in this book.

Thus far the lake dwelling abominations we have examined are the classic serpentine, hump-backed horrors. However some lake monsters have taken on other, stranger forms.

THE LOCH NESS CAMELS
In June of 1990, Lieutenant Mc P Fordyce, related to The *Scotts Magazine* a strange sighting he and his fiancée had some 58 years earlier in 1932, the year before the spate of modern Loch Ness Monster sightings occurred.

> *"In April 1932 while living in Kent, my fiancée and 1 travelled to Aberdeen to attend a family wedding. At the conclusion of the function, rather than return south straight way, I decided to show her a little of my native land. We cut across to Inverness where preparations appeared to be in full swing for a Highland Gathering.*
>
> *After dinner in the late evening we took a stroll through the town and saw men in shop doorways and at street corners practising on their bagpipes. The air was full of sublime music (my version), rent with strange, wild sounds (her description although she never expressed it at the time!).*

The following morning we set off on our journey back to England. The weather was fine, a beautiful spring day, and we had a lovely run by the side of Loch Ness as far as Foyers where we spent a short while admiring the famous waterfall. Shortly after leaving Foyers, the road to Fort William turns away from the loch side and runs through well-wooded country with the ground falling slightly towards the loch.

Travelling at about 25 mph in this wooded section, we were startled to see an enormous animal coming out of the woods on our left and making its way over the road about 150 yards ahead of us towards the loch. It had the gait of an elephant, but looked like a cross between a very large horse and a camel, with a hump on its back and a small head on a long neck. I stopped the car and followed the creature on foot for a short distance.

From the rear it looked grey and shaggy. Its long, thin neck gave it the appearance of an elephant with its trunk raised. Unfortunately, I had left my camera in the car, but in any case I quickly thought discretion the better part of valour and returned to the vehicle. This strange animal occupied our thoughts and conversation for many, many miles and we came to the conclusion that it was an escaped freak from a menagerie or zoo. We felt that a beast of such tremendous proportions would soon be tracked down and captured.

Apart from scanning the national papers for some time in search of mention of the creature, we let the matter rest. I told other members of my family what we had seen and they urged me to publish the story, but I have not done so until now.

At the time of the sighting we were quite unaware of there being anything strange in Loch Ness, but in the autumn of that year, stories started appearing in the Press of an unusual animal being seen in and around the loch. It was the spring of 1933 that the term "Loch Ness Monster" came into general use."

There are and were no zoos in the area, the closest being Edinburgh Zoo and the now defunct Glasgow Zoo. The former is 156 miles south of Loch Ness, and the latter 168 miles south. Neither reported any missing camels. Additionally, there were no circuses in the area at the time and highland folk in the 1930s were not in the habit of keeping pet camels!

Fordyce was not the first to report the Loch Ness camel. In September 1919, a boy called William McGruer and his siblings saw a beast in Inchnacardoch Bay that they described as brownish-yellow, with long legs and a long neck that looked like a camel. When it saw the children it walked into the loch and vanished.

THE CASE OF THE IRISH WATER RHINO

If one can have a camel at Loch Ness, then why not a rhino in an Irish Lough? In March of 1962, a school master bearing the wonderful name of Alphonsus Mullaney, had taken his son, who shared his rather magnificent name, on a fishing expedition to Lough Dubh, in County Galway, Ireland. Mullaney senior felt a tugging at his fishing line. He later told a reporter from the *Sunday Review* what happened next.

> *"Suddenly there was a tugging on the line. I thought it might be caught on a root, so I took it gently. It did not give. I hauled it slowly ashore, and the line snapped. I was examining the line when the lad screamed. Then I saw the animal. It was not a seal or anything I had ever seen. It had for instance short thick legs, and a hippo face. It was as big as a cow or an ass, square faced, with small ears and a white pointed horn on its snout. It was dark grey in colour, and covered with bristles or short hair, like a pig."*

The school teacher and his son fled from the strange beast. Later a group of armed locals searched the lough, but could not find the creature. Apparently the beast had been seen before, on one occasion with two young. But to my knowledge this was the last time the Irish water rhino was reported.

We don't usually associate rhinos with water, but the Indian rhinoceros *(Rhinoceros unicornis)* spends much of its time in water, unlike it's African relations. One group of rhinos, the *Amynodontids,* that flourished in the Oligocene epoch, 38- 25 million years ago, and finally vanishing 10 million years ago, were adapted to an aquatic lifestyle much like hippos. However, even if they did still exist, what the hell would one be doing in an Irish lough? Some have tried to identify the Lough Duhb monster with a feral pig or even a walrus, neither of which match the description of the beast. This water based weirdo seems to have appeared for a short while then dissipated just as mysteriously.

DEATH DEALING DRAGONS IN AFRICA

Lest the reader think that such strange stories are confined to Europe, I will detail some of my findings from an investigation into aquatic monsters in West Africa. Here a beast known as Ninki-Nanka is blamed for human deaths. Back in 2006, I hunted for this beast in the swamps of Gambia. The level of fear the creature engendered was almost unbelievable. Our native guide refused point blank to enter the swamps where it was thought to lurk, leaving us to trek on alone to an abandoned village. The inhabitants had left the village decades before, after one of them had reported a Ninki-Nanka. It was an eerie feeling walking through the abandoned village, knowing that the people had fled in fear, from a dragon, right here in the modern age not in some distant medieval time.

One man told us his grandfather had seen it in some lakes near Banjul in the 1940s. It was a massive serpent with shining scales and a crest or fin on the head. The man, Papa Jinda, died soon after. Afterwards huge mirrors were erected beside the lake in the belief that the beast would be scared away by it's own reflection. There is a strong belief that to look on the dragon means death. A guide at Abuko National Park said that the Ninki-Nanka was like a python, but large enough to swallow a whole cow. It had a crested head, small bat like wings and four small legs.

We were interviewing a man called Bula, whose uncle had seen the beast at a swamp called Kiang West. Bula's fear of the monster was so great that he dared not even look in the direction of the swamp. We had to interview him from behind a bush as he hid, shaking in fear! Our guide Bakary walked a little way into the swamp, but then fear consumed him too, and he turned back as we went on alone.

Another witness, Momounadou claimed to have seen a Ninki-Nanka in Kiang West swamps a few years before. He described a horse like head, with a crest and a vast serpentine body, covered with scales that shone like mirrors. Afterwards he fell ill and his body was covered in lesions. He went to see a local Iman, or holy man, who said instantly *"you have seen a dragon haven't you?"* The Iman gave him a potion that cured him. These cases may simply be psychosomatic, but the creature has supposedly killed people with physical attacks as well. One story tells of a recently erected bridge that was built near a hole where a Ninki-Nanka was supposed to live. The angry monster was said to have emerged and smashed the bridge to matchwood, tossing two people fatally into the waters.

THE DRAGON HAUNTED SEA
Large unknown animals are much more likely to exist in the oceans and seas of our world, than in lakes. Space is the main factor here. Even the most sceptical of scientists would not be surprised by the discovery of large, new life forms in the sea. Indeed statistical biologist, Dr Charles Paxton, at the University of St Andrews, Scotland, has used rates of discovery of large marine creatures to predict that a number of large bodied creatures remain unknown to science in the world's oceans.

However there are still a few very odd cases, that hint that a few sea serpent sightings may fall into the high strangeness category.

Cornwall's Falmouth Bay has had sea serpent reports that reach back to 1876. The beast was dubbed 'Morgawr', meaning 'sea-giant' in old Cornish. The 1970s saw a global monster 'flap' with all kinds of creatures being reported from all over the world. Around Cornwall there were encounters with the Owlman, a winged humanoid that we will examine more closely later in the chapter. There were also sightings of glowing fireballs and weird lights in the sky, as well as animal mutilations. In March 1976, a woman calling herself 'Mary F', sent two photographs purporting to be of Morgawr, to *The Falmouth Packet*. The pictures were taken off Terfusis Point. They showed a large, dark shape, with two distinct humps and a long neck. In one shot the neck is stretched out, in another it is arched downwards. She wrote an accompanying letter.

"I'd say it was fifteen to eighteen feet long (I mean the part showing above the water). It looked like an elephant waving its trunk, but the trunk was a long neck with a small head on the end, like a snake's head. It had humps on the back which moved in a funny way. The colour was black or very dark brown, and the skin seemed to be like a sea-lion's. I'm glad to know that other people have seen the great Cornish sea serpent. As a matter of fact the animal frightened me. I would not like to see it any closer. I did not like the way it moved when swimming."

So again we see, even from the safety of the shore, the animal seemed to generate a deep fear within the witness. Just the year before, a Mrs Scott of Falmouth, had seen the creature with a friend Mr Riley. The creature had surfaced of Pendennis Point. It had a long neck, humped back and stubby horns on the head. It appeared to have stiff bristles running down its back. It submerged and rose again, holding a large conger eel in its jaws. Mrs Scott said that as long as she lived, she would never forget the face on that thing.

Tim Dinsdale, a man who spent decades in search of the Loch Ness Monster, was once told a truly weird tale of a gigantic sea serpent, that the witnesses believed was a manifestation of death itself! Dinsdale was working for the Loch Ness phenomena Investigation Bureau, and was in a mobile exhibition at Acheahannet, in 1970. A man entered and asked to be introduced to him. It turned out that during WWII the man, Mr Ratcliffe, had been petty officer on board the H.M.S *Galatea*. He told Dinsdale that the ship had met with a gigantic sea serpent, in November of 1940. It had a huge, ugly head on a long neck, with a mane-like appendage hanging down. Behind the head and neck, Mr Ratcliffe said there were huge loops or coils of a long body. Dinsdale recounted that the dimensions of the beast, as reported by the witness were 'incredible'. An order came down from the bridge to open fire on the monster, with the forr'd 0.5 inch machine gun. Ratcliffe said he saw the tracer arc and watched bullets strike the monster's neck. The creature's head jerked and shivered and it dived beneath the water.

Soon after a rumour went around the crew that the sea serpent was not a flesh and blood beast, but a manifestation of death and that firing upon it had doomed the ship. A year later the *Galatea* was sunk by a torpedo off Alexandria in the Mediterranean in December, 1941.

A whole year is a long time for curse to hit home, but the crew certainly believed that their attack on the beast had been tantamount to suicide.

Dr Bernard Heuvelmans, the Belgian zoologist who is credited as being the 'father of cryptozoology' formed a classification for sea serpents in his book *In The Wake of the Sea Serpents* in 1968. Heuvalmans, after examining 587 sea serpent reports from 1639 to 1966, came up with nine types or classifications of large marine creatures, unknown to science. These classifications have been criticised as many are vague, with only a handful of sightings involved. In reality, only two groups stand up with large amounts of encounters reported within them. These are the long necked and the many humped. However, there are detailed reports of much stranger monsters encountered at sea. Take the case of the *Tresco* for example.

The *Tresco* was a British cargo steamer travelling between Philadelphia, Pennsylvania and Santiago de Cuba in May, 1903. Second Officer Joseph Ostens Grey recounts what they saw two days into the trip and 90 miles off Cape Hatteras, North Carolina. His account was printed in the October 1903 issue of *World Wide Magazine*.

"About ten o'clock I saw, on our port bow, something creating a vast amount of disturbance in the water. The commotion was so great that I judged it to be a school of porpoises, which herd together and play, jumping above the water like great Newfoundland dogs. It is not at all uncommon to see a school of them in those waters; but, somehow, the approaching school seemed different. I watched them closely as they neared the vessel from the south-east. Whatever was approaching the vessel, the water was surging about some large fish which presently I discovered were not porpoises, but sharks. Now sharks are common enough, but not in solid masses as was the school I now beheld travelling at such great speed. It seemed to me a phenomenal departure from anything I had heretofore observed in regard to these voracious and savage creatures. They were not attracted to the vessel by anything thrown overboard, but held steadily on their way.

They seemed to be some maritime express, bound for Cape Hatteras; for, from the time we sighted them until they disappeared, they kept to their course, as if making all speed. What impelled them to travel at such a rate I could not imagine; nor could I offer any explanation for their assembly in such a solid mass.

Sharks differ in size and there are several varieties. So far as I could tell these were the usual bottle-nosed shark. They were swimming shoulder to shoulder, closely packed together, their dorsal fins cutting the water steadily. Occasionally their snouts appeared. It was a curious spectacle, and, while in no way alarmed, I watched them until they were out of sight. In all, as nearly as I could count them as they passed, their number was about forty.

I saw no more sharks. The time went by uneventfully. My mind reverted several times to that rushing herd of sea-tigers, and no reason for such swift, steady pursuit of an unchanging course occurred to me. My wonder rather increased than diminished.

The passing of the sharks had made me unusually on the alert. About an hour later I espied a fresh object in the water on our port bow. It was some distance away, due south-east exactly the direction from which the sharks had appeared. It was floating low, and it looked black. I thought it must be a derelict one of those wandering, drifting hulks, so desolate to see, so dangerous to encounter.

I instantly gave orders to the man at the wheel to steer for the derelict. The Tresco was steaming along due south; but now she swung gradually about until she was going exactly south-east. The sea was still calm and smooth. We sped easily on our way, with little said except, "It is a derelict; steer for her."

The man at the wheel beside me on the bridge thought so too as we headed for it, wondering how much of a hulk it would prove to be, or what we should ascertain of its history. We always steer for derelicts in the hope of possibly rescuing survivors; or some poor bodies may remain that need decent Christian consignment to the sea. It is, besides, an important duty resting upon the masters of all vessels to report to the Hydrographic Office the name of every derelict met with.

During the twenty minutes we were steering toward it I was decidedly puzzled. It seemed to me that this low-lying, dark object was moving toward us, as well as we toward it. It did not look like the hull of a vessel; nor could it be a raft. Neither would move so swiftly toward us. What could it be? The puzzle grew stranger. I stared intently, as every moment brought us nearer. We would soon know, at all events. The powerful engines were driving us onward so rapidly that the solution would be now a matter of but a few minutes. And yet the time seemed long. Nearer and nearer we drew and at last we were but two ships' lengths away. With a conviction that grew ever deeper, and ever more disquieting, we came to know that this thing could be no derelict, no object the hand of man had fashioned, no object, probably, the eyes of man had ever seen.

Now, swiftly, with a terrible uprising, a mighty and horrible head came out of the water, surmounting a tall, powerful neck that had the thickness and strength of a cathedral pillar, yet spindly in proportion to the huge and awful head it supported. I felt that I must run somewhere, anywhere, to get away; and yet the weird and awful thing, there before us, held my gaze in the one direction.

At length I recovered some measure of my self-possession. "Jump, Leon; jump down into the wheel-house!" I shouted. "Steer down there. Let's get out of this fellow's road!"

The man obeyed with alacrity; and I, only too gladly, followed him.

There were seven steps to be descended; and I felt like a child afraid of the dark does when it runs upstairs to bed, thinking a bogey is after it in the hallways. I was frightened; there is no use to deny the fact.

Once inside the wheel-house, I flung the door to and locked it, thankful for even this frail barrier thankful for the slight protection of the wheel house, a mere nothing to such an adversary. There we were, silent both of us. Leon took his place at the wheel. We waited for what was to come next, still with the same sense of awe and huge, overwhelming dread upon us.

The wheel-house and chart-room adjoin, being one compartment with a partition. In front there are four windows, commanding a wide range; but, unluckily, from his position at the wheel Leon could no longer see the object. It was too near. He stayed at his post, needing no orders. I stepped into the chart-room to his left, where I could obtain a full view of the serpent as it faced us.

I could see it steadily and well from the chart-room port-hole. I looked and tried to notice every possible thing about it, yet wondering anxiously all the while how we should escape. The man at the wheel, and I with my face close to the port-hole, were stricken too dumb with astonishment and fear combined to say a word to each other. We did not say, "What is it? What shall we do if it comes nearer?" Nor did we discuss its appearance and actions. To me it was sickening and horrifying, and Leon had seen quite enough before he fled from the bridge.

Out of the formless horror within me a dread arose which shaped itself into a distinct, dismaying apprehension. What if the thing should attack the steamer? The consequences loomed up, fearfully appalling, to my swiftly realizing imagination. The creature, assuredly, was enraged. So enormous was its size, so vast its strength that even a steamer like the Tresco would be in danger of some kind perhaps of many kinds. The rail of the ship, it was true, was twenty feet above the water; but the head and neck of the serpent were already elevated to a height of fifteen feet. It could easily come aboard. The whole deck, all the upper works, in fact, would be at the mercy of its rage!

But far more serious to contemplate was the problem of its mere weight. That alone was a menace to the ship's safety. As I have said, we were going out in ballast, very light. Such a weight on one side would inevitably list the vessel, for the centre of gravity was so high that any heavy, ill-placed burden meant the gravest danger.

There that evil thing remained, the body motionless, the tail undulating vertically. As it lashed the water with the long, snake-like tail the head all the time was reared high, regarding the Tresco as if waiting to see what such a thing as a ship might be and, until it should decide, determined to maintain its watchful position. It looked for all the world like some fantastic Chinese dragon become a living reality; or a page from a scientific work picturing some ancient saurian monster, neither reptile nor beast wholly, but both in part.

When I first saw it, lying so low as to appear like a derelict, I must have seen only the back and body. The head was probably resting on the shoulders, as a swan sometimes rests, until, coming within two ships' lengths, we alarmed it by our unfaltering approach to the position of defensive attention.

We needed no binoculars. A sailor sees as no landsman sees; his eyes are trained to watch sky and sea and every object which may affect the welfare of the ship. And, indeed, the serpent was so near that even untrained eyes could have distinguished the most minute details of its appearance.
I estimated the length of the creature at about one-third that of the Tresco, or one hundred feet. We saw it only in perspective up to this time, for it remained facing us, neither wheeling nor changing position.

I judged it to be about eight feet in diameter in the widest part of its body, and so about twenty feet in circumference. The body was not cylindrical at all. It had a noticeable arch toward the top, and the hump of the back sloped downwards to the neck as well as toward the tail. It was widest at the forward end, rapidly tapering backward from the hump above the shoulders.

There was something unspeakably loathsome about the head, which was five feet long from nose to upper extremity. Such a head I never saw on any denizen of the sea. The neck, eighteen inches in diameter, was slender by comparison. Underneath the jaw there seemed to be a sort of pouch, or drooping skin; there may have been a slight bulge there. The neck was smallest half-way between the head and where it joined the body.

The nose, like a snout upturned, was somewhat recurved. It was rather pointed in its general formation, but blunt at the end. I can remember no nostrils or blow-holes. The lower jaw was prognathous, and the lower lip was half projecting, half pendulous. Presently I noticed something

dripping from the ugly lower jaw. Watching, I saw that it was saliva, of a dirty drab colour, which dropped from the corners of the mouth.
While it displayed no teeth, it did possess very long and formidable molars. There were two and they curved backward like walrus' tusks. If it had teeth or tongue I did not see, it's mouth was red. Its eyes were of a decided reddish colour. They were set high in the head, like a water-fowl. They were elongated vertically. They carried in their dull depths a sombre, baleful glow, as if within them was concentrated all the fierce menacing spirit that raged in the huge bulk behind.

Below the eyes some scales appeared, which dragged backward, becoming larger and larger until, on the body, they were great plates, or protuberances like the denticulated ridges of an alligator's hide. They did not glisten like the scales of a fish. The smallest of the scales, near the eyes, measured about three inches in diameter, and were so little oval as to appear completely round. The largest of the scales, or indurations, located upon the shoulders, presented a form more pronouncedly oval, and these were some eight inches long, five inches wide, and four inches high, their apex being a distinct ridge.

The hide, in the general tone of its colour, could be compared to nothing but antique bronze, showing the distinct light green hue of the oxidized metal. The tone of the colour was lightest upon the back and sides. As it shaded toward the almost wholly submerged belly it became a dull, dark green, deepening its hue with the decrease in the size of the plates or indurations constituting the creature's defensive armour.

It held itself in the same relative position to the ship during all the time the impressions I have enumerated were photographed indelibly on my brain. Its side fins, extending one-third of the way from the shoulder to the beginning of the tail, and broadest about a foot near the shoulder, worked like fans in swift agitation of the water.

As I gazed, fascinated with the horror of the thing, it raised its dorsal fin, obviously in wrath. And then a thing happened which, strange as it may appear after the recounting of the fearsomeness of the serpent's dreadful front, was more appalling, more sickeningly terrifying, than anything I had yet beheld. Suddenly, at the back of the head, a great webbed crest uprose, and from the eyes, hitherto so dull save for the glow smouldering in their depths, a scintillating glare appeared, as if the creature felt the moment had come for attack. The crest was a foot in height at its forward extremity, where it was supported by a sharp-pointed spine.
The undulations of its tail increased in violence. It lashed the water in fury. Its reddish eyes were fixed upon us; but, threatening as it appeared,

it came no nearer. The novelty of our appearance, and our size, seemed to make it hesitate. In what way it would have attacked us I can only imagine.

This hesitation and anger, combined, kept it at a standstill, and our fear and helplessness for resistance kept us quiet. The creature remained in this fashion, glaring at us, for a few moments more. Then I saw it was about to act.

It was going to turn away from us. I could scarcely credit my senses. I watched its new tactics carefully. Yes, it was moving and turning; it was about to go from us. I felt an infinite, deep-breathed sense of relief.

Its great body turned, as if on a pivot, inward in a circle, followed by its long tail. With astonishing ease for so huge a bulk it made the sweeping evolution. And only then did it lower its ugly head, that had so long confronted us in open antagonism. I began to breathe more steadily. I was certain now. It was afraid, and would go peaceably.

Only at that last moment did I think of Captain Bartlett. I must call him, now that I dared to venture out. I wanted him to see the monster. I unlocked the door and flung it wide, and ran aft along the starboard side as fast as I could. I burst in upon the captain in his state-room. He was lying down, but was fully dressed. The noise of my entrance startled him.

"Come on, captain, quick!" I exclaimed. "Come up and see this animal!"

Springing up instantly he was ready to follow. He comprehended that something unusual was near, yet he was astonished at such a report from an excited mate. Five seconds more and we two stood together on the poop, where we could have a clear view and, as I knew now, a safe place from which to gaze upon our gruesome visitant. I was half glad, half worried to find it was still in sight. The captain would not think me demented.

Captain Bartlett stood transfixed. A moment and he found his voice:

"Good heavens! What's that?"

"I take it, sir," I replied, "to be a sea-serpent."

"I believe you're right," he rejoined. We stood there waiting to see whether it would go or return.

The serpent, or whatever else it may have been, was on our port quarter, for the engines had been driving us steadily ahead. The distance at which it was then removed was about a quarter of a mile. Its tail was now toward us. The back of its head, sunk upon the shoulders, was visible, together with the twenty- five feet of the body which I have hitherto characterized as the hump of the back... As the terrifying thing was gone we could talk and compare our observations and ideas concerning it.

As I have said, I did not notice any nostrils; but I believe it was a breathing animal, endowed with lungs. While no sound reached my ears as we approached it, and while Leon and I were hidden in the chart-room, Captain Bartlett thought he heard distinctly, as we stood side by side on the poop, a noise which came from the creature that was in the nature of a snort or, to be exactly correct, a hoot. The sound, according to the recollection of Captain Bartlett, might be compared to the noise of a shrill tug-boat whistle. For myself, I must frankly say I can recall absolutely no sound. The coincidence of the appearance of the sharks and of the great lizard during the same hour is something I can affirm but cannot attempt to explain. An inference that would seem obvious is that the sharks were fleeing from the monster. But, in the absence of definite knowledge, it must remain coincidence, and nothing more.

After the exchange of these few observations Captain Bartlett turned to me and said:

"I have had many strange experiences, as you know; and I have seen many strange sights. But I confess this thing is, without doubt, the most horrible and blood-curdling that I have ever looked on. Grey," he continued, "words cannot describe its loathsomeness, or the horror and terror with which I gazed upon it."

All this time none of the crew had dared come on deck. Our chief officer, Mr. Griffiths, was asleep in his cabin. The men who had fled so hastily, and the others who came at their call, looked out fearfully at the serpent from the forecastle ports. The steward, John Jackson, a coloured man from Baltimore, who saw it, was greatly terrified. He has since left the Tresco, having been engaged only for the voyage. Those who did not see it, like Chief-officer Griffiths, can testify to the general excitement and the facts elicited by the subsequent discussion among the men who did. When the danger was over the men cautiously returned to the deck. Faces appeared at the hatches, and, after a little reconnoitring, up the companion-way they came, looking carefully astern, to assure themselves that the monster was really gone. Gradually, as they regained courage, they resumed their work, although they were careful to remain in groups, still talking over the astonishing event. After a long time had elapsed they were hardy enough to joke about it, although they had been so scared; and they repeated the story to the men in the engine-room, who had, of course, not even caught a glimpse of the stranger.

All this time the sea had remained quiet and the weather the same, so the conditions throughout were most favourable for viewing the monster.

I now ordered the vessel to be put on her course again—due south. The incident was over; our work was before us. Whatever danger had existed was passed. Santiago was to be reached, and we made that port on the fifth day afterward. As I watched through the port and, later, on the bridge, when, my fear abating, I could collect my thoughts better, I wished we possessed powerful guns which could tear a hole in that appalling head or through the armoured body, so that we could secure the carcass as a trophy and settle once for all the controversy concerning the sea-serpent. And I clenched my hands with annoyance, as I have clenched them many times since, when I thought of that camera of mine, ashore and useless, awaiting my next trip to St. Thomas. Why had I left it there, when now, for the first time in my life, I really needed it?

During the five days that were required for the remainder of the voyage our conversation naturally reverted to the exciting morning and to the experience we never expect will be ours again. I, for one, sincerely hope it will not be repeated, unless for the corroboration of this statement and to assist science by delivering to some learned body the carcass of another such monster.

We have carefully collated all the facts. Our conclusion is that the creature was, without doubt, a mammal, like porpoises and whales, although more like a reptile in appearance.

At Santiago I prepared a report for the Press of Philadelphia, to be presented on my return. Although I made it out carefully, it drew forth the usual jests in several quarters, but it was credited in others. How bitterly I have regretted that I had no photographs to settle the doubts of those who questioned the accuracy of the drawings I have since made from memory! I have but to shut my eyes, and that ineffaceable picture rises before my mind in all its horrible detail.

The Second Officer is clearly mistaken in his assessment, as this monster, if it existed, was transparently a reptile. It bears more of a resemblance to a dragon of medieval legend than any other sea serpent report and this is reflected in the write up from *The Chicago Daily Tribune* on June 17th. This piece seems to be related more from the view of the Captain, one W. H. Bartlett.

"The sea serpent of 1903 at last has put in its appearance. It was observed by truthful Capt. Bartlett of the good steamer Tresco. When ninety miles south of Cape Hatteras the captain noticed a peculiar disturbance in the distance and that the disturbance was headed his way. In a short time it arrived and turned out to be a great number of sharks tearing through the water "like all possessed" on their way towards shore. An hour afterwards Capt. Bartlett noticed a dark object in the distance, and, thinking it a derelict, steered for it. When a short distance from it the "derelict" slowly lifted itself above the water. Instantly the crew went below and barricaded the doors, and through the port- hole in the wheelhouse the captain saw the sea serpent. 'Half dragon and half serpent, it was the most hideous and loathsome reptile, with its gaping jaws and bloodshot eyes. From each side of its horrible mouth two large tusks protruded, similar to those of a walrus, and its lips were dripping with a discoloured saliva which emitted a most offensive smell.' The captain was not in a condition of mind to make accurate estimates of measurement, but he thinks the serpent was about 100 feet long. Fortunately, instead of making an attack 'it turned tail, and, with a swish and a swirl of the water, sank in the depths,' probably much to the relief of Capt. Bartlett and his crew, not to mention the panic stricken bunch of sharks. Now, this is a sea serpent worth having. The veracious captain's statement shows it has not suffered 'a sea change into something rich and strange,' but retains all its old horrible and terrifying aspects. It is no ordinary cephalopod but a worthy descendent of the serpent which Regulus and his army made war on with catapults, and of the monster Olaus Magnus saw, which not only ate calves, sheep, and swine but also 'disturbs ships, rising up like a mast and sometimes snaps men from the deck,' or even of Fafnir, spurting and smoke from his cavern. The sea serpent has been an unconscionable time putting in an appearance, but better late than never. For what should we do in these days of strikes, and floods, and cyclones, and cloudburst, and droughts, and fires, and barbaric regicides had we not our old friend the sea serpent for pleasurable entertainment?

The encounter is recorded in the log of the *Tresco*. Falsifying a ship's log was a grave offence and it is unlikely that a Captain and First Officer would have conspired together to make up this strange yarn. The nameless dread the monster brings to both officers and crew is apparent. Recall Joseph Ostens Grey's words *"I felt like a child afraid of the dark."*

CULT OF THE SEA DRAGON

Sometimes it is not the beast itself that is strange but the circumstances that grow about it. This was the case in a strange story from the northeast of England. Marsden Grotto is a very strange pub. It is located at the bottom off the sheer cliffs of Marsden Bay, South Shields. The pub is accessed via a lift inside a huge tower. The pub itself consists of caves and is built into the cliff face itself. In 1782, a man called Jack Bates, a lead miner from Allendale, Northumbria, and his wife Jessie moved into a small cave in the bay. Using blasting powder from a local quarry, Jack extended the cave vastly, earning himself the name 'Blaster Jack' in the process. Thus, he made a rent-free home for himself. He also created a flight of steps down the long cliff. Mr. and Mrs. Bates opened a tea shop in the caves and it was a great success. It is also said that he aided smugglers to hide contraband in his caves.

After the death of Blaster Jack, Peter Allen took over the grotto and turned it into an inn, extending the caves and even building a ballroom. Marsden Grotto has remained an inn ever since. At some point, a pillar was installed from the now old Lambton Castle, demolished in 1797 and not to be confused with the manor house of the same name built in the 1820s. The pillar features a carving of the famous Lambton Worm, a dragon-like, serpentine beast that terrorized the area in the 14th century.

The pub is, unsurprisingly, supposed to be haunted and the area was investigated by the former policeman turned Fortean Investigator, Mike Hallowell. It was Mike who uncovered the strange story of the cult of the sea dragon.

The North Sea onto which the Grotto faces is said to be the home of a sea dragon called the Shoney. As far back as the time of the Danelaw, the Vikings feared this monstrous beast. They would offer human sacrifice, so the story goes, to placate the Shoney. The victim would be trussed up, have their throat slit, and be tossed into the sea. The hope was that the sea dragon would eat the sacrifice and not attack their longships. Apparently, this became a sort of veneration of the beast and carried on long after the time of the Vikings. It is said that Scandinavian sailors continued the practice. Bodies, tied up, with throats slit, would wash ashore, sometimes half eaten, as far north as Lindisfarne and in particular around Marsden Bay. The beer cellar of Marsden Grotto was said to have been used as a temporary morgue for the bodies. To Mike's astonishment, he was told that the last body washed up in 1928! Think about that for a moment. If the story is true then there was a dragon worship cult carrying out human sacrifice in England well into the 20th century!

Mike tried to verify the claims by checking police records in Tyne and Wear only to be told that their records did not stretch that far back due to boundary changes in the county. He then tried to check records in County Durham only to be told that their records did not stretch back to the 1920s either.

Whilst trying to unearth more information, Mike received several threatening phone calls telling him to drop the investigation if he knew what was good for him. He was unable to trace the calls. Could some vestige of the cult still exist in the remote, windswept northeast of England? The whole story sounds like a particularly scary episode of *Doctor Who* from its 1970s heyday or the plot to a Hammer horror film, but Mike claims to have seen the beast for himself. He was driving along a cliffside road when he and his family saw a huge animal rise from the waters. Mike said it was not a whale, dolphin, or basking shark. The back portion showing above the surface was as large as a car and nut brown with a darker stripe along the back. He and his family could see a long head and neck and a long tail under the water, giving the whole creature a spindle shape. Other cars stopped to look at the beast before it submerged.

Dragon cults and human sacrifice in 20th century England; that really is *beyond* high strangeness.

THE HOLOGRAM SERPENTS

Fortean researcher and author Barry Fitzgerald related to me a very weird tale he had heard from a lake in County Roscommon in Ireland. He was kind enough to allow me to use it in this book and it ticks all the boxes for high strangeness.

"During the Christmas week of 2021, a couple, who wish to remain anonymous, whom for the sake of the report we shall call Sean and Ann McNulty, were visiting some islands on Lough Key in County Roscommon. Leaving Green Island, they headed west toward Church Island; this was a little more than a mile, and Sean had been a boatman for over 30 years and could handle the row without even breaking a sweat. Shortly after leaving the small island, their journey ground to a halt, with no progress made in the open water. They heard something rub against the hull of their small craft; thinking he had strayed onto some rocks, he dropped the oar to check the depth, which far surpassed the oar's length. Confused, he tried again to proceed, but instead, the boat turned on the spot several times, no matter how much effort he applied to the oars. He thought they might be caught in some strange current; he lifted the oars and rested, but it remained still instead of the boat drifting with the current. The sun was starting to get low in the sky, and they knew they had to get to shore. In the distance, Ann noticed what looked like two ducks in the water. She commented to Sean that the ducks acted strangely and got closer to them. Glancing in their direction, they both observed as the ducks got closer that there were, in fact, the heads of two larger animals they described as serpents. A wave of fear hit them hard, and dropping the oars, he broke the boat's invisible hold and headed for shore. They could see these two animals clearly and estimated their length to be somewhere around 8 feet with a thick girth. He said there was something very odd with their appearance, as it looked like they were some type of hologram, but on their heads, they had what looked like crystal crowns. The creatures vanished under the water and Sean managed to reach Doon shore, where he drove the boat into the thick reeds, and both of them fled to shore and clambered through the trees and undergrowth for nearly two hours."

CYBORGS OF THE DEEP

Japanese kaiju are fantastically large monsters in film, TV, comics, and literature. Kaiju are usually titanic biological entities. Godzilla, Rodan, and Gamera are prime examples. But one form of kaiju, the mecha, are vast robots or in some cases cyborgs. These too have a long pedigree in Japanese pop culture with Mecha-Godzilla, a robot replica of Godzilla, being the best known. However, there are a couple of high strangeness sea monster encounters that sound like hybrids of beast and machine that wouldn't look out of place in a battle with the King of the Monsters himself.

The *Tacoma Daily Ledger* of 3rd July 1893 ran a story of a very weird maritime encounter near to Puget Sound on the northwest coast of Washington. The witnesses were William Fitzhenry, H.L. Beal, W.L. McDonald, J.K.Bell, Henry Blackwood, and two 'eastern gentlemen' who did not want to be identified. The group had been on a fishing expedition in Puget Sound in the sloop *Marion*. The wind having got up, they weighed anchor at Black Fish Bay. They made camp on shore not far from the camp of a survey group. The men turned in at 11 pm. A little later the strangeness began.

> *"It was, I guess, about midnight before I fell asleep, but exactly how long I slept I cannot say, for when I woke it was with such startling suddenness that it never entered my mind to look at my watch, and when after a while I did look at my watch, as well as every watch belonging to the party, it was stopped.*
>
> *I am afraid you will fail to comprehend how suddenly that camp was awake.*
>
> *Since the creation of the world I doubt that sounds and sights more horrible were ever seen or heard by mortal man. I was in the midst of a pleasant dream, when in an instant a most horrible noise rang out in the clear morning air and instantly the whole air was filled with a storm current of electricity that caused every nerve in the body to sting with pain and a light as bright as that created by the concentration of many arc lights kept constantly flashing. At first I thought it was a thunderstorm but as no rain accompanied it, and both light and sound came from off the bay, I turned my head in that direction, and if it is possible for fright to turn one's hair white, then mine ought to be snow white, for right before my eyes was a most horrible-looking monster.*
>
> *By this time every man in our camp as well as the men from the camp of the surveyors, was gathered on the bank of the stream; and as soon as we could gather our wits together we began to question if what we were looking at was not a creation of the mind. But we were soon disburdened of this idea, for the monster drew slowly towards the shore and as it approached, from its head poured out a stream of water that looked like blue fire. All the while the air seemed filled with electricity, and the*

sensation experienced was as if each man had on a suit of clothes formed from the fine points of needles...

One of the men from the surveyors' camp incautiously took a few steps in the direction of the water that reached the man, and he instantly fell to the ground and lay as though dead.

Mr McDonald attempted to reach the man's body to pull it back to a place of safety but he was struck with some of the water the monster was throwing and fell senseless to the earth. By then every man in both parties was panic-stricken, and we rushed to the woods for a place of safety, leaving the fallen men lying on the beach.

As we reached the woods the 'demon of the deep' sent out flashes of light that illuminated the surrounding country for miles, and his roar - which sounded like a roar of thunder-became terrific. When we reached the woods we looked round and saw the monster was making off in the direction of the sound, and in an instant it disappeared beneath the waters of the bay, but for some time we were able to trace its course by a bright, luminous light that was on the surface of the water. As the fish disappeared total darkness surrounded us, and it took us some time to find our way back to the beach where our comrades lay. We were unable to tell the time as the powerful electric force had stopped our watches. We eventually found McDonald and the other man and were greatly relieved to find they were alive, though unconscious. So we sat down to await the coming of daylight. It came, I should judge, in about half an hour, and by this time, by constant work on the two men, both were able to stand.

The monster fish, or whatever you want to call it, was fully 150 feet long and at its thickest part I should judge, about 30 feet in circumference. Its shape was somewhat out of the ordinary insofar that the body was neither round nor flat but oval, and from what we could see the upper part of the body was covered in very course hair. The head was shaped very much like the head of a walrus, though of course very much larger. Its eyes, of which it apparently had six, were as large around as a dinner plate and exceedingly dull, and it was about the only spot on the monster that at one time or another was not illuminated. At intervals of about every eight feet from its head to its tail a substance that had the appearance of a copper band encircled the body and it was from these many bands that the powerful electric current seemed to come. The bands nearest the head seemed to have the strongest electrical force, and it was from the first six bands that the most brilliant light was emitted. Near the centre of the head were two large horn like substances, though they could not have been horns for it was through them that the electrically charged water was thrown.

Its tail, from what I could see was shaped like a propeller and seemed to revolve, and it may be possible that the strange monster pushes himself through the water with a propeller-like tail. At will this strange monstrosity seemed to be able to emit strong waves of electrical current, giving off electro-motive forces, which causes a person coming within the radius of this force to revive an electric shock."

It may be pertinent at this moment to remind ourselves that the term robot was not coined until 1920 and cyborg until 1960.

The whole story is fantastical and weird, the epitome of high strangeness. If someone were going to fabricate a story about a sea serpent, why would they make it half machine when such a concept was not even dreamt of in the relatively new genre of science fiction?

But this mechanical, maritime monstrosity was not the only one of its kind reported.

The *Daily Mail* reported on a strange story that was said to have occurred on October 28th 1902. The steamship S.S Fort Sailsbury encountered a seemingly mechanical sea monster in the Atlantic off the coast of Central West Africa. The lookout saw a huge, dark object in the sea in front of them at 3 a.m. It bore two lights. The thing passed forty to fifty yards on the ship's port side. According to the second officer A.H Raymer, it was five to six hundred feet long. It seemed to have an armoured back and was propelled by fins that caused a commotion in the water. Phosphorescence was being churned up as the object passed.

In his own words taken from the ships log...

"October 28, 1902, 3:05 a.m. dark object, with long, luminous trailing wake, thrown in relief by a phosphorescent sea, seen ahead, a little on starboard bow. Look-out reported two masthead lights ahead. These two lights, almost as bright as a steamer's lights, appeared to shine from two points in line on the upper surface of the dark mass. Concluded dark mass was a whale, and lights phosphorescent. On drawing nearer, dark mass and lights sank below the surface. Prepared to examine the wake in passing with binoculars. Passed about forty to fifty yards on port side of wake, and discovered it was the scaled back of some huge monster slowly disappearing below the surface. Darkness of the night prevented determining its exact nature, but scales of apparently 1 ft. diameter, and dotted in places with barnacle growth, were plainly discernible. The breadth of the body showing above water tapered from about 80 ft. close abaft, where the dark mass had appeared to about 5 ft. at the extreme end visible. Length roughly about 500 ft. to 600 ft. Concluded that the dark mass first seen must have been the creature's head. The swirl

caused by the monster's progress could be distinctly heard, and a strong odour like that of a low-tide beach on a summer day pervaded the air. Twice along its length the disturbance of the water and a broadening of the surrounding belt of phosphorus indicated the presence of huge fins in motion below the surface. The wet, shiny back of the monster was dotted with twinkling phosphorescent lights, and was encircled with a band of white phosphorescent sea. Such are the bare facts of the passing of the Sea Serpent in latitude 5 deg. 81 min. S., longitude 4 deg. 42 min. W., as seen by myself, being officer of the watch, and by the helmsman and look -out man."

The sheer size of this object, together with the lights and strange propulsion make it sound like some kind of weird, titanic machine rather than a biological entity.

These mecha-monsters are not always huge. The *Pittsburgh Daily Post* of April 4th 1889 ran a story of a seemingly part mechanical creature encountered in the Delaware River that runs in to the Delaware Bay in the Atlantic Ocean. Charles Wooden and Charles Adams were rowing on the river when they were attacked by a weird creature. The beast, which had a strange, musky odour and let out a cry that was half hiss and half bark, savagely bit through one of the oars. It was about six feet long, had a large head shaped like a bulldog, and jaws furnished with two rows of teeth. The eyes were sunken deep into the head and had long lashes. The creature had a long neck and the witnesses saw it had two, duck-like, webbed feet; it was covered in short black hair, but strangest of all, its tail terminated in four metallic blades like the screws of a propeller.

Perhaps the strangest marine case of all seems to involve a sea serpent fighting a robot! Reported in the June 2006 issue of the Ukrainian magazine *Riviera*, a story written by Anatoly Tavrajeski tells of just such an alleged occurrence. An unnamed Russian family were on the cliffs overlooking the Black Sea in the Sevastopol region of the Crimea. There was a rocky outcrop separating Laspi and Cape Aya and it was off this area that the father noticed something odd whilst looking out to sea with his binoculars. What looked like a silver, humanoid figure was in the water. It seemed to be armoured and far bigger than a man, about ten feet tall. The man's family ran over and began to watch as did other people in the area. As the group watched, the silver figure seemed to be attacked by a light brown, serpentine creature that coiled around it. The 'robot' and the sea serpent seemed to struggle with each other in a fight that lasted twelve minutes. Finally, the combatants split apart. The robot like figure swam away close to shore and vanished around the coast where the view was obscured by vegetation. The serpent swam back out to sea. Neither seemed harmed.

The family wrote to Professor Anton A. Anfalov, a Ukrainian paranormal researcher from Sevastopol and head of the awkwardly named Society of Observers of Anomalous Phenomena in the Environment. The Professor wrote back telling them of his theories that the whole area was riddled with extensive cave systems that had secret UFO bases and colonies of sea serpents in them. Make of that what you will. Now it is time to shake off the water and head onto dry land.

NIGHT OF THE BLOODY APES

There can be little doubt that there are still undiscovered species of large primate inhabiting our planet. The fossil record makes it abundantly clear that the ancestral tree of mankind is a truly tangled bush. At one time it was thought that our ancestor *Homo erectus* left Africa and speciated into *Homo ergaster*, *Homo heidelbergensis*, *Homo neanderthalensis* and others. However, we now know that the situation was much more complex. *Homo floresiensis* was discovered in strata some 50,000 years old on the Indonesian island of Flores in 2005. The cave in which the remains were found were so perfect for preservation that the bones were sub-fossil. The tiny species was only 3 feet 7 inches tall. None the less it made tools and fire. It was, at first, thought to be a dwarf, island form of *Homo erectus*. Examination of the bones, however, showed that it was a sister species to *Homo habilis*, an African hominin that died out 1.9 million years ago and until now had not been known to have left Africa. Five years later, on the island of Luzon in the Philippines, a related species, equally as tiny and named *Homo luzonensis,* was unearthed. It too was closely related to *Homo habilis* and also dated to around 50, 000 years ago. Now it seems that a hominin only known from Africa left its home continent and travelled halfway around the world to create its own lineage quite distinct from our own ancestors. Again, we may ask the question: what else is out there?

Another recent discovery was of a large fossil hominin species called the Denisovans. Initially, they were identified from finger and toe bones found in a Siberian cave in 2010. Later, a jawbone was found in China. This species may have lingered until only 14,500 years ago. It is thought that they were related to Neanderthals, but were even more robust in build. Both Denisovans and Neanderthals interbred with anatomically modern humans and indeed have left their DNA in various populations of modern man.

One as yet unnamed species of hominin, whose remains have been found at Red Deer Cave in China, lived only 10,000 years ago, less than an eye blink in evolutionary terms.

Interestingly, some populations of modern humans around the world have genetic markers in their DNA that suggest their ancestors mated with an as yet unknown hominin that we do not even have fossil remains for, just genetic markers.

There are other types of data that support the existence of unknown primates.

The Centre for Fortean Zoology brought back hair samples from the jungles of western Sumatra. The hairs were found close to orang-pendek tracks. Lars Thomas, an expert in animal hair associated with Copenhagen University, examined the hair. He concluded that it came from a primate related to but distinct from the Sumatran orang-utan. He said that he was forced to conclude that there was a large unknown species of primate on Sumatra.

In 1977 the Chinese government backed a large expedition to the Shennongjia Forestry District in search of the yeren, the local name for the yeti. As well as turning up possible nests and 16 inch footprints, a number of hair samples were found. Microscopic, keratin content, and particle induced x-ray emission tests indicated that the hairs were primate, but matched nothing known. Also, the copper/zinc ratio in the hair was unlike that of any other animal.

Dr Eva Bellamain found the eDNA of a higher primate in a pool in Bhutan, one that shared 99% of its genetic make-up with man. Chimpanzees (*Pan troglodytes*) and their relatives pygmy chimps (*Pan paniscus)* share 98.7% of their DNA with man. Whatever left the eDNA in that Bhutanese pool was more closely related to us. Both chimps and pygmy chimps are native to Africa. Gorillas share 98% of their DNA with us, but once again are African. The only Asian great apes, the three known species of orang-utan, have 97% of DNA in common with humans. Orang-utans lived on mainland Asia until the late Pleistocene/early Holocene. As far as we know, they live only on the islands of Borneo and Sumatra today. All this points to a higher primate, unknown to science, and as closely related to man as is the gorilla.

I tried to contact expedition leader Mark Evans with zero success, but managed to contact Dr Bellemain via e-mail. Eva turned out to be a very helpful lady. She explained that she could do no further work with the samples as she no longer worked for the company that had them, a wonderfully Ian Flemming-esque organization called Spy-Gen. On enquiring further, we found out that Spy-Gen had destroyed the samples for reasons unknown.

So, it seems that there are several large, unknown higher primates, be they great apes or relic hominins, still living in the world today. But things are muddied by other, strange reports that seem of a quite different type. Monstrous, sometimes aggressive, ape-like creatures are sighted in places that could not possibly support them as biological entities. They appear in tandem with other Fortean phenomena like UFO sightings, sightings of other cryptids, and even poltergeist phenomena.

Strangely, these weird reports seem to come from certain areas such as the USA and Australia. They are notably absent from Asia, which is the continent most likely to support unknown, large primates. Aside from a few superstitions about bad luck if one talks about the yeti or almasty, I have only been able to turn up two 'paranormal' type story relating to a possible Asian man-beast.

Scottish scientist Bill Grant was approaching a small lake near the Tibet/Nepal border in 1983. Suddenly he heard what he describes as a voice that was not a voice, warning him to go no further. A profound feeling of fear crept over him and, despite being a veteran explorer; he felt he could not take another step forward. He fled, but returned hours later to find the feeling of fear had gone. But, beside the lake he found huge, man

-like tracks in the mud.

In 1981 in the Pamir Mountains of Kyrgyzstan, several campers saw, at around midnight, a hair covered bipedal, ape-like monster wander past their camp. The thing did not seem to notice the campers and lumbered off into the night. The witnesses said it was holding a glowing sphere.

For truly weird tales we must go Stateside.

SASQUATCH HORROR

In March of 1973, three teens Richard Engels, Brian Goldojarb, and Willie Rosermann went camping in Big Rock, in the Angeles National Forest, Southern California. They had been there for two days. On the third day, Rosermann became increasingly uneasy. He felt that something was watching him and wanted to leave. He became more insistent as the day drew on and, as it was his car, the three drove out of the park in the evening. Rosermann felt that whatever had been making him uneasy would be waiting for them near the entrance of the park. As they drove down the canyon road and passed Sycamore Flats campsite, Rosermann saw a gigantic, dark, humanoid figure lurking in the trees beside the road. The huge creature emerged from the trees and began to lope along the road after Rosermann's '53 Chevy pickup. All three saw it silhouetted against the full moon. It had a conical head, no visible neck, and seemed lean and spare for its great height. The creature chased the terrified teens for a while until they lost it around bends in the road.

The trio reported the encounter to the local Sheriff's Office, fully expecting to be laughed at. Instead, the officer told them they had already received five Bigfoot reports in the past week.

Next day the lads, feeling braver in daylight, returned to Sycamore Flats and tried to estimate the creature's height. It had run under some tree branches and using this as a guide they created a cardboard cut-out of the figure, finding it was some eleven feet tall! They also found 18 inch long tracks, that unlike standard Sasquatch prints, bore only three toes.

The witness formed the Angeles Sasquatch Association (ASA) in order to investigate sightings and track the creatures. They were to have a number of very strange encounters.

In 1974 Neil Forn and Rich Engles of the ASA visited Antelope Valley Southern California to investigate Sasquatch sightings. Whilst using torches to search for tracks Forn came across a small pyramid of stacked rocks on top of a flat boulder. As he approached them an eerie feeling came over him.

> *"I could sense a hostility, like something didn't want us up there! It was like we were just barely being tolerated. And I had this feeling that if I touched those rocks it was going to kill me!"*

The two men saw a seven-foot creature loom up from a ledge in the beams of their torches. It had a conical head and was covered in black hair. They said it seemed to project cold, raw hatred towards them. Both men fled.

The nameless dread is far from the only strange factor in these North American sightings. For example, it seems that they can have a weird effect on machines. One case from West Virginia took place in October of 1960. WC 'Doc' Priestley' was driving through the Monongahela National Forest. He was following a bus that had some friends of his in it. Suddenly the car started to splutter and then stopped. Priestley then saw a hairy monster standing at the side of the road. Its long hair was standing on end as if electrically charged. The witness's friends realized he was no longer following them and backed the bus up. The creature fled at the sight of the bus, its hair dropping back down as it went. Priestley decided not to tell his mates about the beast. The car started up again and he once more drove after the bus. But once more the car spluttered and this time sparks and smoke came from under the bonnet. There again on the side of the road stood the Bigfoot, its hair standing on end once more. As the bus backed up again, the monster retreated, and this time did not re-emerge. The witness found that his car had suffered a bad short.

An almost identical story occurred decades later. Cheryl Boyer was driving one night near Chickies Rock, Lancaster Country, Pennsylvania, when her car stalled. The car had just come back from the garage the day before so she was annoyed. She got out and opened the bonnet. As she did so she saw something in the tree line. It was covered with hair, walked on two legs, and had glowing eyes. She slammed the bonnet hoping to scare the creature. When she looked up again, the thing had vanished. Her car started once more.

Another case where the creature seems to have its hair effected by electrics occurred in September of 1969 at Lost Trail Pass, Montana. The witness was cutting poles with a power saw when he was overcome with the creepy sensation that he was being watched. He switched off the motor and, on turning around, he saw a seven foot, black haired 'ape-man' watching him from only 25 feet away. Fearing an attack, he switched the saw on again. The hair on the creature's neck stood up. When he turned it off again the beast's hair fell down flat. The Bigfoot then moved away, never taking its eyes off the man.

In some cases, it is claimed that the creatures can mentally effect witnesses. One hot summer evening in 1972, Ron Bailey, a thirty year old from Palmdale, California, drove into the car park of a housing tract where he and his family rented a small house. Bailey, a Vietnam vet, worked for Rockwell International, a manufacturing conglomerate involved in aircraft, defence, the space industry, and commercial electronics. He was a happy, well respected family man, but his life was about to change.

As he exited his car he noticed a weird silence. None of the neighbourhood dogs where barking and even the crickets were silent. The light from the house next door lit up a large bush behind the car park. The bush seemed to sway as if in the wind, but the night was dead still. A smell like sewage reached his nostrils and he was overcome by fear. He ran to the house, locked the door behind him, switched on the kitchen lights, and drew up the blinds.

> *"So, I thought to myself, it's as plain as day, I'm losing my marbles. I started to let the blinds down when something grabbed me mentally. I don't know what it was-it's beyond description, but it was literally hypnotizing me. I'm not an dramatic person or an imaginative one. You could say I'm about as down-to-earth as a doorknob, but I felt like something alien and outside myself was trying to control me."*

Unable to move, he started into the courtyard where a thing emerged from the shadows about fifty feet away. It was black, hairy, stood on two legs, and seemed to be about nine feet tall. The shadows obscured the shaggy giant's face. Bailey felt that the thing was trying to control him mentally and draw him outside to it. He felt his resistance to the mental attack was weakening with every passing second. Then suddenly the monster turned and shambled off into the night. Bailey collapsed, exhausted and sweating, to the floor.

In early 1973 the beast was back. As he climbed into bed beside his wife, an awful thumping noise began from the rear of the house. It sounded like somebody stomping heavily in one position. On investigating, Bailey saw a huge, dark figure running in the direction of the desert. Next morning, three-toed tracks were found behind the house.

Becoming interested in what the creature was, Bailey began taking trips into the mountains searching for it. He found other sets of three-toed prints like those left behind the house.

One summer night, he was out with his wife Margaret and his friends John and Joyce Baylor. The women remained in one of the cars parked at the base of the mountain and the men trekked up a narrow canyon. As they climbed higher, they were bombarded by rocks hurled by the monsters (there now seemed to be a number of them). As they beat a retreat, one rock hit Bailey's ankle causing him to fall over. Looking back, he experienced what he called the 'mind grab' again. This time he said that the thing was mentally telling him to put his gun down and come to it. He found himself obeying until his military training kicked in and he pulled free of the mental grasp.

The men ran back down the mountain to find only John's car remained. Their wives had driven off. Once they were back home, the women explained that as they had waited in the car a huge creature had emerged from a stand of sycamore that were ten to twelve feet tall. The monster's head was taller than the trees. It had walked over to examine a mobile home parked on the other side of the road. The sight had scared the women so badly that they had driven home in a panic.

The ASA investigated another seeming case of a hairy giant communicating with a human via thought. It began one Saturday in October 1974 when seventeen year old Jim Mangono and six friends hiked into the Angeles National Forest. They followed a stream into the canyon and set up camp, ate, and his friends lay about. Mangono decided to walk upstream for a bit. He sat quietly upon a large rock and meditated. About an hour later he came out of his meditation, but something felt wrong.

> *"I was like in a blank trance. I couldn't pull myself out of it. I kept waiting...for something to enter my mind...waiting for something which I couldn't figure out."*

When he returned to camp the others noticed that Jim simply sat quietly staring off into space. He rose again and retraced his steps to the rock, feeling as if something was calling him back. The surroundings now felt weird, and he began to feel afraid. He returned to camp after what he thought had been fifteen minutes, but it turned out that he had been gone for a full two hours! This kind of missing time is often associated with so called alien abduction cases. What had happened to the boy in the two lost hours?

That night he was awoken by a series of ghastly screams that his friends slept through. The following morning he found an 18-inch, three-toed track in the sand close to the rock he had meditated on.

He reported these weird events to the ASA who decided to investigate. On November 16th 1974, three SA trackers, Val Stemler, and Jim Mangono backpacked to the area with the idea of using Jim as bait. That evening Jim meditated some way from the campfire, then he rose and began to walk downstream like a somnambulist.

"They want me to come to them," he said.

The ASA trackers refused to let him go into the night alone.

"No, they want me to come alone." Insisted Jim.

Not knowing the intentions of the huge creatures, the ASA members escorted Jim back to camp. Rich Engels and Neil Forn walked down to the rock but saw nothing.

The group was later awoken by dreadful screaming and the sound of something huge crashing through the bushes. Then a barrage of rocks came flying towards the camp, some of them narrowly missing the team's heads. The attack continued until dawn.

The ASA decided to try hypnotic regression on Jim Mangono. The session took place on January 15, 1975 in the Palos Veres offices of psychiatrist Dr Robert Jorden. The hypnotist was Donna Welke, director of the School of Applied Hypnosis.

During the hypnosis Jim told of being called to the rock and meeting a huge Bigfoot that put a hand on his shoulder and spoke to him with its mind. Apparently this one was a leader and Jim saw others looking down at him from the canyon. The creature communicated that Jim should tell people that the creatures were here first, before man and that man was ruining their planet.

These words will ring a bell with UFOlogists. Supposed aliens have often given environmental messages to abductees. Everything from pollution to nuclear arms have been mentioned. This may tell us more about the abductee's subconscious mind rather than about their abductors.

Jim did not take part in any follow up sessions saying it was *"too much hassle."* He lived a long way from the offices and was busy with work and education. Maybe, in reality, he was just too freaked out. It would be interesting to track him down today and see what he recalls of his weird adventures.

Further north in the USA and further back in time, a deeply weird case of Bigfoot telepathy took place. Conser Lake lies in Linn County, Oregon. In 1959 a series of weird events began. A man driving a truck full of mint claimed that a white, gorilla-like beast had run alongside his vehicle, keeping pace at 35 mph, looking in at him. The thing only gave up when the driver approached the mint distillery.

The media at the time spoke of a 'flying saucer' having crashed in the lake and a creature that screamed like a cat, left web-footed prints, and seemed to be unaffected by bullets.

The following year a group of seven teenagers visited the lake looking for the monster, and found it.

It was on the evening of July 31st. The seven foot tall, white haired creature made an odd squishing sound as it moved and let out a cry of *"Fleep! Fleeoweep!"* The scared kids high-tailed it out of the area and one of them, George Hess, reported the encounter to the sheriff. In the meantime, his friends had returned with an armed posse. In the coming weeks would-be monster hunters frequented the lake and a fisherman narrowly avoided getting a bullet in the head when an overzealous hunter mistook him for the creature. At the same time more webbed tracks were found and one farmer claimed that the monster had trampled over most of his mint harvest.

A Mrs Penning claimed that the monster had left a handprint on her bedroom window. She had heard a sound like water dripping and heard a strange cry that rose in pitch. Looking up she saw the pale ape-thing gazing in through the window at her. Two weeks later she and her sister heard rappings at the two windows of the house and found another hand print.
In mid-October Marilyn Samard, one of the teens that had seen the monster back in

July, had a second encounter. Together with her mother, aunt, and two boys she saw the creature emerge from a rubbish tip and begin circling the car. It even looked in at the windows. Weirdly, she could make out no facial features. We do not know if this simply meant it had hair falling down over the face or that it had no face at all. This detail is mentioned in other areas of Forteana such as with ghosts. Indeed, the thing had become known locally as The Ghost of Conser Lake.

So far, so weird, but things got stranger when the monster began to communicate telepathically.

Local journalist Betty Westby, who had written widely on the monster, decided to go into the area looking for the creature, taking a 'telepath' with her. This telepath claimed that the entity didn't like being called a monster and preferred to be referred to as a 'visitor' or 'alien'. The being continued, via the telepath to say he meant no harm and came only with affection for the creatures of Earth. The creature said he was one of many and his name was 'Flix'.

Another person, Alvin Hammock claimed he went looking for Flix and sent out a mental message that he meant the Bigfoot no harm and could trust him. He said that he received a psychic reply from Flix saying *"How can I know I can?"* Hammock thought that was a reference to the hunting knife he had and so he left it in his car. He also claimed that he was pushed into some water by invisible hands. On his way back to the car, he glanced backwards and saw a twenty-inch-wide section of grass flatten down into a circle with nothing visible touching it. So, we have what is ostensibly a Bigfoot case that has overtones of UFOlogy, ghosts, telepathy, and a mini crop circle!

The next report is just as strange and suggests these entities can shift shape and/or are not fully corporeal. The witness spoke to researcher Leonard Stringfield some ten years after the event. In 1964 the witness was an 18 year old girl who had been parked in a car with her future husband, Lew Lister. They were at Point Isabel, Clermont County, Ohio. It was 11.pm and were about a mile from Lew's family farm. Then both of them saw a figure walking towards them through a field. Lew shone the car headlights on the interloper. The thing was no man. It had a horrible head, pointed at the top and narrow at the chin. Its brow was wrinkled, and the nose looked like a pig's snout. The large ears looked pig-like too. It was six feet tall with broad shoulders and was covered with a 'yellow fuzz'. The creature seemed to move in a series of weird leaps. It passed through a three-strand barbed wire fence like a ghost. This is in direct conflict with many cases where 'normal' sasquatches have been seen tangled on barb wire fences, even leaving hair behind. The horror tried to grab Lew as he frantically wound up the windows. Mrs Lister screamed but then felt hypnotized by the monsters glowing eyes. She described feeling as if she were in a time-lapse but then the encounter got even stranger. The beast seemed to change form. The witness said it looked like watching a slowed down film. The creature's hands became paws and it fell onto all fours becoming a quadrupedal creature, then vanished utterly.

The slowed down time effect and the shape changing are factors seen in other Fortean cases and maybe give us a clue to the nature of the events.

In the same area of Ohio, some four years later, the Abbott family were terrorized by a Bigfoot that seemed equally ghost-like in nature. One evening in 1968, a sound like banging metal caused fifteen-year-old Larry Abbot, his father, and a relative, Arnold Hubbard, to investigate. After training their torches on some rustling bushes, they saw a huge beast emerge. It was ten feet tall with shoulders four feet wide. It had long arms that swung at its sides and was covered in light brown fur. The monster's ears were pointed and it had prominent teeth. Larry said that its glowing eyes seemed to put him into a trance much like Mrs Lister. He found himself unable to talk. The creature dropped to the ground and the trio lost sight of it.

Shortly later they heard it again and Arnold got a .22 rifle and the three of them began to search the property. As they crossed a field, the Bigfoot rose up only fifty feet away. Larry caught it in the beam of his torch and Arnold shot and hit the thing three times at close range. The hairy giant made a horrid scream then seemed to be enveloped by a white mist. Seconds later, the mist dissipated and the monster had vanished. A search of the area uncovered no blood or footprints.

Bigfoot creatures have also been seen glowing as if covered in phosphorous. A young woman driving along a lonely stretch of highway outside of Brooksville, Florida on November 30th 1966. She stopped to change a tire when she saw a hulking, hairy monster beside the road. It had glowing green eyes and it seemed that its torso was glowing green as well. It stood watching her before the approach of another car made it shamble away.

A year later in Elfer, Florida, four teenagers were parked in a 'lovers' lane area one January night. A foul stench suddenly pervaded the area. The ape-like beast leapt onto the bonnet of one of their cars. One witnessed spoke to investigator Joan Whritenour.

> *"Then we panicked. The thing looked like a big chimp but it was greenish in colour, with glowing green eyes. I started the motor and the thing jumped off and ran back into the woods. We tore like the blazes back to the dance we were supposed to be attending."*

Later, a police officer from New Port Richley visited the area and found a sticky green substance that was never identified.

Back in Ohio and in 1968, the Allison family seemed to have been victim not only to Bigfoot but to a whole Fortean flap. The events unfolded in Salem near Youngstown. The family were haunted by a huge, shadowy, man-like figure that seemed to be watching their house from the woods. The thing was also encountered in the house's driveway. The family's son, Bruce, said that as it ran it didn't seem to touch the ground. They were also visited by a cat-like beast three feet tall that left tracks around the house. The puma (*Puma concolor*) has long been extinct in the area and when the family contacted local zoos no big cats were missing from the collections. To top it off Mrs Allison saw a UFO! She witnessed a black, unlit craft that made a noise like a helicopter but looked like a plane with no wings. The object had a transparent dome on top in which there was a humanoid figure. The occupant seemed to be humanoid with olive skin and slanted eyes. We will see a number of these multi phenomena cases as we go through the book.

Another case involving Bigfoot and big cats occurred on the 23rd of August 1976 in Paris Township, Ohio. The Cayton family lived near Minerva and all nine members witnessed the weird events. The family were out on their porch at 10.30 pm when they heard noises from an old, disused chicken coop. They saw two pairs of glowing eyes reflected in the beams of the torches they shone towards the sounds. Scott Patterson drove over to investigate and saw that the eyes belonged to two big cats resembling pumas. Then a Bigfoot strode in front of the cats as if to protect them. The monster then began to lurch towards Patterson's car. At this point the family who were watching from the house phoned the police. As they waited in the kitchen the creature approached the house and looked through the window. It watched them for ten minutes. The family had guns but decided not to shoot as the creature did not seem aggressive. When deputy sheriff James Shannon arrived he noted an ammonia or sulphur-like smell.

Randy and Lou Rodgers lived in a farmhouse outside of Rochdale, Indiana. In August of 1972 the couple became aware of a strange presence around the house. It would bang on walls and windows at night. The pair would sometimes catch glimpses of a six foot tall ape-like creature loping off into the cornfields. The creature seemed to visit between 10-11.30 p.m. at night and Mr and Mrs Rodgers began to wait for it. Mrs Rodgers even put out left over food for it. The thing brought a stench like rotting flesh with it when it came. Weirdly, when it ran through mud, it left no tracks. Likewise, it moved through vegetation with no noises. The couple also said that when they looked at it, it seemed as if they could see through it like a ghost!

The phantom monster was seen elsewhere by other people in the neighbourhood. Also a UFO put in an appearance. Mrs Rogers' brothers saw a glowing light over the cornfields into which the creature retreated after visiting the farm. It seemed to explode but no debris or damage could be found.

Some locals got up a posse to hunt the creature. One said it has walked out into a road in front of him. He fired upon the creature, but the bullets had no effect on it whatsoever.

In August 1973 a driver told a police officer that whilst travelling between Hillside and Millwood in Derry, Pennsylvania, a huge animal like an ape had run out in front of the car. The man and his wife both saw the car collide with the beast, but the animal simply vanished like a ghost.

In early October 1973, near Galveston, Indiana, Jeff Martin was enjoying an evening of fishing. Looking over his shoulder he observed, just twenty feet from him, a figure. Thinking it was a person in the shadows he called out to it, but it ran away. Sometime later he felt something touch him and upon turning around he saw a sandy coloured Bigfoot standing behind him. It turned and ran with Jeff bravely giving chase. He saw it leap a ditch and vanish into the trees. Right afterwards he saw a glowing bronze light rise from the woods.

Two days later, Jeff returned with his fiancée, her father and two friends. The car was followed by a strange light in the sky as they drove to the area. The four all saw the creature standing in an area of tall weeds. They estimated it to be eight to nine feet tall. The beams of their torches seemed to grow weaker as they were trained on the monster. Two of them retreated to the car in fear as the remaining witnesses shouted at the creature. Another car came along the track and they had to move their own car. When they looked back, the Bigfoot was gone. It is not known if anyone in the second car saw the creature.

A very strange looking Bigfoot was the centre of a Fortean flap that occurred in Sharpsville, Indiana. In June of 1970, Dale King was visiting his girlfriend one evening when an odd feeling came over both of them. A wave of fear and dizziness came over them. To clear their heads, they drove to Dale's home in the country. They encountered a patch of road that seemed much darker than the surround countryside, even though it was night. Driving into it they found that it seemed that the temperature increased and that visibility became much poorer. The car's headlights could hardly penetrate the blackness. Driving on, they finally got to Dale's home. When getting out of the car, Dale's dog Zipper, normally friendly, savagely attacked the couple and Dale had to beat the dog off before he stopped.

A year later, in June of 1971, Zipper began barking madly one evening and, upon investigation, Dale saw his dog attacking a hairy monster nine feet tall and only 25 feet away from him in the back yard. The monster had a sickening stench like rotting meat and decaying vegetables. It was covered in dirty, stringy, brown hair and had long arms. Weirdest of all, the creature's head looked like a 'hair covered helmet'. Zipper was getting the best of the fight as the thing seemed to swipe at him in slow motion. Dale said the monster also seemed 'confused'. All the time it let out a low rumbling growl. Running inside to grab a shot gun, he found the creature shambling off. He fired two shots at it, seemingly with no effect. On phoning the local sheriff, he was laughed at.

About a month later, Zipper and Dale's other dogs began barking wildly again. Grabbing his gun, he investigated. Zipper had picked up the scent of something and led Dale through a graveyard to a river. Dale again smelt the vile odour of the monster but did not see it on this occasion.

Later that year, Dale was awoken by his dogs barking at four thirty in the morning. Looking out of the window, he saw the monster walking down by a creek. Grabbing his gun once more, he tracked the monster. Watching from the window, Dale's mother saw that the weird creature had doubled back and was now tracking her son. It finally wandered away.

Oddly, Dale found a strange circle of weeds in a dried-up pond. Thirty feet across, it had been flattened in a counterclockwise manner. This is similar to the crop circles that began to appear *en-masse* in the 1980s. Back in the 60s and 70s these were known as 'flying saucer nests' because of the belief that they had been flatted by alien space ships.

The monster was back in the spring of 1972. Both Dale and his brother were now married and on this particular night had gone out leaving their wives at home. Upon getting home they found the women hysterical with fear. They claimed that the thing had attempted to get in through the window. On investigating the men saw that the side of an aluminium storm window had been wrenched out by something with immense strength. The monster's awful odour was thick in the air as well.

The final visitation was in autumn of 1972. The barking dogs again awoke Dale who once more gave chase to the creature but didn't get close enough to shoot it. He tracked it through the graveyard and up into the woods. He waited for it in the cemetery, but it never returned.

APES FROM SPACE.

Renowned Fortean researcher Stan Gordon has investigated and recorded several of these high strangeness Bigfoot cases, often under the auspices of the Westmorland County UFO Study Group (WCUFOSG). Stan was phoned on September 1st 1973 by a man with a strange tale. The man said he knew three women who had been driving on a country road near Penn Township, Pennsylvania when they observed a large, silver, rectangular, metallic object on the ground. A door seemed to appear in the UFO and a set of stairs were lowered down. Two tall, hairy Bigfoot like creatures emerged, ran down the steps, and into the nearby woods. The women, who were frightened by the encounter drove off quickly.

Beginning in August of 1973 strange events began to unfold on a farm owned by a Mr and Mrs Bell in Derry, Pennsylvania. On a warm night Mrs Bell was driving on a country road at 7.30 pm. From a cornfield, she saw a creature emerge that she thought was a gorilla that had escaped from a zoo. However, the animal walked on two legs, crossing the road in two strides. It turned briefly to look at her. It had brown hair, long arms, and a hairless face.

Mrs Bell saw the monster again on September 18th. She was walking to her truck at 10.20pm. She saw a nine foot, tall, broad shouldered, hairy, man-like beast standing next to it. The Bigfoot took a step towards the woman and raised an arm. Mrs Bell screamed and the Bigfoot seemed to imitate her. She ran for the house screaming louder and the beast imitated her screams again. The creature ran off through a field. Mrs Bell said she had noticed a strange silence and stillness around the farm before the encounter. This strange feeling of stillness associated with many Fortean experiences has been dubbed 'The Oz Factor'. The term was coined in 1983 by British UFOlogist Jenny Randles.

Mrs Bell's husband and his friend went out to search for the beast that had badly scared his wife. They did not see the Bigfoot but saw a tubular, red UFO that illuminated the ground with a red light. Both Mr and Mrs Bell also reported another UFO that would hover over their farm at about ten o'clock at night. The star shaped object was silver and would turn to red and back to silver again.

Stan Gordon, George Lutz and Barry Clark from WCUFOSG investigated the farm on several occasions. Barry heard the creature scream one night. He described the noise as earth shattering and scary. It started as a low rattle and grew into a high scream that terrified the farm dogs as much as it did Barry. On the same night Barry saw the star shaped UFO and watched as it changed colour and moved jerkily from a 45-degree to a 90-degree position. Barry contacted some other researchers and together they drove up to the top of a hill for a clearer view. The four researchers observed the night sky lit up by five flashes like those from a welder's arc, one after the other with about a second between each flash. The flashes formed red rings. As these descended, they grew into a tubular section of light. But the show wasn't over. Suddenly the sky was filled with little silver lights that moved towards the red tubular light turning red themselves as they drew closer. The observers likened them to bees swarming to a hive. A female investigator began to cry and ran back to her car. Three men were left to watch the whole object seem to drop towards the ground. The investigators left the area.

The night's events scared Barry so much that he stopped his investigations on the farm.

Another strange case occurred in Beaver County, Pennsylvania in September 27th 1973. Two girls waiting for a lift in a country area at 9.30pm saw an eight-foot tall, white haired Bigfoot. The creature was holding a luminous sphere in its hand. The girls ran home in terror, and later the father of one of them went into the woods to investigate. He was gone for over an hour.

WCUFOSG investigator Bob Boyde went to investigate and had the father confirm that she and her friend had seen a strange creature but said that he saw nothing in the woods. He forbade anybody looking into the woods saying 'some things are better left alone'. Those who knew him said that his personality had changed and he would talk about the end of times that he said would occur in just six months' time. This kind of talk has been recorded in UFO witnesses and 'abductees'. At the same time the man was in the woods, other witnesses said they saw a craft shooting beams of light into the forest. Just what did that man see that scared him so much?

Stan Gordon was involved in what was surely the strangest Bigfoot case of all time.

On the 21st of October 1973 (a great year for weirdness) at around 9.pm, fifteen people on a farm near Greensburg, Pennsylvania, saw a large red ball of light descending to a pasture. Stephen (a pseudonym) a twenty-two-year-old man from the farm set out to investigate with two ten-year-old twin boys. Upon reaching the area, they saw a vast, dome-like object some one hundred feet across.

It was now glowing white rather than red and lighting up the area around it. The object made a noise like a lawn mower. Strange screams were also heard. Then one of the twins shouted a warning that something was moving along the edge of the field. By the light of the UFO the three could see two strange creatures approaching. They were ape-like creatures covered in long hair. One seemed to be eight feet tall and its companion a little smaller at seven feet. They had glowing green eyes and were making whining noises and sounds like a crying baby. One of the twins became so scared he ran back to the farm. Stephen let off a warning shot with his rifle, but the creatures continued to shamble towards him. He fired three rounds into the chest of the larger creature. It let out a whining sound and raised its hand towards the other creature. The glowing dome and its accompanying sound vanished, and the two Bigfoots lumbered off into the forest.

Stephen and the remaining twin ran home and phoned the state police. At 9.45 pm the state trooper who came out and Stephen returned to the pasture. Where the UFO had rested was an area of illumination bright enough to read a newspaper by. The farm cows and horses refused to enter the area. The men heard something big crashing about in the woods. Shining his torch into the woods, the trooper illuminated one of the beasts just ten feet away. Stephen shot at the beast hitting its chest and causing it to sway. Then it charged at the fence and both men fled.

The trooper returned to his barracks and phoned Stan Gordon who arrived at 1.30 in the morning together with other investigators from WCUFOSG. He found no unusual radiation readings and the glow had now vanished; the hard ground showed no marks in it. Suddenly Stephen began to shake violently and collapsed. Investigator George Lutz held him up but then he began to thrash about causing him and George to fall to the ground. He began to snarl and make wild, animal-like noises.

His own dog attacked him, but when he growled at it, the hound fled. Stephen ran on all fours around the field whilst continuing to snarl. Suddenly, he fell face first into a mass of manure. At the same time Stan and the other investigators began to have breathing problems and a sulphur-like smell permeated the area. Stephen seemed to wake up from his strange trance and was helped by the others back down the hill. Stephen seemed confused and shouted out phrases like "*Stay away from the corner*" and "*It's in the corner.*" He then began to rant about the end of the world and the doom of mankind, much like the man in the previous account. The investigators took him home and later Stephen said he could not recall making animal noises and running around but he recalled seeing a hooded figure with a scythe, looking much like the traditional image of the personification of death. It was this figure that told him about the impending doom. Stephan eventually received psychological counselling from Dr Berthold Eric Schwarz, a psychiatrist who had investigated both UFOs and paranormal phenomena. Dr Schwarz concluded that the witnesses including Stephen were telling the truth and that Stephen had been in a 'fugue' state, a disturbed state of consciousness where the victim cannot recall the actions they took whilst suffering the attack.

In later years Stephen was haunted by weird dreams. In one he was visited by a dead relative whom he had never met. Another was about a plane crash that was confirmed on the news a few days later to have happened.

Later Stephen told Stan and George that two weeks after the event he was visited by two men. One in uniform who claimed to be from the Air Force and another man in a suit. They asked him to describe the encounter in detail. They told him he was not crazy and took a collection of photographs out of a briefcase. The shots were of Bigfoot creatures, including one holding a pig under its arm as it stepped over a fence. They told him the pictures had been taken all over the country. The men said that they wanted to hypnotize him and Stephen agreed. After the session, they thanked him and said they would be back in touch. They left and he never heard from them again.

Stan stayed in touch with him and visited him on a regular basis until he passed away.

It later emerged that on the same night, the owners of a neighbouring farm also saw the strange object heading towards Stephen's family's farm at the same time his cows were making strange noises.

The weird encounters in the area continued into 1974. On February 6th in Uniontown, Pennsylvania, a woman heard noises at night on her porch. Suspecting the neighbours dogs, she picked up a shotgun thinking to scare them away. Instead she was confronted by a seven foot, hairy ape-like monster standing just six feet away. Thinking the ape was about to attack she raised the gun and hit the monster from point blank range. The hairy beast vanished in a flash of light.

As if this wasn't enough, the woman's son-in-law, who lived close by heard the shot and came to investigate. On the way he saw five, tall ape-like monsters with glowing red eyes standing on the edge of an area of woodland. As he watched, he saw the woods lit up by red flashing lights.

A case from Sandusky, Ohio seems to have Bigfoot and aliens tag teaming to terrorize the witness.

At about 1.30 in the afternoon of November 7th 1977, Milard Faber was walking through the woods near Monroeville on the west branch of the Huron River. Faber was collecting algae that he sold to local pet shops for aquariums. He stumbled across a massive creature crouching in the undergrowth. It was about eight feet tall when it stood up. Its muscular arms seemed long and the thing was covered in dark hair save for the leathery skinned face. The Bigfoot looked at him with glowing orange eyes then leapt into the river. Thoughtfully, it left behind a branch with a claw mark, massive tracks, an area of flattened grass, and a vile stench.

As the area was on a county line, Faber contacted the newspapers and sheriff departments of both Huron and Erie counties. Later he met deputies and reporters from both counties. The nasty smell lingered but Huron County's Captain Silcox said that the flattened grass was probably the work of pot smokers and went on to tear down the branch with claw marks. Suspecting a cover-up was in the making, Faber did not show the officials the monster's tracks. A small story was run in the *Sandusky Register.* The day after, six helicopters were seen in the area of the sighting. Two days later another four visited the place. Faber's friends claimed to have been chased away from the woods by both Huron and Erie Counties Sheriff's Departments when they tried to investigate for themselves.

Things got even weirder when Faber had gone to bed for the night. Becoming aware of something, he looked up and saw five weird entities glide through his bedroom door. They were about five feet tall, glowing pink, and had bulbous heads with luminous eyes. They wore what looked like skin-diving suits with belts and all seemed to have a stooped posture. Though they seemed to float without touching the floor, their legs still moved as if they were walking. This odd little detail is seen in some ghost reports. Four of the creatures sat themselves in chairs and the fifth approached his bed, standing about a foot from it. Faber said he felt hatred and rage emanating from the 'aliens'. He felt they were angry with him for stumbling upon the Sasquatch. He felt the entities wanted to take him back with them from wherever they came from. Faber switched on his lamp, but that had zero effect on the beings. However, when he began angrily shouting and swearing at the supposed extra-terrestrials they quickly rose and fled from the room, never to return.

Why would 'aliens' be interested in a person seeing a Bigfoot? What are the chances of somebody seeing both a relic hominin and then, just three days later, seeing five little pink spacemen? The latter seem to be a form of 'bedroom invader', a class of entity we will look at in a later chapter. Both Milard and his brother Robert returned to the sighting area several times and were beset by feelings of intense anxiety that seemed very localized. Again, this is a feature that turns up in other cases, an often very restricted area in which powerful feelings of fear are experienced.

Remaining in Ohio, a case from 1981 has some remarkable details in it. It was investigated by Don Worley, a prominent UFO researcher. The events centred around a farm in the northeast of the state, owned by a Vietnam war veteran called 'Ben'. Ben had seen a banana boat shaped craft hovering above his farm in early 1981. Then in May and June something began attacking his livestock and his ducks had their heads bitten off. One night he caught a huge beast in the beam of his torch. Measuring it against a winch he estimated the black haired, red eyed monster to be nine feet tall. His dogs seemed terrified of the thing. As the days went on Ben and his family saw weird dancing lights over nearby woods. The lights seemed to shine beams onto the ground. They also saw the red eyes of creatures moving in the woods. They also saw human like, shadowy forms that seemed to grow out of bluish white lights. Ben noted that his teenaged son Andy seemed to be a catalyst. When the light phenomena were at their peak, Andy would lapse into a deep sleep from which he could not be roused. When driven away from the farm to his relatives, Andy would awake with no memory of the events.

One time, Ben tried to photograph a Bigfoot, but when he examined the photograph, it didn't show a hulking hominin, but a glowing reddish orange ball of light trailing streaks behind it. Could the light have been somehow projecting the Bigfoot image into the mind of the witnesses? We must remember, though, that these things seemed to be physically real enough to kill livestock. Andy's trances seem very interesting. They resemble the trances mediums go into at seances and the strange mesmeric states into which some UFO and fairy witnesses slip.

In some cases, Bigfoot type creatures have been seen inside UFOs apparently operating the controls!

As far back as 1957, hairy monsters were reported as piloting flying saucers. Multiple witnesses at Wingdale, New York saw a strange craft with two tan coloured Bigfoots at the helm. Witnesses said they wore clear goggles and had bright blue eyes. One described them as looking like 'super intelligent orang-utans'.

UFOlogist Len Stringfield invested a case that occurred to a 'Mrs H' on October 21st 1973 (what is it with that year!) Mrs H lived in a trailer park in the Coverdale area of Western Cincinnati, Ohio. Mrs H woke up at 2.30am and went to get a drink of water. She was puzzled by an intense light shining through the curtains. Looking out, she saw a row of six silvery blue lights forming an arc about six feet from her window, She saw another light further away in the car park beyond her trailer, about thirty five feet away. The light seemed to be emanating from a dome shape like a huge, transparent umbrella. Inside this clear dome was an ape-like creature. The grey coloured creature seemed to be looking at a warehouse. It moved its arms stiffly, without bending them, and it reminded the witness of a robot. She could make out no facial features other than a downward sloping snout. The thing looked like it was operating a leaver of some kind that Mrs H could not see. She woke up her son Carl who also saw the strange sight.

Mrs H phoned the police, but the officer on the other end of the phone didn't believe her. Whilst she was talking, she heard a booming noise and looking up she saw the Fortean sceptical had vanished.

The following year in Frederic, Wisconsin, 69 year old dairy farmer William Bosack was driving home at 10.30pm. Beside the road he saw a disc shaped object the lower half of which was obscured by mist. Looking inside the illuminated, curved window of the object, Bosack could see a creature covered in tan hair save for its face. It had a flat nose and mouth, but the ears stuck out about three inches from its head and reminded the farmer of a calf. The creature's eyes were large and bulging and it had an expression of surprise on its face. Its arms were raised above its head. The witness could not see it's lower half as it was hidden by the mist. Bosack looked at it for ten seconds before accelerating past. As he did so, his car headlights dimmed and he heard a whooshing noise.

HAUNTED BY BIGFOOT

If Bigfoots in flying saucers is weird, then how about Bigfoots that turn up in people's houses?

In February of 1925, in Forest County, Mississippi, a family had a frightening encounter with something that entered their home. The father was away on a business trip, leaving his wife and three daughters aged five, seven and ten at home. During the night they heard their guard dog barking madly at something then retreating and seeming to go to run away. The family had wedged a chair against the front door but could hear something fumbling with the handle. Eventually that something pushed the chair aside and entered their house. The frightened family could hear its heavy breathing as it wandered around the house. Finally, it entered the bedroom. The family, paralysed with fear, saw it walked on two legs, was covered in long, course hair, and smelled of sulphur. The beast lay down across the bottom of the bed across the legs of the mother and children.

The thing's arms and legs reached down over the sides of the bed. It lay there for some time causing the legs of the family to have cramps due to its weight. Finally, it stood up and left. As it went, it left the door open and even though the night was cold, none of them dared to get up and close the door for the fear that the monster was awaiting them on the porch. The encounter was recounted by the youngest girl to the researcher Albert Rosales.

Wild animals do enter houses. Bears, tigers, leopards, crocodiles, deer, snakes, and kangaroos are among the species that have been known to wander into people's homes. Could this just have been a flesh and blood Bigfoot come in from the cold? Well, yes, but its behaviour is very weird. It ignores the family, does not try to steal food and simply lies across their legs. Then there is the odd sulphurous smell too. This has been associated with a number of Fortean phenomena.

A woman calling herself 'JJ' wrote to the Beyond Creepy YouTube channel with one such tale. She had moved into her new home in Oklahoma City in 2012 along with her boyfriend and a dog and cat. The house was not old, being built in 1982. It was a single level house with a vaulted ceiling backing onto wooded hills. The first two years were fine but then, after JJ had been away for a couple of weeks and returned, at 3.00am she woke up and went to the kitchen for a drink of water.

She was overcome by a creepy feeling and after her drink walked back to the bedroom. Glancing down the hall to the living room, she noticed that the light that usually came in was mostly blocked by something in the doorway. The figure seemed to stoop slightly and JJ could now see a shoulder and head that looked at her with dull red eyes. JJ said it looked like a 'man-gorilla' and seemed 'ghostly' but was solid, as it blocked the light. She was terrified but seemed to have a mental message 'go to bed'. She went to her bedroom, shut the door, and pulled a clothes hamper in front of it. Looking back she knew the action was absurd as it would have proved no hindrance to the thing if it had wanted to enter. She sat in bed watching the door for an hour, unable to move. Oddly, she did not wake her husband but sat as if in a trance; she finally fell asleep and awoke at about 10 am.

She heard somebody moving about the house and assumed it was her husband. On looking, she found that he wasn't in the house. JJ burnt sage around the house, a ritual supposed to drive away spirits. Later that day, whilst she was having a barbecue with her husband in the garden, her dog began franticly barking at something neither of them could see in a corner of the garden. Later JJ burned sage in the garden as well. Their cat hid all day. JJ left the house in 2016 and has never had any other experiences.

John Lockwell took a job as a night watchman at a brewery in the northwest of the USA. It was a large complex with several processing plants and a bar and shop. Soon Lockwell became uneasy in the job, thinking that there was someone watching him. One night after inspecting the plants, he returned to the bar area. As the automatic lights came on, he felt the feeling of being watched again.

Turning around, he saw a huge, muscular, humanoid figure covered in long hair. It was sanding inside the bar and seemed to have followed him through the doors. As he looked at it, the thing vanished. Running outside, Lockwell found no trace of the creature. When trying to discuss this with colleagues he found that nobody wanted to talk about it. Finally, one of them admitted that there was indeed something strange on the site but it seemed harmless.

About a week later, he was conducting another walk-through of the bar area. This had two sets of automatic doors forming an 'airlock' between them. He used his swipe card to open the first set and walked through into the 'airlock'. Staring into the glass, he was terrified to see the Bigfoot reflected in the window. The monster was standing right behind him and was in the 'airlock' with him. He could see its chest rise and fall as it stared down at him with an intense gaze. He tried to remain calm as he swiped his security card to open the outer doors. Walking quickly through, he spun round to see that the 'airlock' was empty and the beast had vanished. The incident scared Lockwell so badly he left his job.

The ASA had cases with a ghostly feel to them. One time, Willie Rosermann and Bruce Morgan, a science major, drove to Sycamore Flats with a tape recorder and mic. Bruce had the recorder on his lap and held the mic out of the open window. Suddenly, what

seemed to be a huge, ice cold and utterly invisible hand, wrapped itself around his. On playing back the tape later, they heard the footsteps of something approaching the car, heavy breathing and a clicking sound, as if something had touched the mic.
Another time, the ASA put a running tape recorder with a microphone, on the ground at Big Rock.

At the time all the team could hear was crickets chirping. When played back they could hear a sound like a generator or hydroelectric plant. It seemed to be coming from beneath the ground. Suddenly, the noise stopped and was replaced by a harsh metallic voice that seemed to say *Keep out! We don't want..."*and then the voice faded.

In March of 1975, Dick Milett wrote to the ASA from Cedar City, Utah, with a very spooky story.

> *"My wife and I moved into a mountain cabin 70 miles from Cedar, high in the mountains. Every night at about 11 o'clock, we hear a motor running, but it sounds far away and runs through the night. There is also a mine tunnel close by that has recently been dug further in.*
>
> *I entered it with a kerosene lamp, and when I came to the part that was the new diggings, the walls lit up like little diamonds. But the owner of the mine had not dug there for some time. He refuses to stay there alone anymore because of strange happenings. But he stayed there for years.*
>
> *One morning I heard heavy footsteps coming toward our cabin, then I saw a huge dark form move across the window. I grabbed my gun and threw a cartridge into the chamber as our door knob turned. Whatever was out there never entered. I slowly opened the door. There was nothing there and I heard no footsteps leaving. Our two dogs were just outside the door and never barked. What puzzles me is that there were no footprints of any kind and we have lots of snow here. My Indian friends tell me it was a Newputz, meaning ghost in their language and only a few will dare stay in the area overnight.*
>
> *The sound of the motor we hear runs on the weekends. If the sound was coming from a mining company, it would be shut down on weekends."*

These stories have the overtones of a haunting, as if one phenomena is bleeding into another. Cryptozoology has melded with UFOlogy in some of the above cases and now it melds in with parapsychology. Could all three phenomena be parts of a greater whole? The story also has parallels with the demons and bedroom invaders we shall be looking at in a later chapter, some of which are monkey or ape like in appearance.

MAN BEAST TERROR DOWN UNDER
Though the USA may have the most high strangeness accounts of big hairy man beasts,

it is not the only country to produce such stories.

Tony Healy interviewed a 17-year-old youth called 'Len' who, with some friends, had encountered a yowie, an Australian ape-man, near Cotter Dam, Australian Capital Territory, one night in 1990.

After hearing strange cries and footsteps in the scrub, the youths retreated to their campsite in the park below the dam. As they walked, their stalker overtook them, running faster than any human possibly could, and vanished into the shadows ahead. It reappeared under a street light, some 50 meters from them. It looked man shaped but covered in grey hair and seemed to move with a blurring speed. It emitted a weird moaning sound. The youths became very frightened and huddled together on the steps of a locked and deserted pub, clutching pen knives and literally crying with fear.

Len phoned his mum from a call box next to the pub, begging her to pick them up. The thing had vanished by the time she arrived, but the atmosphere of terror remained. As they drove back home they all got the feeling that the thing was somehow following them. Even Len's mother started to cry with fear. Len suffered nightmares for some time afterwards. This sounds very like the effect the Lough Fadda monster had on Georgina Carberry. We will encounter this loathsome feeling of pursuit in other cases as well.

Another witness, 'R', who encountered a yowie whilst parked on the Mount Kembla Road, New South Wales, gave a graphic description of The Nameless Dread. Having a 'bad feeling' he looked over his shoulder to see that his car was being approached by a man-like, hair covered beast with glowing red eyes.

> *"My immediate reaction was to flee. I had this feeling I was being hunted-an immense feeling of dread...like someone had drained all the life blood out of you...an awful, awe-inspiring feeling of like...all your sense come alive. It is very hard to explain."*

He lost no time in driving away as fast as he could.

In 1996, Dean Harrison was jogging one night near Ormeau, Queensland, when he began to get the feeling he was being followed by something that stayed hidden by the bush. The thing brought a horrid sensation with it.

> *"I just had an indescribable chill that ran from my head to my toes. The unfamiliar and hugely terrifying sensation just overtook my entire body...I simply could not understand why I suddenly felt so uneasy and so vulnerable as I did but I knew something was terribly wrong and it had something to do with whoever or whatever was behind me."*

Turning round he saw a seven foot, hulking silhouette, behind the trees. Dean felt paralysed and had to force himself to move. The creature pursued him with a blood chilling roar. He narrowly escaped its grasp as it lunged for him from the bush line and he dodged into the light of a street lamp.

Jerry and Sue O'Connor, whose property backs onto the forested wilderness of the Blue Mountains, have had many yowie sightings. They too speak of a weird effect the creatures seem to have. They describe a 'sickening electrical feeling' in the kidneys, that they liken to liquid nitrogen. It induces paralysis and can even waken them from sleep, it is so intense. They feel that the yowies somehow 'project' the feeling.

In one case, the yowie seems to be linked to a poltergeist case. Tony Healy and Paul Cropper recorded it in their excellent book *Australian Poltergeist*.

Farmer, George Nott, told historian Martin McAdoo, a weird tale about what happened when he moved his family to a remote, long abandoned homestead near Wilcannia, New South Wales. Soon after they moved in the family found huge man-like prints around their paddocks. The horses all seemed spooked. Then they began to hear something stomping around in the attic. The heavy steps caused dust to fall down on The Nott's bed. George took a look into the attic in the morning, but found it was empty. He thought they had a ghost in the house.

A couple of days later, Mrs Nott was putting their youngest child to bed, when the family heard her scream. Rushing into the bedroom they saw a big, hairy monster standing upright like a man, rush out of the door. It was covered in brown hair and had broad shoulders. They thought at first it was a gorilla, as at the time they had not heard of the yowie. Mrs Nott had been bending over the bed, when the creature came through the door and grabbed her by the neck. It tried to drag her away.

To get away from the thing, the family re-located to an out-station, but the monster followed them. Mrs Nott awoke one night to find it standing over the bed. Her screams awoke George and the monster fled. He chased it outside and could hear it bellowing and the sound of its crashing footsteps.

Another time the eldest daughter saw it lurking on the porch at three in the afternoon. George went to investigate, but then from inside the house, that was empty at the time, came banging on the inside of the kitchen window. Rushing back inside he found the salt cellars tossed to the floor. Poltergeist phenomena continued at the out-station with doors opening on their own and showers of pebbles hitting the roof, like heavy showers of rain. This is a phenomenon that occurs again and again in poltergeist outbreaks. You will recall the North American reports of Bigfoot turning up inside houses like a hairy ghost.

An ape-man in a haunted house? It is as much of a coincidence as ape-men seen with UFOS.

One of the downright weirdest yowie accounts was recorded by the Australian Yowie

Research group. It occurred in Inglaba State Forest, New South Wales. The witness was with his girlfriend Lea and a friend called Tom. Tom was teaching Lea how to drive and they were along a track in the forest. The witness and his girlfriend had walked their dogs there many times. The witness was sitting on the bonnet of the car, as his girlfriend drove it slowly up the track. The area was called Burnt Bridge road. The year was 1994 and the witness was 17 at the time.

Suddenly a huge animal emerged from the trees and lumbered across the road, glancing briefly at the car, but scaring the witness. It had long arms that it swung as it walked. It had deep set eyes and a brow ridge. The witness said it had 'intelligent looking eyes' and seemed to project an aura of power. The creature was muscular, with large hands and was covered in brown hair, with a copper or bronze tint. The yowie stood eight to nine-feet-tall. As they stopped the car, the beast crossed the road. It had frightened the witness so much that he ran to get his gun out of the boot of the car to shoot it. Tom and Lea jumped in the way, to prevent him shooting the yowie. Then things got really weird.

Tom says the animal just vanished, it didn't slip into the woods, it simply disappeared in front of his eyes. This happened instantly. However both Tom and Lea saw something different. Tom says the creature walked behind a tree stump and emerged as a man in a blue tracksuit, that jogged off up the road. Lea said it turned into a goanna (a generic name for monitor lizards in Australia), crawled up the tree stump and vanished into a hole. All three observed something different. Tom ran after the jogger that nobody else could see and shouted out to him. Afterwards all three argued about what they had seen in the last moments of the encounter. All saw the yowie initially but all diverged in their accounts at the end of the encounter.

Is this a case of the yowie having some kind of mental effect on the witnesses, that each interpreted it differently? Or maybe it was something that the human mind could not fully comprehend and each witness's brain processed what they were seeing into something they could understand. It is interesting that all saw the yowie as a yowie at first. Is this the true form of whatever the phenomena is? We have already heard of a Bigfoot transforming into a four legged beast in front of a couple in Ohio and we shall see other cases of strange transformations as we proceed. All of this seems to support the idea that these kinds of monsters and indeed other Fortean phenomena are malleable, or at the very least they can affect the human mind.

BIGFOOT IN BRITAN!?

Great Britain is a small country. Its land area is 80,823 square miles, about as large as the state of Oregon. It has a population of 68,369,309 as of November 2021. That's a lot of people on a small island. Most of the vast ancient forests were cut down hundreds of years ago. Most of our countryside is now farmland or pine plantation, neither conducive to wildlife. The largest forest is the Galloway Forest in Scotland with 297 square miles of coverage. The largest in England is the Kielder Forest at 235 square miles. These are tiny in comparison to the huge swathes of forest in the US and

Canada. The idea of a race of eight-foot-tall ape-men living unnoticed in Britain, is utterly absurd. Yet still there are reports of Bigfoot type entities here.

The Abbot's Bromley Horn Dance is the oldest folk dance still performed in the UK. It is performed on 'Wakes Monday', the first Monday after the first Sunday after September 4th. Its origins date back to at least the 13th century. A lease written then grants right to villagers for foraging and collecting wood, on the condition that they do the Horn Dance. So the dance must have been in existence before the lease was written. The dance is held in the Staffordshire village of Abbot's Bromley and consists of six men with harnesses bearing the antlers of reindeer, villagers playing Robin Hood and Maid Marion, a hobby horse, a fool and a musician. The antlers themselves have been dated to the 11th century. Interesting, but you may ask 'what has this got to do with Bigfoot sightings in the UK?'

Mick Dodds ran a bric-a-brac shop in Lichfield. In 1986 he decided to drive his family to Abbot's Bromley to watch the ancient ritual of the Horn Dance. At 10.45pm the family was driving home.

As they passed the 11th century Chartley Castle, Dodds had to slam on the brakes, as a huge stag walked out in front of his car. As the creature crossed the road, Mrs Dodds screamed in horror as what looked like a chimpanzee lurched out of the darkness and followed the stag. Halfway across the road the creature turned to stare at the family, before charging at the car. At the last moment it veered off to the left and vanished, taking the same route as the stag.

The creature sounds remarkable, like one seen over a century before, also in Staffordshire. It is recorded in Charlotte Sophia Brune's book *Shropshire Folklore*. The events in question happened on the night of 21st of January, 1879. A man was carrying luggage by horse and cart from Ranton in Staffordshire, to Woodstock in Shropshire. It was about ten at night, when his route led him across the Birmingham to Liverpool Canal, at bridge 39. Just before he reached the bridge, a black creature leapt from the plantation beside the road and onto his horse. It resembled an ape and had glowing white eyes. The horse panicked and broke into a run. The man lashed at the creature with his horse whip, but it passed clean through the beast as if it were a ghost. The horse galloped on, with the thing still clinging to it, until it eventually leapt off.

The man was so scared that he took to his bed 'prostrated' with fear for several days. A few days later the man's master was visited by a policeman, enquiring about the 'robbery' on the bridge. The master explained that there had been no robbery, but his man had been badly scared by a strange creature. The policeman seemed unsurprised and answered:

"Oh, was that all sir? Oh I know what that was. That was the man monkey, sir, as does come again ever since a man was drowned in the cut."

Cut is a slang term for canal in Britain. It is interesting to note, the ghost-like nature of the man monkey. The witness's horse whip passes through it, as if it has no substance, yet it can hang onto a galloping horse's back and frighten the animal. The policeman seems to indicate that the entity is some sort of ghost, linking it to a drowned man. In a later chapter we will look at cases of Bigfoot like beasts manifesting at seances.

One witness that I interviewed, was the owner of an articulated-lorry. He had worked for The Forestry Commission, transporting logs. He was sleeping in the cabin of his lorry one night in November. The location was Fixton Woods, in Sussex. At around 2.30am, he went outside to relieve himself.

Close to his vehicle was a large piece of equipment used for loading logs onto lorries. The machine had a large red light, so the drivers could avoid backing into it at night. The witness noticed a huge figure standing next to the machine, apparently fascinated by the light (attraction to electrical equipment again). The man, a former solider, was trained to estimate size accurately. He reckoned that the creature was eight-feet tall. He returned to his cabin to get a torch. When he shone the beam on the thing, it turned and ran into the forest. He described it as black and ape-like.

Fortean researcher and writer, Nick Redfern, has interviewed several witnesses to the British Bigfoot. One such was Bob Caroll. Bob was a former lorry driver for a paint manufacturing company. It was the early hours of January or February of 1972/73. Bob was due to deliver a pallet of paint a 6.00am to a depot in Leicester. As he approached bridge 39, he slowed down and a powerfully built, upright creature, covered in bluish-black hair came out of the trees, bounded across the road and ran down towards the canal. The thing stood about five feet tall. Bob pulled over and showing some guts, ran back to where the creature had crossed the road. He looked under the canal bridge but it was too dark to see anything. Then he heard a strange noise.

> *"I would say it was like a baby crying, but it sounded louder, like it was evil or not right: and I got a funny feeling hearing it."*

A noise like a baby crying accompanies many Fortean phenomena, including Bigfoot sightings and ghosts. Bob's quote also makes it sound like he was beginning to feel the effects of the nameless dread.

Another witness, Peggy Baker, was driving with her daughter Kathleen near the village of Ranton in 1997. It was 11.30p.m. when a shambling ape-man lumbered out of the trees and raising its hands, roared at them.

Cannock Chase is a large area of woodland, or rather, large for the UK in Staffordshire. It seems to be another hotspot for weird encounters. Nick Redfern interviewed a man named 'Gavin' who, in 1997 had driven to the Chase in order to get romantic with his girlfriend. They had parked near a large rock called Glacial Boulder. Suddenly Gavin's girlfriend screamed. Standing atop the boulder was a huge, hairy man-like thing waving its arms at the night sky. Gavin jumped back into the front seat and started the car. As he put his foot on the accelerator, the creature leapt onto the bonnet of his car. He claims the thing held on for five minutes, before being thrown to the ground and bounding back into the forest. Though the amount of time it clung to the car may have been overestimated, the encounter sounds like a modern updating of the 1879 case. Indeed Nick heard several more cases of big dark shapes lunging at cars in the area. Fortean phenomena from UFOs to monsters seem to approach courting couples in cars. In some areas of the USA they have even been dubbed 'lovers lane monsters'.

Bolam Lake is a small country park in Northumberland, in the northeast of England. It has a lake surrounded by woodlands and is about 800 meters across. Back in 2002 there was a spate of sightings of an ape-like creature. In January, 2003, the Centre for Fortean Zoology undertook an expedition to investigate these strange sightings. We teamed up with Fortean investigator Mike Hallowell and his now defunct Twilight Worlds group.

Oddly, our electronic equipment started to play up at the location. Camera, mobile phones, tape recorders and laptop batteries seemed to fail, despite having been fully charged.

The first witnesses we spoke to were a woman called Naomi and her fourteen-year-old son. She told us how, after hearing the stories of 'bigfoot' at Bolam Lake, they had driven over to investigate. When she parked her car, Naomi and her son hear a strange, eerie call from the woods. They got out of the car and saw a tall, hairy creature standing in the trees, about a hundred feet away. Though the creature was standing totally still, both mother and son felt that it was somehow rushing towards them. They were both overcome with fear. Jumping back into the car, they drove off as fast as they could. Naomi said that she had the horrible feeling that the figure was somehow chasing the car, even though she could see nothing in the car mirror. You will recall this effect from other encounters with strange creatures.

Another witness, Neil, had been fishing with two friends one night. As they were walking back to their cars, Neil stopped to tie his shoelace and the other two men looked back to see why he had paused. To their horror they saw a huge, dark figure with white glowing eyes, standing right behind Neil. Neil said it appeared to be bent over looking at him. He could see no features in the tall, dark, man shaped thing other than the glowing eyes. All the men fled back to their cars and drove away.

Splitting up into several groups, the team patrolled the park in the evening. The group I was in saw nothing odd whatsoever. However CFZ director Jon Downes, and several of the Twilight Worlds group saw something very strange indeed.

At about five in the evening, Jon Downes, saw a black, tall, man-shaped thing run through the woods, right to left at high speed and then run back again, before vanishing. Jon said it looked to be about seven feet tall and muscular. It was a black silhouette that seemed two dimensional and darker than the night around it. He described it as looking like a '*man shaped hole in reality'*.

Another man we spoke to, who lived in the suburbs of Newcastle, told us that whilst walking his dog at the park, he saw a huge, man-shaped figure standing next to a tree. The thing scared the witness so badly that he never went back to the park.

Obviously a race of 8 foot ape-men cannot live undetected in a small country park. The sightings sound more like a ghost than an animal. As far as I know the strange sighting stopped as quickly as they started, leaving only a mystery in its wake.

Like their US counterparts, these UK Bigfoots are also seen in tandem with UFOs from time to time.

In her book *Mind Monsters,* UFO researcher, Jenny Randles, records such a case. It took place in Wollaton, near Nottingham, a place we shall visit again later in the book. Frank Earp was a 15-year-old in 1966, when he and two friends went searching for UFOs near a canal. At dusk a misty vapour rose up from the water and formed into a glowing doughnut shaped cloud, about the size of a dodgem car, and began to float towards the boys. They became afraid and walked quickly away. As they approached their homes, Frank looked back and was horrified to see a creature standing behind them where the cloud once was. It was six feet tall and covered in hair. In its hands it clutched pencil shaped, glowing, red objects. The thing's legs seemed to fade away into nothingness like a ghost. Frank and one of the other boys ran off in terror. The third boy, who had been close enough to touch the thing, shouted after them, confused at their fear as he could see nothing at all after the glowing cloud!

This is an interesting case for several reasons. Firstly only two of the three boys could see the creature. We have seen this in other Fortean phenomena, like ghosts and UFOs. Indeed, this case seems to combine elements of the ghostly, the UFOlogical and cryptozoology. The boys, who were UFO hunting at the time, saw a glowing object that then became a hairy monster, that in itself seemed semi-corporeal like a ghost. This case seems to involve something that shifts from one form to another, encompassing three classic Fortean tropes.

Jenny records a second British Bigfoot/UFO case. In April of 1977, Robert Holmes and Sally Jensen, a couple of courting teenagers, were parked up in Myers Lane, Worrall near Sheffield, Yorkshire. The area is close to a golf course and a city dump (who says romance is dead?). Sally heard a noise like the crackling of somebody walking on leaves. Looking round they saw a light approaching. At first they thought it was another car wanting to pass them on the narrow road. But as it drew closer they saw it was an orange glowing dome. In front of it strode a huge, hairy creature that seemed to be surrounded by a white haze. They likened the thing to the Bigfoot that had turned up in an episode of *The Six Million Dollar Man*. The creature was 10-feet-tall. The car radio was full of interference.

Panicking, they drove off at 70 miles per hour, but the UFO and its hairy owner seemed to keep pace with them, never getting any further away. They passed another car on the way. Looking back they could see the headlights of the car between them and their weird pursuers. They thought that the other car had seen the monster and were reversing to get away from it. Robert and Sally reached the end of the lane, but did not see the other car emerge and neither did the Bigfoot or UFO. Robert drove Sally home. They were both scared and later reported the events to a local newspaper, who declined to print the story. They were later interviewed by UFO researcher, Nigel Watson, who found them to be honest and pleasant.

In 2017, user 'Min Bannister' wrote of a strange sighting she had, on the Forteana Forums website. She saw a huge, black figure from a train, as it passed some woods.

> *"A few months ago my husband and I went to Cardiff on holiday. We went by train and of course spent a fair amount of time looking out of the window. At some point, somewhere in England between Scotland and Wales (annoyingly I can't remember where) the train left a station and soon passed a wood with a field next to it. I saw something. I saw something weird. I only caught a glimpse as the train passed by. After a short space of time suitable between two people who don't quite know whether to say anything to each other or not, my husband looked at me. His expression suggested he had seen something weird but wasn't sure how to express it. I asked him if he had seen something weird and he said he had. I had seen a humanoid figure at the edge of the wood. I got the impression it was facing away from me but I am not sure how as it was entirely black. Not like someone wearing black clothes but like a total absence of that figure in the landscape. And it was freakishly large. I can't really say how big but just too big.*
>
> *The really annoying part is – my husband now has absolutely no recollection of having seen this!*
>
> *The blackness puts me in mind of the black stick men but mine wasn't very stick like and in fact pretty broad. Could the UK version of Bigfoot be the same or a similar phenomenon to the BSM?*
>
> *Most of my Fortean experiences have been either really cool or interesting without being scary but this has rather disturbed me."*

There seems to be two distinct kinds of Bigfoot. The first seems to live in remote, forested mountain areas, where it tries to avoid man and moves around at night. It seems to be some unknown form of hominin. The second appears in areas that could not possibly support them. They generate extreme fear and appear alongside other Fortean phenomena, such as UFOs, out of place big cats and hauntings. This second kind seems to have a more flexible physicality and a less tangible form. They seem to be a whole different kind of 'real' to the flesh and blood Bigfoot.

OH WHAT BIG TEETH YOU HAVE

All over the world there are legends of humans transforming into wild animals, usually large, powerful, predatory ones. There are stories of men transforming into crocodiles, bears, tigers, lions, leopards, snakes and wild boar. But the most famous is the werewolf. This beast seems to have gripped the imagination of story tellers and film makers. The first such movie was *The Werewolf* back in 1913, which is now lost, but it was followed by a legion of movies. Indeed, the werewolf seems to be the monster that most films have been made about, challenged only by vampires.

But the idea of a man transforming into a beast, must be one of the most outlandish in legend. Nevertheless, people claim to have seen wolfmen in reality.

THE CASE OF THE HEXHAM HEADS

Nationwide was a British current affairs show that ran from 1969 to 1983. It was very popular and is still fondly remembered. In 1976 it featured an item on modern day sightings of a wolfman in the small town of Hexham, in Northumberland. The idea of a national news show tackling such a topic today may seem strange, but back in the 1970s such things were not unknown. Indeed, I can recall seeing this program as a six-year-old. My friend, Dave Archer, who accompanied me on many of my cryptozoological expeditions, also watched this item and recalls being very badly frightened by it.

The story began in 1971, when two brothers, Colin and Leslie Robson, dug up two stone heads in their garden. The heads, which were about the size of satsumas, resembled a skull and a witch. After they were taken indoors, the Robson family noticed that the heads seemed to move about, without anybody touching them. Objects like bottles were flung around by unseen hands. The family next door, the Dodds also reported poltergeist like phenomena, with one of the children's hair being pulled at night. Mrs Dodds who had handled the heads, said she saw a creature like a werewolf walking past her open bedroom door at night. Her screams awoke her husband, who ran out of his room in time to hear something padding down the stairs and out of the front door.

The heads were finally sent to Dr Anne Ross, an expert in Celtic mythology, at Southampton University. She believed them to be ancient and connected to the 'Celtic head cult'. Anne Ross had asked Professor Hodgson, at Southampton University, to look at them and he concluded *"both heads are made from the same material; a very coarse sandstone with rounded quartz grains."*

But another expert she showed them to had totally different conclusions. Dr Douglas Robson of Newcastle University concluded *"The material from which the heads have been formed is an artificial cement, and the material is unlike any natural sandstone."*

Dr Ross took the heads from the University to her own home. As it turns out, this was a mistake. Dr Ross soon had her own encounter with the wolfman. She woke in panic one night to see:

> *"this... thing... going out of the door... It was about six feet high, slightly stooping, and was black against the white door. It was half-animal and half-man. The upper part, I would have said, was a wolf, and the lower part was human... It was covered with a very dark fur. It went out and I just saw it clearly, and then it disappeared. Something made me run after it, a thing that I wouldn't normally have done, but I felt compelled to run after it. I got out of bed and ran, and I could hear it going down the stairs. Then it vanished towards the back of the house."*

A few days after this, her daughter, Bernice, returned home from school to see the wolfman on the stairs. It leap over the banister and padded into a corridor, the girl felt compelled to follow it as it entered the hall and vanished. The disturbances ceased as soon as the heads were removed from the house.

In 1977, the heads passed into the care of Don Robins, a chemist and author of the book *The Secret Language of Stone*. He became convinced that the Hexham Heads could help prove his stone-tape theory (that stone could record events and human emotions and then could play them back. This theory is used as an explanation for some hauntings, but it was actually invented by the science fiction writer Nigel Kneale of *Quatermass* fame. We will look at this theory more closely in a later chapter.) Robins had no real paranormal experiences with the heads and after his dog chewed one of them in 1978, he passed them on to the dowser Frank Hyde, who wanted to conduct dowsing experiments with them. And there the trail goes cold. Nobody knows the current whereabouts of these strange items.

To top it all, a man called Desmond Craigie, reported that he was the creator of the heads. He had lived in the house before the Robson's. He claimed to have made the heads for his children in 1956 and they were supposed to be Hitler and Mussolini! Craigie worked for a concrete company at the time. His attempts to recreate them were not convincing.

Maybe the heads are locked away in a draw or attic, long forgotten but awaiting rediscovery.

But the haunting horror unleashed by the Hexham Heads is not the only wolfman to have stalked the British countryside.

OLD STINKER AND OTHERS
Barnstorm Drain was a water channel built in 1798, to dry out salt marshes around the city of Hull, in Yorkshire. In 2016 there was a spate of sightings of a werewolf dubbed

'Old Stinker'. The name, apparently, was first given to a creature that haunted the area in the 10th century, feeding on human corpses. Its foul breath gave it its charming moniker.

One woman who saw it said:

> *"It was stood upright one moment. The next it was down on all fours running like a dog. I was terrified. It bounded along on all fours, then stopped and reared up on to its back legs, before running down the embankment towards the water. It vaulted 30ft over to the other side and vanished up the embankment and over a wall into some allotments."*

Other witnesses, a couple, saw the beast eating an Alsatian dog. It bounded over an 8-foot-fence, still holding its prey in its jaws and vanished into the night.

A bipedal, grey furred, wolf-like beast was reported by a woman delivering pizzas in her car.

Local councillor, Steve Wilson, kept a log of sightings and reported them to the town council. After a citizen named Wayne Carr filed a freedom of information request, to inquire if the city council had prepared any written policy, or performed any risk assessments for the purpose of dealing with the werewolf effectively, he received a formal response through the appropriate channels, the council didn't elaborate further.

My good friend, the Reverend Peter Laws, a vicar, horror novel writer, horror movie buff, and reviewer for *Fortean Times*, bravely set out to search for the beast, like the hero of a Hammer horror movie of the 1960s. Using a large piece of steak for bait, he spent a night in the monster's haunt, but failed to see the beast.

Some sixty miles north of Hull lies the small village of Bempton. Paul Sinclair, Fortean investigator, has been recording strange happenings in the area and the surrounding countryside. On February 7th, 2019, 'Rob' an ex-paratrooper, told him that he had been wild camping with a friend. One night they had walked down to Bempton Cliffs to an RSPB reserve. As they approached, they saw a crouching animal about sixty feet away. At first they thought it was a hyena, until it stood up as they trained their torches on it. They said the thing had long arms and human like hands. The creature stood some seven feet tall. It's hair was silvery brown. Rob and his friend backed away and returned home. Rob was so scared that he told Paul he would only return to the area with a group of men.

CULT OF THE MOON BEAST

The strangest werewolf story in the UK, was unearthed by Nick Redfern, whilst researching the Man Monkey of Ranton. Nick was, at the time, summer of 2000, co-writing a column on the paranormal with Irene Bott, the then president of the now defunct Staffordshire UFO Group. The column appeared in a local paper the *Chase*

Post.

After writing a piece on the Man-Monkey, he appealed for witnesses. Around the same time Nick was a guest on a number of local radio chat shows, speaking about the strange sightings in Cannock. Nick was contacted by a man named Rob Lea, who claimed to have knowledge that tied in with the Man-Monkey case. Nick agreed to meet him in a pub, in the village of Milford.

Rob, as it turned out, was, like me, a goth. Over drinks Rob explained his research. As it turned out his family owned a farm in Newport. Early one morning, in August of 1989, Rob's father had found five of his sheep slaughtered in a strange manner. The sheep all had their throats slit and were arranged in a circle. Their internal organs had been removed and piled up in the centre of the circle. Rob's father had instantly thought of devil worship and phoned the police. As it turned out, the local constabulary didn't do a whole lot other than advice the family not to speak too much about the event. Rob had brought a briefcase with him that had notes and photographs from his research, including pictures of the slaughtered sheep. Rob had travelled around the UK researching animal mutilations. During the course of his investigations he stumbled upon an occult group based in Bristol, in the southwest of England. Rob dubbed them 'The Cult of the Moon-Beast', though as far as he knew, the group itself had no name. The group sacrificed animals in order to call up strange creatures, including things that resembled wolfmen, that were then sent out to perform their master's will.

Rob produced an audio tape interview with an elderly man from Newport, called Sam. Sam said, that as a child he had heard of a spate of sightings of werewolves in the 1910s, around the same bridge on the Shropshire Union Canal where the Man-Monkey had been seen.

Rob had discovered that the group numbered about fifteen and had members across England and as far north as Scotland but their base was Bristol. Rob claimed to have seen them performing rights near the Ingrestre Golf Club in Cannock in 2000.

According to Rob, the Man-Monkey and the werewolves had the same point of genesis. The Cult of the Moon Beast could summon creatures that looked like giant serpents, werewolves, apes, huge bird-like creatures, huge phantom hounds and panther-like big cats. The creatures, he said, came from another dimension that coexists with our own and can intersect with it in certain areas. The rituals were performed at these areas. In their true form the entities looked like balls of light, but took on their various monster forms once they had 'come through to our world'.

And the point of this summoning was murder. Rob asserted that the cult charged huge amounts of money to use the creatures to kill people. The beasts rarely attacked physically, but mostly frightened their victims to death. If somebody wanted somebody dead, the Cult of the Moon Beast would send out a monster to scare them to death, looking like the victim died of a heart attack. This, he explained, was cleaner than a knife or bullet.

The whole story is as hard to swallow as a bowling ball. It is tempting to think that Rob had been watching too many episodes of *The X-Files,* which was at its peak at the time. The fact that the phone number Rob had given to Nick for future contact turned out to be bogus, does not inspire confidence.

However there are a few things that make me pause from total rejection of the story. Firstly, the description of the types of monsters raised by The Cult of the Moon Beast closely resemble the creature archetypes from my own 'global monster template' theory, something that will be explained in the final chapter of this book. Secondly, the idea that these creatures took the form of balls of light. We will see balls of light appearing alongside all kinds of Fortean phenomena during the course of this book, even cases of monsters and other entities changing into balls of light. Thirdly, what did seem to be ritually slaughtered animals, with innards removed and piled up, were discovered around the Cannock area, some had candles and other signs of ritual magick found along with them.

Fourthly, there are other cases of alleged occult activity involving monsters.

We have already covered Alistair Crowley's rituals at Loch Ness and the cult of the sea dragon around Marsden Bay, but another weird cult story comes from the other side of the world, in the Russian Far East. Russian explorer, Alexander Remple, was told a fantastical story by Vladimir Semyonovich Kuzetsov, a seasoned hunter, having spent many a day wandering the inhospitable Russian taiga in search of game. He was seventy-one in 1991, when Alexander Remple came to interview him about a bizarre experience he had had 'some years before the Second World War'. He was pushing through the wilderness one day, when he heard strange singing coming from a nearby clearing, in which he could also make out a blazing bonfire. Curious, he quietly approached the clearing and was intrigued (and doubtless frightened) to see a semicircle of people congregated around the fire singing in a strange language. In the dusky light of the setting sun, he could see the cultists performing 'incomprehensible gestures' with their hands before all starting to bow down. Kuzetsov was horror-stricken as he watched a huge shape come crawling out of the forest, from the direction of the sunset.

As he stared at the bizarre scene, he slowly came to the realisation that the shape was an enormous black snake. He estimated its length as being approximately 10 meters (33ft). He said that he thought he could see small front legs on the creature's body, but he added that he was unsure of this. The cultists all raised their voices in a 'guttural chant' and Kuzetsov felt that he had seen enough - panicking and madly fleeing back through the trees without being able to see the trail. He had no concept of how long he was running for, but when he finally felt safe enough to stop, his face and hands were covered in scratches.

Finally, there was the case of a werewolf reported from the very area.

Simon had been studying for a zoology degree back in 1982. He had taken a fellow student out bird watching, close to the bridge on the Shropshire Union Canal, where the Man-Monkey was reported. It was broad daylight and a fine day. Suddenly the birds in the area took flight, chirping in panic, as an awful scream enveloped the area. The pair saw a hair covered figure emerge from the bushes on the opposite side of the canal. It was about five and a half feet tall. The face and hands were free of hair and the body was highly muscular. The creature's head, seen side on, had a long, dog-like muzzle and pointed ears. Simon described it as a werewolf. The wolfman bounded back into the bushes. The female student that Simon had taken with him, was screaming hysterically at the sight of the wolfman. They both beat a hasty retreat.

Germany too has its werewolf stories. In 1988, there was a werewolf sighting at the Morbach munitions base, near the village of Wittlich, in Germany. Wittlich has a long werewolf history. As the supposed location of the last werewolf killed in Germany, there's a shrine just outside the village with an eternal flame. According to local legend, if the flame goes out, the werewolf will return. One night in 1988, a group of policemen were on their way to the munitions base, noticed the candle was out, made a few jokes and returned to their posts. Later that night, alarms started going off at the base's perimeter fence. Investigating, several men saw a huge animal standing on its hind legs. *"The creature that we saw was definitely an animal and definitely dog or wolf-like. It was about seven to eight feet tall, and jumped a twelve-foot security fence after taking three long leaping steps,"* one of the witnesses stated. A guard dog was brought to the scene of the sighting to track the animal, but it *"went nuts"* and wouldn't peruse the wolfman.

WOLFMEN OVER AMERICA
Back across the Atlantic, lupine horrors have been stalking the backwoods and lonely roads for many years.

St. Coletta School for Exceptional Children, in Jefferson, Wisconsin, was a hospital for the mentally disabled. Long since abandoned, its most famous inmate was Rosemary Kennedy, sister of former US President, John F. Kennedy. In 1941, Rosemary's father, Joseph P. Kennedy, had her lobotomized at the age of 23, for nothing more than developing mood swings and a somewhat tearaway personality. Indeed modern experts think she may have been suffering from depression. This young woman, who had been presented to King George VI and Queen Elizabeth and who had attended the coronation of Pope Pius XII, was reduced to the mental age of a two year old, incontinent and unable to talk coherently. She remained in care till her death in 2005.

In 1936, Mark Shackleman, a husband and father in his thirties, was working as a night-watchman at St. Coletta School. The grounds covered fields, orchards and a Native American burial mound, as well as the asylum. During his rounds, Mr Shackleman saw something that seemed to be digging into the mound. The crouching thing bounded away as the watchman approached. He found claw marks in the earth when he examined the mound.

The next night Mark got a better look at the creature in his torch light. This time he saw it stand up on two legs. As it turned to face him, he saw it was about six feet tall and covered in black hair. The beast had a vile stench about it, like rotting meat. He noticed that the thumbs and little fingers on its hands seemed shrivelled. The muscular creature has a long, wolf-like muzzle and pointed ears. It made eye contact with the witness and spoke what seemed to be a three syllable word 'gad-arr-ah with emphasis on the second syllable. Its snarling voice sounded half human. Mark was convinced the creature could easily kill him and despite having been a heavyweight boxer, became very afraid.

He silently prayed to God to save him. The wolfman seemed to sneer at him and walked away from him, leaving behind a foul smell that lingered for some time.

For years Mark kept the story to himself, but finally told his mother and in later years his son Joe, who related it to journalist and author, Linda S. Godfrey.

THE BEAST OF BRAY ROAD

It was in the same state that America's best known werewolf emerged and the case was broken by Linda Godfrey, who was writing for a newspaper called *The Week* in Elkhorn, Wisconsin. Linda was contacted by a freelancer who had a story. She didn't want to write it up herself, as she knew some of the people involved, so she passed it on to Linda. As it turned out the story was about werewolf sightings along a stretch of rural highway, sparsely scattered with farmhouses, known as Bray Road. It is lined with woodlands, cornfields and marshy meadows.

The first witness she spoke to was Lorri Endrizzi. Lorri was driving home from work one night via Bray Road. It was an autumn morning in 1989. At about 1.30am Lorri noticed a strange creature that seemed to be kneeling in a ditch beside the road. The brownish-grey creature had pointed ears and a dog-like face. It was clutching some road kill in its claws and as Lorri drove past it, turned its head to look at her, its eyes reflecting in the headlights. It was about the size of an average man. She said it was so man-like, that she was convinced she had seen something supernatural and somehow connected to cult activity.

Trying to find out what she had seen, Lorri later looked through books in her library. She came across an image of a werewolf in Jane Wermer and Sol Chaneles' *The Golden Book of the Mysterious*' published in 1976.

She contacted Jon Fredrickson, animal control officer for Walworth County, in the hope he could tell her what the creature was. Whilst she was talking to Fredrickson in his office, several large books seemed to leap off a shelf and crashed to the ground, as if pushed. Nobody was near the shelf which had bookends. Poltergeist activity in a werewolf case?

Linda visited Fredrickson herself and found to her surprise, he had a large file on sightings of weird creatures around the area, labelled 'werewolf'.

The next witness Linda interviewed was Doris Gipson who had, fittingly, seen the beast on October 31st, 1991. The night had been foggy and Doris hit something whilst driving. She stopped the car thinking she had hit an animal. As she got out of the car she saw a huge, upright, dog-like animal in the fog. It had a muscular chest and long brown hair. The creature was charging towards her and she could hear its feet pounding the ground. Doris barely made it back into the car, as the beast grabbed the boot. She stepped on the gas and drove quickly away. She later found claw marks on the boot of her car.

After Linda's article was published in *The Week,* more people contacted her with stories of sightings. Linda ended up writing a book on the subject *The Beast of Bray Road Trailing Wisconsin's Werewolf.* It was a tremendously good book and Linda S. Godfrey went on to write many other books on werewolves and other monster sightings across the USA. All are well researched and written and several accounts in this book are taken from Linda's work.

One of the most dramatic sightings, occurred to Steve Krueger, a Department of Natural Resources worker. Steve was called in to remove the carcass of an 85lb doe from a road in Holy Hill Wisconsin one night, in November 2006. As Krueger sat in the cab of his truck filling out the paperwork for the removal, he felt his truck shake. Thinking it was simply the wind, he ignored it. A second, harder, shake caught his attention and he looked out the back window of his truck, to see a shadowy figure standing at the tailgate of his truck. Krueger shone his torch through the back window to get a better look and saw a seven foot tall animal, with a wolf-like face, reaching into the bed for the deer carcass he had just removed from the road. The beast had a wolf-like head, but a broad chest and muscular arms. It reached for the carcass with clawed hands and hauled the carcass off the truck, Terrified Steve drove away leaving the monster to its supper.

Some have argued that the Beast of Bray Road is nothing more than a bear with mange. When a mammal loses its hair, it can look radically different. Hairless bears do indeed resemble the Hollywood idea of a werewolf. However bears do not run whilst standing erect. Also bears with mange are usually sickly creatures, unable to run at speed. The next case is clearly not a bear, or any other natural creature and it seems to show the entity in a state of transformation, from one form to another. Oddly, this case occurred in an urban setting, the city of Madison, Wisconsin, in May of 2004. Linda Godfrey was approached by a man who worked in a bookshop, who had read *The Beast of Bray Road.* The man 'Preston' claimed that he had been walking back from his ex-girlfriend's house just a week before, at 1.30 in the morning. He saw what he thought was a man walking on all fours. As it drew closer it seemed to change into a huge dog. As Preston watched the 'dog' began to flail its legs wildly, in a way that he described as reminding him of someone breakdancing. Then the thing stopped and turned its head to look at him. It now had a face like a gorilla! Preston ran back to his ex's house and banged on the door until she let him back in.

It sounds as if the entity was changing, from a wolf like form into a Bigfoot like one. Could the two creatures in fact be one kind of creature, that is capable of changing its shape?

The sightings in Wisconsin continue, sporadically to this day.

WEREWOLVES ELSEWERE

Other parts of the States, werewolves have been reported. Indeed it seems that the USA is wolfman central. Trying to cover them all is beyond the scope of this book, so we will look at a few of the stranger or more dramatic reports.

Like other Fortean phenomena such as Bigfoot, road ghosts and UFOS, werewolves seem to have an unhealthy interest in cars. A man wrote to Linda about his co-worker's experience in Livermore, California, in 1990. She was driving home from a late night shift in foggy conditions. She stopped the car at a four way crossing. Suddenly, a creature leapt from the fog and landed on the bonnet of her car, glaring in at her. It gripped the bonnet with hands, but its head and face was that of a wolf. Its fur was grey and the car shook as the beast leapt onto it. The witness floored the accelerator and the creature was thrown off the bonnet. The Witness was so scared that she never drove along that rural road again.

Another witness 'Jill' had an encounter in central Illinois. She was living in a small town between Pretoria and Bloomington. It was 10.30 at night in the autumn of 1983 or 1984. Jill was driving to meet up with some friends in Pekin, along a rural road. The moonlight on the clear night lit up a pasture about ten feet below the level of the road. Jill saw what she thought was a big dog sitting in the field. Suddenly it jerked its head towards the car and stood up, some seven feet tall on two legs. Jill remembers that the legs looked more canine than human and the thing had pointed ears on its head. It charged aggressively at the car and the witness felt an overwhelming sense that the beast was enraged and meant her harm. She drove away as fast as she could and was badly frightened by the encounter. Even twenty years later driving along that stretch of road in broad daylight made her nervous.

Susan, a retired woman from northern Michigan, wrote to Linda with a very strange story that occurred in 2013. She and her husband were driving home from an auction in northern Indiana. They were heading north on State Highway 66, about ten miles south of Sturgis, Michigan. They saw a large, fast-moving creature lit by the headlights of an oncoming car as it rushed across the road. When they reached the area where the beast had crossed, their headlights lit a wolf-like figure, standing at the edge of a cornfield. Susan recalled it had ears like an Alsatian dog, but the head seemed to be wider. Its eyes glowed an orange/red and seemed to be actively glowing by themselves and not reflecting light. And here the story gets very strange. In fact they eyes seemed to radiate so much inner light they blurred together into one bar of light as it turned to look at them.

> *"I swear I got such a strong impression of this thing letting me know it was superior. I felt it telling me I would be prey if I were outside with it, and if I thought I was safe in the dark, I was wrong. Yes I know how crazy that sounds and I thought, get a grip! But the message just seemed to slam into my mind, and I didn't forget it."*

This is far from the only case of witnesses to Fortean phenomena getting telepathic messages. It seemed like the entity was purposely trying to create fear in the witness. Maybe there is a good reason for this.

Mostly, monsters just seem to frighten witnesses, rather than causing physical harm but this is not always the case. The land between the lakes is a national recreation area, formed in 1963, by President John F. Kennedy. The forced movement of people from the land was, and still is, a source of anger. The land itself is located in rural Kentucky and Tennessee, between Lake Barkley and Kentucky Lake, hence the name. The area, known as LBL for short, has a long history of ghosts, weird balls of light and strange wild beasts, unlike any native fauna.

One story from the area cements our latter day werewolves as having in interest in motor vehicles, or at least their contents. A group of students from Murry State University, were camping in the LBL in 1973. One of the group went into the trees for a pee and came back ashen faced, saying that he felt that something was watching him. Soon after, something could be heard prowling around the camp just out of sight, sniffing the air. As they shone their torches into the night, they picked up red, glowing eyes and a nerve shattering howl rang out. The panicked students ran for their Volkswagen mini bus. As they drove off, a hulking beast lunged from the shadows and grabbed the back of the bus with such strength, that it stopped the vehicle in it's tracks. The terrified driver stamped on the accelerator and the bus broke free from the monsters grasp and drove off at speed. Back at the university, the friends examined the bus and found deep claw marks in the bodywork at the back end.

Kentucky Fortean researcher and blogger, Jan Thompson, has a couple of hair raising stories to tell, about a werewolf type beast said to roam the LBL. One of these occurred to her in the summer of 1978. She was sitting by her house with her cousin, whilst her other cousin Joe, had been out riding a dirt bike. Suddenly Joe burst out from some nearby woods peddling his bike frantically and came to a halt, skidding outside of the house. He was looking back with a frightened expression towards the woods. Jan takes up the story.

> *"We followed his gaze not understanding what this escapade was all about. In silence we watched with him for about 30 seconds and then the dogs started barking. Growling. And then whining, trying to get out of*

the pen in a frantic panic of digging and gnawing at the fencing. IT GRABBED ME!! LOOK AT MY LEG! Joe screamed, making us jump with alarm at the sound of his voice. We looked down at his Levis and saw scratch marks going across his right thigh, scratches that tore through the tough denim and left small bloody marks on his skin. The marks were like a bears-claw-rake, not those caused by branches or sticky bushes, but a definite wide pattern of a paw print. IT WALKED ON TWO LEGS! His voice startled us again, as he was trying to tell his story in between huge gulps of air. He was frightened beyond belief, and the bits and pieces of what he was striving with extreme effort to tell us was coming out in loud syllables that filled us both with the same dread. It was following me through the woods along the path from the old sawmill. Hairy, it was so hairy and its' snout was so long and it walked on two legs, it ran on two legs, his voice was sputtering, slowing, his eyes were still wide, and I could see the pulse of his heartbeat throbbing under the skin of his temples."

Soon his strange story was vindicated, as a huge, hairy, bipedal creature emerged from the woods. It had a long, wolf-like snout and let out an eerie howling call that drove the family dogs to cower in terror. The three cousins ran and locked themselves in the house. The security lights illuminated the massive wolfman as it stalked up to the house. The kids grabbed kitchen knives and hid under a bed like the three little pigs, as the beast circled the house looking for an entrance. It even broke a window at one point. Luckily Jan's aunt arrived home in her car and the lights seemed to drive the beast away.

According to Jan, not everybody was so lucky. In the early 1980s police were called to a camp ground in the LBL, close to some abandoned military bunkers. So the story goes, a family of campers, a couple and their child, had been attacked and killed by some powerful wild animal. Their bodies, half eaten, were scattered on the ground. The sight shocked even the hardened officers, but things only got worse. The family apparently had a younger child whose body could not be located. An officer called Adam began to search the surrounding woods. Jan tells what happened next.

"From somewhere in the nearby woods, about 50 yards from the campfire, a scream was heard. A mad shriek that turned into a long wail and then to whimpering. As others arrived they could see by the gleam of several flash-lights that the cop was holding his hat in one hand and his light in the other. There was blood on his face, the front of his shirt and on the brim of his hat. More blood could be seen dripping on him. It was coming from above. High in the trees the flash lights swung, searching for the source of the mysterious bleeding. A very small hand could be seen dangling down from a tree limb way up high, as well as a slender lifeless leg that still had a white sock still on the foot. The missing child had been located. It had been Adam that the

blood had trickled upon, hitting his hat first, making him look up, and then feeling the thick cold fluid sprinkling his face then sliding down to his neatly buttoned shirt. It had been Adam that had screamed. The little girl had apparently been carried up the tree and leisurely eaten upon while carefully laid across a large tree branch. More of the same long grey and brown hair was found sticking in the bark of the tree near her body."

Saliva and hair was sent away for analysis, and though not fully identified, it was apparently that of a wolf. Jan says the local government hushed up the story for fear of it effecting tourism. There is no way of corroborating this gruesome tale, like many in this book, so it remains in the nebulous world between fact and folklore. Some have even accused Jan of fabricating the story to create a modern myth. There are, however, other stories of strange attacks and killings in the woods of the USA. For now the killer wolfman of the LBL must remain a mystery.

These cases are just the tip of the iceberg. There are many, many cases of werewolf sightings in the USA, but here is the thing, the descriptions of modern day werewolves in no way resemble the original werewolf stories of Europe. True werewolves in stories reaching back centuries in Europe, were people who transformed themselves utterly into wolves, indistinguishable from ordinary wolves, except perhaps, in odd behaviour. They were usually granted this power by the devil, a sorcerer or witch, or a 'forest lord' who gave them a magical belt or jacket, that when donned allowed them to become a wolf. These stories may have been based on attacks by rabid wolves. Rabies would have made the animals bolder and more aggressive and able to pass on the 'curse' through their bites. The man wolf hybrid, the beast with a wolf-like head and a hairy, humanoid body, is a creation of 20th century film. So why then are these fictional wolfmen being reported? It is almost as if they are taking the form we expect them to, due to cultural influence. We will see this elsewhere and we will return to this idea, but for now just squirrel it away in your noggin.

MONSTER CATS

We have all heard of the big cats said to stalk the British countryside. They have been filmed, photographed and on occasion even captured. There are other accounts of mysterious cats from around the world, most seem to indicate escaped exotics or unknown species. But some are a lot stranger.

On April 10th, 1970, near Olive Branch, Illinois, Mike Busby was driving along a little used road alongside the Shawnee National Forest. Suddenly his car engine faltered and stopped. As Busby got out to check beneath the bonnet, he noticed a pair of green, cat-like eyes glowing in the shadows. A black panther-like creature emerged from the forest *walking on its hind legs!* The monster attacked Busby, dragging him to the ground and slashing his arm and abdomen with its claws. Luckily, a passing lorry frightened the creature off, allowing the victim to scramble back into his car, that strangely started once more. Once he reached Olive Branch, he was able to find the lorry driver who confirmed his story.

Needless to say, black panthers do not walk on their hind legs and they are not native to the USA. Furthermore, we must note the strange effect the creature had on the car.

Herman Belyea was attacked by an identical beast 19 years earlier, in Queens County, New Brunswick. On November 22nd, 1951, a black cat-like monster, six-feet-tall and walking on its hind legs, forced him to run for his life. The creature dodged his axe blows and pursued him until he reached the edge of town.

Earlier still in 1948, at the Rice Hope Plantation, Santee River, South Carolina, Sam Lee and Troy Rodgers were looking for poachers. At about 10.30pm they heard a noise and observed a cat-like, maneless lion, that stood up on its hind legs like a man. The pair were so frightened that they ran back to their car.

On other occasions, the cat creatures seem bullet proof, like the 'panther' shot at by police near Atlanta, Georgia, in April 1958. The patrol men were searching woods after reports of the beast. Their torch beams illuminated glowing yellow eyes and the thing charged. The men opened fire but the creature sped past them unharmed. Officer J.F. Porter recalled… *"Both of us were firing at point blank range. I don't see how we could have missed it."*

Edward Banks of Bushlease Farm, at Crondall in Hampshire, had his property haunted by a phantom big cat in the 1960s. The creature savaged a heifer and left claw marks and a stink of ammonia on trees. Strangely, prior to each appearance of the creature, Banks had noticed weird lights on the roofs of his farm buildings. It was as if a beam of light was being shone on them, but the beam had no source. On each occasion the big cat would return to the farm. Take note of the weird animal, seen in tandem with a strange light!

Now we turn our attention upwards to the winged monsters of the skies.

MONSTERS ON THE WING
Why is it, that on a summer's day, when the wind cuts up rough and sends clouds across the sun, making vast shadows sweep across the land, we freeze. If only for a moment, it's as if someone has walked over your grave. This same freezing behaviour has been recorded in monkeys and lemurs as a defence against birds of prey. So why do we have it as well. Surely there are no flying predators big enough to constitute a threat to full grown humans? Legend and Forteana says otherwise. The skies of our world are haunted.

DRAGONS: KINGS OF THE MONSTERS
The dragon is the great, great grandfather of all monsters. Before the daemon, before the vampire, before the werewolf, before the giant, before them all, was the original uber-monster, the dragon. The dragon's image has crawled across cave paintings

25,000 years old, dwarfing mammoths. It has slithered across Chinese rock art in Shanxi province, 8000 years before Christ. It haunted the Sumerians and the Babylonians, was worshipped by the Aztecs and feared by the Celts. Its image graced the prows of Viking longships. In the east, a glittering rain god, in the west a flame spewing, maiden devouring monster. It is found in every culture on earth. In the UK alone, there are around one hundred dragon legends. Recently Michael Wetzel, a Harvard University linguist and philologist, used phylogenetic analysis of legends to trace back their origins to a far more distant time than anybody had previously thought. In his book *The Origin of the World's Mythologies* he dates the first dragon legends to 75,000 years ago, the immortal dragon has its fangs and claws deep in the psyche of mankind. And it is still seen today.

Sightings of winged dragons still occur in the modern age. Picture a ghostly, ethereal globe of light bobbing and flitting in the inky night sky. In the west this would doubtless be called a U.F.O, and some would deem it to be an alien spacecraft. To the Namaqua people of Namibia, however, this light would mean something infinitely more terrifying - a latter-day dragon. The Namaqua have been reporting such creatures for decades. Their flying-snake is described as being the size of a large python, yellow, speckled with brown. From its cranium two horns sprout, and a pair of bat-like wings grow from behind the head. Strangest of all, a glowing ball of light is said to shine on its forehead. This is strikingly reminiscent of the magickal pearls or jewels that were said to be embedded in the heads of Asian dragons. It would be easy to dismiss this as native folklore, but European settlers have seen them as well. In January 1942, 16 year old Michael Esteruise was tending sheep, when something emerged from a cave on top of a nearby hill and launched an attack.

> *"I heard a sound like wind blowing through a pipe, and suddenly the snake came flying through the air at me....it landed with a thud and I threw myself out of its path. The snake skidded, throwing gravel in all directions. Then it shot up in the air again, passing right over a small tree, and returned to a hill top close by."*

Michael had been sent out by his father - the owner of a vast farm in Keetmanshoop - to dispel the native mumbo-jumbo that had been costing him both men and money. All of his farm workers had left after he ignored their stories of a giant flying-snake, that laired in the mountains where his sheep grazed. He finally deputised his boy to show the ignorant savages the folly of such beliefs. The boy did not return. He was later found unconscious and when he came to, related his dramatic tale. Oddly he related that the snake smelt of "burned brass." Police and farmer investigated in time to see the winged serpent crawl back into its cave. Lighted sticks of dynamite were hurled in after it. After the explosions they heard a low moaning for a while, that gradually died away.

This incident was investigated by no less an authority than Dr Marjorie Courtenay-Latimer, a woman forever immortalised in the annals of cryptozoology as the discoverer of the coelacanth (*Latimeria chalumnae*) - an archaic fish believed extinct for 65 million years. She interviewed the boy, who took her to the spot of the attack. She did not see the beast but observed the great furrow that it had made in the dust. She also noticed the lack of small animals such as birds and rats.

In 1988 Professor Roy Mackal (better known for his Congo and Loch Ness excursions) investigated fantastic claims on a remote property, owned by German settlers. Locals described a massive featherless creature, with a nine metre (thirty foot) wing-span, that glided between two hills about a mile apart at dusk. The thing seemed to have lairs in crevices in the hills. Team members discovered the remains of ostrich carcasses in highly inaccessible areas, and believed that the creatures had preyed on them, before taking their kills back to the nest area. Mackal returned to the U.S.A without having spied the animal, but shortly afterwards, one of his team members got lucky. James Kosi - who had stayed on in Namibia for a while - saw the monster from a distance of around a thousand feet. He described it as a giant glider, black, with white markings.

Another dramatic sighting was in 1978, where a French farmer who was tending his cattle in the Karas region, when he saw a bright white light heading straight towards him. Then afterwards he heard a massive thud and he heard one of his cattle cry out in pain, as he rushed to check what attacked the cow, he saw a dragon, He told the local police:

> *"I saw what looked like... as the best matching description I can give you is that it looked like a dragon, it had a white bright light on its head, which was blinding me, the colour of it was brown and yellow, it had green eyes, there was a tar-like smell coming from it and it had smoke coming out of its nostrils."*

Again we see strange light associated with monsters.

Elsewhere airborne, reptilian monsters have been seen over America.

In 1783, a silversmith and his daughter were chased off Rattlesnake Hill near Silver Run, Maryland. They described it as 'a fiery dragon with gaping jaws'.

In a letter to the *Fredrick News* (a paper in Maryland), a man who signed himself "R.B" claimed he had seen a dragon. At 6.30 one morning, in 1883, whilst standing on a hilltop, he saw a monstrous dragon with glaring eyeballs, a wide mouth, and a tongue that hung like flame from its jaws. The creature was rearing and plunging above Catotcin Mountain. It sounds like the same type of creature described above.

In 1873, in the skies over Bonham, Texas, a flying serpent manifested itself. Men working on a farm saw a yellow striped snake, the size of a telegraph pole, floating in the sky. The creature would coil itself up, then lunge forwards as if striking at something. Shortly after, the beast was seen over Fort Scott, Kansas. This may have been the same animal that western historian, Mari Sandoz recorded earlier:

"Back in the hard times of 1857-58 there were stories of a flying serpent that hovered over a Missouri river steamboat slowing for a landing... In the late dusk it was like a great undulating serpent, in and out of the lowering clouds, breathing fire, it seemed, with lighted streaks along the sides."

The action switched to Darlington County, North Carolina in 1888. Witnesses claimed to see a winged reptile fifteen feet long that flew as fast as a hawk and made a hissing noise. Reports also came in from three neighbouring counties.

In 1891, the skies over Crawfordsville, Indiana were haunted by a winged reptile twenty feet long and eight feet wide, that swooped down from a hundred feet to ground level, on violently flapping wings. It scared those who saw it with searing hot breath. The dragon manifested for two nights.

In a letter printed in *Occult Review* of December 1917, by a man who called himself a "philosophical aviator"— the writer had been told of an encounter between a First World War air pilot and a dragon. Whilst at a considerable height, the pilot had seen a dragon rapidly approaching him. He rapidly descended to avoid the colourful reptile. On reaching earth, he said nothing to his colleagues for fear they would think him drunk. This account is suspect because the details are so scant. The narrator and the witness are nameless, and the location is not mentioned. The sighting may have been caused by oxygen deprivation at a great height. Even so, another account that is so alike to this may give weight to the pilot's story.

Again, the *Occult Review* took up the story, with a letter from Georges Lajuzan-Vigneau, printed in April of 1918. He claims to have seen a letter in a French newspaper in 1909. The story involved three aviators who encountered a huge bluish dragon, that caught up with their plane, and kept pace with it easily. The men panicked, and began to descend, whereupon the dragon seized one of the trio, and flew away with him. A grim story, but one with little evidence to back it up.

Further north in the early 1990s, the Rocky Mountains of British Columbia were the setting for a bizarre encounter. A woman was out hiking when she came across a dragon in the wilderness. The report ran thus.

"The creature was in a beautiful shade of dark green and could easily blend with trees as he had been standing by them but the witness reported that he was perched on a rocky outcropping on the side of the mountain. He was fanning his wings slightly, looking quite calmly into the valley below. I had been hiking up this mountain, when the movement of his head caught my eye. I had been this way before, and there was a group of trees on the cliff where there had been none before. I did not believe what I had seen at first, but the shape was too obvious, and he was parallel to me, about seven bus lengths away. I was climbing

up one rock outface, he was on another. He was the most beautiful creature I had ever seen. His head was long, with a large eye ridge and two smaller bumps with a triceratops-like horn on his nose. At the back of his head were two large horns, jutting out backwards, and two smaller horns below them. They were a greyish-white and caught the light like dull silver. His forelegs were slightly smaller than his hind legs and were gripping the edge of the cliff. He looked as though he were a quadruped. He had slightly darker dorsal ridges running from between the longest horns to about halfway down his tail. As I stood there, gaping like a fish out of water, the dragon turned and looked at me. He cocked his head to the side, almost like a bird, then spread his enormous wings and vaulted off the cliff. He was absolutely elegant in the air, flapping his wings several times before banking into a glide and disappearing around the side of the mountain. My legs felt so weak that I had to sit down. I have been camping in those mountains for over ten years, and I had never seen anything to suggest that dragons might actually exist there. But after that encounter I began to think about it. What better place for a dragon to live than in the mountains? There are places in Banff and Jasper that nobody has ever been to, and there are many elk and deer and possibly even bears for it to feed on. Plenty of lakes, and the mountains themselves have many hidden caves and the like."

In October 2001, Stevenson Fisher of Camden, Minnesota claimed to have seen a flying monster with a twenty-four foot wingspan. The grey creature had leather wings of skin. Fisher said he could see light through them. It flew close to telegraph wires and the roof of his house. He used these to gauge the monster's size. He saw the same creature (or another of the same species) that winter.

Cryptozoological researcher, Linda. S. Godfrey, was contacted by witnesses who claimed to have seen dragons in Wisconsin, in 2007. One young man she calls 'Jim' (a pseudonym) wrote the following...

"On October 7, 2007, in Oconto Falls, Wisconsin, some of my friends and I were at a now closed down arcade. It was the same as always, a few of the best local bands, me and some of my friends joining them on stage and hanging out afterwards. While we were outside on the cold, partly cloudy night, one of the guys that was there from Green Bay said the thought he saw something in the sky. Most people were sceptical, but we just decided to lay down on the grass and on top of vans and trucks and maybe we could see what it was.

After about fifteen minutes of talking and laughter, those emotions changed to surprise and astonishment as we watched a massive white / tan dragon fly over the clouds. We knew it had to be a dragon, because how else would you describe something flying over that was almost silent, larger than a plane, had a tail, bat like wings, long neck, and a narrow, pointed head and scales?

I remember noticing the scales because they dimly reflected the street lights below. We thought we were all seeing things but five minutes later it flew over again, this time in the opposite direction. The eight people from Green Bay wanted to stay but they had to leave. They hoped to see more back home.

Me and my friend K, however, decided to go to my house and lay in the backyard and watch. My mom joined us, not really believing any of it until as soon as she was about to go back in, another big one flew over the house. If I remember right her words were "I'm tired and I'm going to bed. I doubt that it even-holy (exclamation)."

We saw a few smaller ones after that but I haven't seen them since that night. I believe that they were migrating during that month. Hopefully I will see them again next October. When talking about the indecent at school, one of the girls around us claimed to have seen the same thing with her cousin two years earlier."

Linda was later in contact with Jim's mother, whom she referred to a 'Janet'. Jim had phoned his mother for a lift and told her what he had seen. Janet thought her son was joking and went to pick him up. She noticed a shadow crossing a parking lot, then saw a fire ball, blue with orange around it shooting from east to west in the sky. Jim told her that he and his friends had seen the same thing before the dragon had arrived. Janet then drove her son home.

Linda also spoke to Jim's sister Jill. Jim had talked her into staying up. The pair lay on a trampoline in their back yard, looking at the sky. They saw a fire ball shooting east to west, followed quickly by one moving west to east. Then they heard a loud screech that caused all the local dogs to bark. Jill said...

"Then as we were watching the sky, coming from the west, from the river to over our yard, we saw what appeared to be-and this is the only thing we could think of to describe it as-a dragon. It was making gliding movements up and down, never flapping its wings, kind of like how a mermaid is...

It never flapped its wings at all. And it looked almost cream. The drawing is showing it as we were looking up at the underside. Its stomach reminded me of the underside of a cow, barrel-chested, and from where we were it looked as large as a cow. I remember the moon was very bright and full that night. It had a snake-shaped head and a long- pointed tail."

The creature was flying only twenty feet above their two-story house. After looking at pictures of pterosaurs they saw that the wing structure was very different. The family also saw smaller dragons that they took to be the young of the larger ones. The smaller creatures seemed to be playing by circling each other. She continued her statement.

> *"I don't care what the scientists say, it was NOT a pterodactyl. I could see it had pearly, pale scales. And the fireball came from its mouth. They were beautiful and flew gracefully. I could see they had four legs, too. They were tucked up underneath them like when a bird flies."*

On March 18, 2012, in the southern part of Fayette County in Pennsylvania. A man was walking his dog in a rural location at about 11:45pm. He was in the front yard and away from any lights when his attention was drawn to look upwards after hearing a whooshing sound coming from overhead.

Flying above him, at a distance of about 55 feet, was a large flying creature that, *"looked like a dragon."* As the flying creature passed, he was able to get a good look at the strange flying animal. The body was about 22-feet-long, with a wingspan of about 18-feet-wide, and looked to be shiny with an almost reflective body, with no scales. Its colour was dark, auburn brown. At the end tip of the wings there appeared to be talon-like fingers, about 3 to 4 in number. The arms of the wing structure appeared muscular. The wings were quite thick, not like skin. There appeared to be a rear fin on both sides of its body, and the creature displayed an arrowhead shaped tail. The witness also saw what appeared to be two extended rear legs. The creature had a cone shape around the head that stopped flat on the base of the neck. The oddest physical feature that the witness mentioned to me, was that the mouth and eyes were illuminated with, *"a very ominous orange glow."* As the creature flew over a tree at the bottom of the yard and moved off in the distance, the fellow heard a deep-throaty sound, similar to the fog horn on a boat. The entire observation lasted about 20 seconds.

In 2016, a witness calling himself 'Malcolm' wrote into Lon Strickler's *Phantoms and Monsters* website with a story of his own dragon sighting in New Jersey.

> *"Hello - last weekend I was driving on Marshallville Rd. by the river near Tuckahoe, NJ. I don't know the time, but it was dusk and the sky was light enough to see anything in the sky.*
>
> *I caught a glimpse of a huge flying creature crossing the roadway approx. 100ft in the air. I swear it was shaped like a dragon - silhouetted against the lighter sky - flying south towards the state forest.*
>
> *I looked at it for almost 10 seconds. The wings were wide and there was a long tail. The head was like that of an unknown reptile. It was dark in colour and flapped its giant wings, which had to be 30 ft or more in*

width. There was another car behind me - they definitely saw it by the expression on their faces. I pulled over and the 2 people in the other car looked over when they passed me, pointing to the sky. I tried to get them to stop but they were moving by too fast. I lowered my window and yelled but they continued on. When I stuck my head out of the window I immediately smelled a foul stench and felt warm down drafts. It was very weird. I have added an image I found that closely mimics what I saw.

I'm not naive and I know that this could not have been an actual animal. I almost feel like I entered a time warp at the time this creature appeared. For the rest of the evening I had a physical reaction similar to jet lag.

I haven't said anything to anyone else - I live alone and retired. I've lived in this area for most of my life and I've never experienced anything like this. There has been no mention of this in the local news - I may have seen something I shouldn't have.

Sir, what are your thoughts? Malcolm."

Lon wrote back to the man who added that the sighting happened at 8.15pm. The dragon was 50-60 feet long and had horn-like projections on its head. It stank like rotting fish.

The letter drew out another alleged witness, who wrote in.

Hi,

I just read your post about the dragon sighted over southern New Jersey. It took me by surprise because I had a similar experience in the mid-eighties when I was still a kid (about ten years old, I think).

It was before noon, in the springtime, and I was biking around a local ballpark by my house. (Nottingham baseball field in Hamilton, NJ). There were no games scheduled that day so I was alone, as far as I could tell.

There was a play area and a pond by the woods that I would frequent, when I was bored, in the hopes of finding other kids to play with.

About halfway there, while looking up at the clouds and sky, I saw what I thought was a plane coming from the south. There's a small air strip by us and there were some very cool bi-planes that would fly by from time to time. My favourite was a red baron style one, and I thought that was it, because of the colour.

I stopped and turned to watch (and wave, it's what kids do). When the plane gave its wings a single, long, flap, that's when I realized it wasn't a plane.

It was a dragon. There is no doubt in my mind. It was dark red and I could see the sun shimmering off its scales as it got closer. Needless to say, I didn't wave. I felt frozen to the spot, dumbstruck, a little scared, and very, very excited.

It was huge, flying at about the same height as the bi-planes do, and it didn't make any sound as it went overhead, just gliding. It was like a classical European dragon, but the proportions were a little different.

When it was right overhead I felt a blast of heat. I assumed it was its fiery breath at the time. I was absolutely vibrating with excitement and looking around frantically for anyone else that might be around. The ballfields were empty, no one was outside in the nearby houses!

It kept going at a steady clip and right before it disappeared over the tree line to the north it gave another slow flap of its massive wings.

I turned right around and sped out of the park hoping to find someone else who saw it too. No one, everyone was inside their homes that early. When I got home I realized no one would believe me. So I never told my parents. They'd think I was making it up. Instead I told my older sister, but she said I probably saw a hang-glider or a fancy plane.

It was such an odd sighting I never perused it further. Nice to see the dragons are still flying over New Jersey though. Thanks for all you do, your blog is awesome. D"

Viking folklore is full of dragon legends, so the next account should come as no surprise.

In 2010, the Icelandic volcano, Eyjafjallajökull, erupted spewing out titanic clouds of ash into the atmosphere and grounding planes for weeks. One local man watching the eruption saw something flying about the volcano. He realized that what he saw must have been huge due to the great distance he was observing it from. At first, he thought it was a plane, until he saw the wings flapping. In an e-mail to Lars Thomas, the Danish biologist and cryptozoologist, he said the following...

"As I see it, there are only two possible explanations, either I was imagining the whole thing, or what I saw was in fact a dragon. I find that

very hard to believe, but then again, I don't think I could imagine something like that."

In May of 2003, David Nardiello was teaching English in Nigshimozu high school in the town of Watagh Shinke-Cho, Osaka, Japan. He was cycling home late one night through heavy rain. The torrent had formed a pool in some nearby rice fields. Nardiello saw a white animal emerge from the water and turn to look at him. It had a long neck and snake-like head with black shark eyes and fangs. The body and tail were akin to a lizard whilst the four legs resembled a cat. The animal had leathery, featherless wings. It flew into the air to a height of thirty metres (100 feet), and Nardiello - increasingly scared - cycled home as fast as he could.

Later that night, he saw it flying through the night sky from his third story flat. He asked his neighbours if they had seen it, but none had. Some however said they had heard weird cries from the fields for a few nights. His co-worker Kato Sensi dubbed it Nekohebitori or 'cat, snake, bird'. Nardiello felt strongly that the animal was a predator and was dangerous.

In the late 20th and early 21st century, Gunung Kancing, a mountain in the province of Lampung, Western Java, was thought to be the lair of a dragon the locals called the 'ular naga' or dragon snake. It was described as resembling a huge snake with a horse-like head and bat's wings. The monster terrorized Peserwaran village for years. It stole and ate children, often attacking in broad daylight. The beast would snatch the children whilst their parents were busy in the fields. One village elder, Darga, said that the dragon was seen almost daily. In the end, the villagers blocked up the entrance to the cave where they thought the creature lurked. After that, the killing stopped but the people were so scared that they mostly abandoned their homes and moved to other villages.

As we can see, far from being relegated to the pages of fantasy novels or to ancient legend, dragons are very much with us today, terrifying, eldritch and unknowable.

THE VAN METER VISITOR
Van Meter is a small city in the state of Iowa with a population of around 1484. It was founded in 1869 and named after Dutch settler Jacob Rhodes Van Meter. From September to October of 1903 the city was visited by one of the strangest flying monsters in the annals of Forteana. The first witness was U.G. Griffith, a salesman who was walking home at 1.00am on September 29th. As he walked down Main Street, he saw a beam of light shining down from the Mathers & Gregg building.

He thought it was an open lantern and that somebody was trying to break into the building. He shouted out a challenge and the light, to his amazement, rose up and seemed to fly onto the roof of another building. Again, he shouted out and again the light rose up and this time flew across the road above him and landed on a building on the opposite side of the street. Then it seemed to rise straight up into the sky and vanish.

Next day, Griffith, who was a respected and prominent citizen of Van Meter spoke to the owners of the building in question, still thinking some sort of illicit activity was afoot. The roofs of said building were checked and no signs of a prowler were found. That night the town doctor, Dr Alcott, was asleep in a room near his office. He was awoken by a bright light shining through his window. Having heard the rumours of a prowler, he grabbed a pistol and headed outside, expecting to find a felon trying to break in. Instead, he saw a bizarre creature clinging to the side of the building. The horror was vaguely humanoid but with huge, clawed, bat-like wings. It had a long beak and a small blunt horn on its forehead that seemed to project the light. The doctor fired his gun at the beast, which swung its head around and shone the beam of light from its horn directly into the man's face dazzling him. He fired a further four bullets at it. The monster let out a shriek and flew off into the night. It seemed utterly unhurt by the bullets.

As night fell the following evening, Clarence Dunn, a teller at the bank, waited with a shotgun in case the premises was the next place targeted. After dozing off he was awoken at around 1.00am by a gargling, rasping sound, like a strange breathing. On looking out of the window he was blinded by an intense light. As he staggered back, he saw the beam sweep the room as if searching for something. He also saw a hideous face with 'dead' eyes and a beak full of sharp teeth. In panic, Dunn shot the beast through the window at point blank range. Again, it seemed unharmed by buckshot and flying glass. It dazzled him once more with the light that shone from its horn before vanishing into the dark.

Next morning, huge, bird-like, three-toed tracks were found on the ground below the window. And that night the Van Meter Visitor returned again.

O.V. White had seen the monster perched atop a telegraph pole near a hardware store. Again, the witness shot at the monster and again it shrugged off the bullets as if the weapon had been a pea shooter. It clambered down the pole using its beak and talons like a parrot climbing down a branch and then released a vile stench towards White. The monster flew off in the direction of an old coal mine. A Sidney Gregg observed the monster in flight as well.

The following day both men told their neighbours about the sighting. Due to the growing alarm, a deputation of business owners confronted the mayor, who summoned the owner of the mine, a Mr Platt. Platt confessed that some of his workers had been hearing weird noises coming out of some of the abandoned shafts in the mine. Two guards were posted at the mine that night with orders to contact the police if they saw anything strange. The guards returned, saying that they had seen two of the creatures emerge from the mine and fly towards town. A posse of armed miners and townsfolk were gathered to ambush the creatures when they returned.

As dawn approached, the pair of flying horrors returned. As they landed, a volley of gunfire was unleashed upon them, but seemed to have no effect other than to make them let out unearthly screams and more of the vile odour. Unharmed by bullets and buckshot, the creatures entered the lair once more. The miners quickly collected some dynamite and blew up the abandoned mineworks. Nothing more was seen of the monsters.

US newspapers in the 19th and early 20th century were well known to concoct wild stories just to fill space and sell copy. Tales of ghostly trains and other stories circulated much like wild tales do on the internet today. But the Van Meter story seems to be something different. Many of the people involved where affluent and influential people, including the mayor. Would they be party to a hoax? Nothing matching the description of the Van Meter Visitors has ever been reported anywhere else.

ENTER THE MOTHMAN

The creature that was to become known as the Mothman made its debut in 1960 or 61. A woman and her father were driving along Route 2 in the Chief Cornstalk Hunting Grounds of West Virginia, USA. They saw a figure taller than a man standing in the road ahead of them. The woman described what happened next.

> *"A pair of wings unfolded from its back and they practically filled the whole road. It looked almost like a small airplane. Then it took off straight up...disappearing out of sight in seconds. We were both terrified. I stepped on the gas and raced out of there. We talked it over and decided not to tell anybody about it. Who would believe us anyway?"*

On November 16th 1963, the Mothman seemed to have had a short holiday in England. Four teenagers, including John Flaxton and Merryn Hitchinson, were walking in a lane by Standing Park, Hythe, Kent. The boys saw what they at first thought was a star, but the light seemed to move and follow them. The teenagers became unnerved and moved away. Suddenly, the light appeared much closer. This time it was only one hundred feet away and ten feet above the ground. It appeared to be a glowing, golden oval. Suddenly, the boys heard a snapping of twigs as something moved in the bushes nearby. Something waddled out of the shadows. It was as tall as a man and stood on two legs, but that's where any resemblance ended. It had black, rubbery skin and bat-like wings folded against its sides. It seemed to have no head, just a rounded torso. The terrified youths fled.

Back in West Virginia after its dummy run, the Mothman was back with a bang in 1966 haunting an area called Point Pleasant. Two couples, Linda and Roger Scarberry and Steve and Mary Mallett, were driving, late one night around an abandoned wartime munitions factory called the TNT Area.

It was November 17th, and as they drove past an old generator plant, they saw two red lights in the darkness. The couples realised that what they were looking at were the eyes of some kind of creature. As the owner of the eyes shambled towards them, they were horrified to see it was a grey-furred figure, seven feet tall, with huge wings folded at its back. It seemed to have no head, the eyes being set in the chest area. The witnesses drove off but soon encountered the thing again, standing by the side if the road. As they sped past, the creature unfolded its bat-like wings and pursued the car.

They hit one hundred miles per hour but the flying monster kept pace with them. They noticed that it didn't seem to flap its wings as it flew. Linda said she could hear it squeaking like a big mouse. As they approached the town, the monster veered off into a cornfield. The witnesses headed straight for the sheriff's office. The deputy, Millard Halstead, could see the kids were scared and drove over to the TNT Area. Halsted found nothing other than some strange dust.

The story made it to the local paper, *The Point Pleasant Registry,* and the journalists dubbed the creature 'Mothman'. For the next eighteen months, the Mothman haunted Point Pleasant and the surrounding areas. The creature seemed, like other monsters, to be able to generate intense fear in witnesses. This is shown in the case of Marcella Bennett. Mrs Bennett had driven over to the house of her friend, Ralph Thomas. As she parked her car, a figure stirred near the automobile. *"It seemed as though it had been lying down,"* she later recalled. *"It rose up slowly from the ground. A big gray thing. Bigger than a man with terrible glowing eyes."*

Marcella was so scared of the creature that she dropped her baby girl. Luckily, Ralph and his family managed to drag the mother and daughter inside. Mothman waddled up to the house and peered in through the window at the terrified humans. The police were called, but the Mothman had vanished by the time the authorities arrived.

Mrs. Bennett would not recover from the incident for months and was, in fact, so scared that she sought medical attention to deal with her fear. She was tormented by frightening dreams and later told investigators that she believed the creature had visited her own home too. She said that she could often hear keening sounds (like a woman screaming) near her isolated home on the edge of Point Pleasant.

Legendary Fortean investigator John Keel came to Point Pleasant in 1966 and found himself in a paranormal ground zero. The town was gripped with fear. Keel recorded his findings and experiences in his classic work *The Mothman Prophecies*.

There were many, many other sightings of the Mothman over the coming months. During this flap (pun intended) other Fortean phenomena were occurring. There were visitations from the sinister men in black, threatening figures associated with UFO sightings (tackled in a later chapter), and, indeed, UFO s themselves. One such craft chased an ambulance taking blood supplies to a hospital.

On December 15th 1967, the Silver Bridge that connected Point Pleasant West Virginia and Gallapolis, Ohio collapsed, killing 46 people. Many thought that the Mothman had appeared as a harbinger of the disaster as the sightings ceased after the disaster. In fact, the bridge failure was due to a defect in a single link, eye-bar 330, on the north of the Ohio subsidiary chain, the first link below the top of the Ohio tower. The crack was formed through wear and internal corrosion. It was only 2.5 millimetres deep, but when it failed, the whole bridge was doomed.

There is small wonder that there were no Mothman reports after the disaster. The media was focusing on the horror that had unfolded on the bridge. People soon forgot about the strange flying creature. Whatever the Mothman was, it had no connection with the collapse of the Silver Bridge.

THE CORNISH OWLMAN, MOTHMAN'S UK COUSIN.

John Keel thought certain places attracted or were more prone to Fortean phenomena. Point Pleasant was one such place. But it seems like certain times or eras have more weird occurrences in them than others. We have already noted that 1973 seemed to be the champion Fortean year for some strange reason. But the 1970s seemed to be a haunted decade. The British writer, broadcaster, and performer Bob Fischer has dubbed those who grew up in the 1970s, 'The Haunted Generation.' The area around Falmouth Bay in Cornwall, England was the focus for a cavalcade of the strange in 1976. A sea serpent was seen multiple times off the coast and named Morgawr by Doc Shiels. UFOs were reported and birds acted strangely, sometimes attacking people.

On Good Friday, 1976, Don Melling took his family on holiday to Cornwall. They were stopping at a caravan site in Truro. They had come all the way from Preston in Lancashire in the North West of England. The next day the family went for a picnic in the woods near the sleepy village of Mawnan Smith, a mile or so southwest of Falmouth. His two daughters, June, 12, and Vicky, 9, went off to explore the 13th century Mawnan Old Church which stands next to the woods. Suddenly, the girls came running through the lychgate screaming hysterically. They would not calm down or explain what was wrong until they were miles away, back in their caravan.

The girls claimed that whilst they were exploring the graveyard, they had heard a strange noise, then seen a 'huge, feathered bird-man' hovering above the church.

Trying to salvage the holiday, Don took his family to a steam fair in the town of Penryn. At the fair was Doc Shiels. As Doc had been investigating the sightings of Morgawr, somebody pointed him out to Don as the 'man involved in hunting monsters'. Don angrily approached Doc and accused him of rigging something up to scare his daughters with. Doc managed to calm him down and make him see common sense, that it would have been impossible for him to have created a flying birdman that hovered over a church. Doc wanted to interview the girls, but their father insisted that they had been frightened enough. He did, however, produce a drawing of the thing by June. It showed a winged, humanoid creature with an owl-like face. The family were so distressed by the affair, they cut their holiday short and returned to Preston.

Doc wrote a letter about the whole affair to *Fortean Times* and thus the Owlman was born.

Soon others claimed to have seen it. On July 3rd, two fourteen-year-old girls Sally Chapman and Barbara Perry, were camping in the woods near the church. They had been brewing tea when a hissing sound alerted them to something strange. Sixty feet away, a freakish figure was standing. They met up with Doc Shiels on Grebe Beach below the church. Once again, they had been told that he was the man to talk to about monsters. Doc interviewed them.

> *"It was like a big owl with pointed ears, as big as a man. The eyes were red and glowing. At first I thought it was someone dressed up, playing a joke, trying to scare us. I laughed at it, we both did. Then it went up in the air and we both screamed. When it went up you could see its claws were like pincers!"*

Barbara added...

"It's true, it was horrible, a nasty owl-face with big ears and big red eyes. It was covered in grey feathers. The claws on its feet were black. It flew up and disappeared in the trees."

Doc separated them and got the girls to draw the thing they had seen. The pictures were alike, but Sally thought that Barbara had drawn the wings wrong. Both girls made additional notes.

Sally said...

> *"I saw this monster bird last night. It stood like a man then flew up through the trees. It is as big as a man. Its eyes are red and shine brightly."*

Barbara said...

> *"Birdman monster, seen on third July, quite late at night but not quite dark. Red eyes, black mouth. It was very big with great big wings and black claws...feather grey."*

Cryptozoologist, Fortean, and director of the Centre for Fortean Zoology, Jon Downes, tried to trace the girls for many years without success. Then, when he was preparing a book on the Owlman in 2000, he received a letter from somebody claiming to be Sally, now 38, married and living in Pembroke, Wales. She had been told about Jon and his work by a friend and found his details online.

She noted how nice Doc had been and how the event still affected her 24 years later.

She said her friend had emigrated to Australia. The event clearly scared her and she recounted it in the detailed letter. The hissing had alerted them to something odd and then...

"I had seen a horror film in Plymouth a few months before, a werewolf film with Peter Cushing. It was the first thing I thought when I saw it. I thought it was a werewolf. The face wasn't really like an owl-thinking back, it was like a frowning sneering black thing. The eyes were burning, glaring and reddish. I don't know if it had fur or feathers, but it was grey and grizzled like the werewolf in the film. I remember hearing Barbara start to laugh but it was a sort of choked, panicked laugh...

I knew right away it was REAL. It wasn't like a monster in the films that look rubbery and fake. It just looked like a very weird, frightening animal, as real as any animal in a zoo. It looked flesh and blood to me, but there is simply no way it could have been. It couldn't have been something that was born and grew. No way. I have no idea what it was. My head hurts even thinking about it.

It was more frightening than I can really describe. I remember blood rushing to my head, making it pound. It just stood there for what might have been a minute. I'm not really sure how long

Barbara was laughing, but it was more like a sort of breathless, hysterical sound by now. I wanted to run but couldn't. It was so EVIL, intensely so. When it moved, that nearly did it, I nearly started running. Its arms or wings or whatever went out, and it just rose up through the trees. Straight up through the evergreens. It didn't flap, it didn't make a sound. Then, weirdly, I thought 'costume' for the first time because the legs looked wrong. They looked like a kind of grey trouser material, certainly unnatural. I can't be entirely sure now. And then the feet. Black, hooking things. I have no idea how it stood up on them. They were like an earwig's tail-piece.

It's difficult trying to remember exactly what happened next. The wood was quiet, but it felt as if it, the thing, could appear again at any second. I think I had nearly fainted at one point. Babs was the same. We were shaking like leaves. I was thinking that someone was going to come out of the woods, laughing at their trick, but really I knew that it couldn't possibly be a trick. On one level, my mind simply wasn't accepting it. It still doesn't in a way. That's how I got over it, I think. By pretending that it hadn't really happened. As I mentioned earlier, in some ways I lost a lot of years to it. Somehow, shaking and crying a bit, we got packed up. That was the worst time, waiting for it to come back. I don't think I could have coped seeing it again. My mind was

POUNDING, ballooning. I don't know how it could have disappeared like it did. The woods weren't that thick. Not thick enough to hide what I can only think of as a monster. I know that sounds silly, but it is perhaps the most apt way to describe it. It seemed to just vanish, like a ghost."

The girls spent a sleepless night in another part of the woods. Sally seemed to know that the encounter was over when the normal sounds of the woodlands came back. In the morning, they walked down to the beach were they met Doc.

Soon after, a letter appeared in *The Falmouth Packet* written by a girl called Jane Greenwood.

"Sir. - I am on holiday in Cornwall with my sister and our mother. I, too have seen a big bird-thing like that pictured in 'Morgawr-The Monster of Falmouth Bay'. It was Sunday morning and the place was in the trees near Mawnan Church, above the rocky beach. It was in the trees, standing like a full grown man, but the legs bent backwards like a bird's. It saw us and quickly jumped up and rose straight up through the trees. My sister and I saw it very closely before it rose up. It has red, slanting eyes and a very large mouth. The feathers are silvery grey and so are his body and legs. The feet are like big, black, crab's claws. We were frightened at the time. It was so strange, like something out of a horror film. After the thing went up, there were crackling sounds in the tree-tops for ages. Our mother thinks we made it all up just because we read about these things, but that is not true, we really saw the bird-man, though it could have been someone playing a trick in a very good costume and make up. But how could it rise up like that? If we imagined it then we both imagined it at the same time"

The publication mentioned, *'Morgawr-The Monster of Falmouth Bay,'* was a booklet published at the time. The author calls themselves 'A. Mawnan-Peller'. Nobody knows who the writer was.

Other witnesses included the sixteen-year-old daughter of one Ken Opie, who said she saw a creature like a devil flying through the trees near the church, and two French girls, students at Cambourne Tech, who saw a big, furry bird with a gaping mouth and big eyes.

The next witness to the owlman was probably the most important for reasons that will become clear as the story moves on. In 1995, Jon Downes bought up the notes and letters sent to a magazine known as *Crypto Chronicles*. The magazine had folded, so Jon had purchased the whole collection of notes and correspondence from the magazine's editor / owner. Among these was a letter from a young man who claimed to have seen the Owlman in 1988/ 89 whilst on holiday with his girlfriend. He was 12/13 at the time and had seen the creature as he walked at night through the woods near Mawnan Old Church.

Jon wrote to the witness who much to his surprise wrote back. He insisted on keeping his real name from the public, so he was referred to as 'Gavin' and his girlfriend as 'Sally'. In his lengthy letter he confirmed the events and confessed they had badly affected him.

> *"The Mawnan Owlman is a ridiculous thing that only a bizarre attention-seeker would even pretend to have seen. I am ANGRY that I have seen this creature. The whole thing is so stupid. But because I have seen this—AND I KNOW WHAT I SAW, I am prepared to discuss the event intelligently. For the above reasons (the obvious old 'laughing stock' excuse of those who have witnessed paranormal things), I am anxious not to be identified, and the idea that what I say might be on TV scared the living shit out of me!"*

The pair saw the creature at about 9.30pm. It was squatted on a branch of a conifer tree, gripping the branch with black, pincer-like claws. It was silvery grey and had glowing eyes. As the torch beam fell on it, the thing drew its head back and leapt off the tree. Despite being scared, the couple returned later with a camera, but did not see the creature again. They photographed the tree that they had seen it in.

Gavin went on to say how the creature had haunted his dreams and one dream forced him to hunt and kill another person with a bow and arrow.

> *"But this isn't all and the rest of my story is more kind of psychological. You cannot possibly imagine what effect an event such as this can have on your life...Well, following our return to Southampton (prepare yourself for a big shock) both 'S' and I kept seeing the Owlman. I saw it in the woods behind my house. And 'S' in the woods she passed on the way home from school."*

Jon asked Gavin if he would be willing to travel down to Devon to be interviewed. Bravely he agreed. Jon and a colleague, the late John Jacques, conducted a long, recorded interview with 'Gavin' about this experience. He expanded on the description.

> *"It was similar to a man in shape, and four or four and a half feet high. It was grey, and with a texture like that on a bird with soft feathers, and the torch made it brighter than I think it was. The torch caught the eyes so they were reflected in the light and were bright yellowy, I think.*
>
> *It had a torso like a man, but had legs which were in proportion to its*

body, again similar to a man. It was actually standing on this branch, which would have been quite a thick branch and virtually all of its feet were wrapped around the branch. The feet were black and looked like pincers...

About half way up the legs was an ankle with the legs bending back, with no visible knee like in a person's knee. The torso would have been spindle shaped' broad at the top and at the bottom, so that it tapers down towards the legs and went straight down.

The head was really odd shaped. A really flat face, without any real outstanding features like an animal with a snout or anything. It had a black mouth curling down quite sharply at the corners, like an inverted 'V', and with no other features on the face apart from the eyes, really...

The top of the head was pointed and it had flaps at the side of the head about the same level as the eyes, that I think were pointed. It had quite large arms, or wings, with long feathers, the same colour as the rest of the body —a light grey— which weren't folded; they were held at the side so that the very tips of the feathers were like hanging around, nearly touching the branch that the animal was standing on. "

The fear he felt lingered with him.

"Still now, even today, if I look at trees, I look for the shape of an Owlman in the trees. For quite a long time afterwards, at least the next year or so, I was very very nervous of going near trees in the dark...

About the freaky dreams, I think of Owlman now as a malevolent entity, which is a thing that isn't just there, but is bad. I'm not going to say 'evil' because 'things' can't be evil. I'm not saying Owlman is an animal. An animal can't be evil right?

I think that Owlman is maybe like other psychic phenomena. It is a kind of projection towards people that it involves. "

I met and became friends with Gavin through Jon Downes. I have known him for well over twenty years. Gavin is now a very well respected and internationally known scientist, a man gifted with exceptional intelligence. We have spoken at length about the Owlman and he still fears it over thirty years later. He told me how he had thought that the creature had somehow followed him home and was watching him in the woods from behind his house and haunting his dreams. Whilst talking about Owlman, he went from a confident expert in his field to a scared child. It was truly alarming.

The Owlman was seen on and off in the following years. The creature itself has entered the Fortean hall of fame.

So it is here we finish with our round up of weird monster encounters. I could have written several books on this one subject. We can conclude that, broadly, there are two types of monsters in the world. The first are unknown animals, some of which are large, frightening, and potentially dangerous, but they are flesh and blood. With luck, you could eventually capture one and put it on display at the zoo. The second type is something very different. They shrug off conventional weapons such as gunfire, like pea shooters. They appear and then vanish quickly. They have an effect on electrical equipment. They manifest alongside other Fortean phenomena such as UFOs, animal mutilations, and poltergeist outbreaks. They provoke intense and lingering feelings of fear and dread in their witnesses. They seem also to be able to project a feeling of pursuit, as if they can somehow follow the witness from the sighting area all the way to their home —a truly creepy thought. In short, this type of monster is not an animal in the conventional sense, but this does not make them any the less real or any the less frightening.

CHAPTER TWO: GHOSTS

"Just when I thought I could not be stopped
When my chance came to be king
The ghosts of my life blow wilder than the wind.

Japan, Ghosts.

A clay tablet held at the British Museum and dating to ancient Babylon, some 3,500 years ago, was recently discovered to be the instructions for dealing with ghosts infesting the home. The ritual instructs the reader to make figurines of a man and a woman, and equip them with specific items, including travel provisions for the man and furniture for the woman. The figurines are then to be buried together at sunrise while the exorcist recites a spell. The spell on the tablet is incomplete, but it begins by calling the sun god Shamash, who was responsible for the movement of ghosts to the underworld. The writer of the ritual instructs the exorcist not to look back.

Like monsters, ghosts have a long pedigree. Everybody thinks they know what a ghost is supposed to be. It's a restless spirit of a dead person. In fact, relatively few ghosts can be linked to known people who have passed on. The field is a broad one and many phenomena are included under its umbrella. To link most of these with souls of the departed is as naïve as thinking that all strange light phenomena are alien spaceships.

There is confusion even inside the field. Parapsychologists often make the distinction between ghosts and poltergeists. A ghost, they say, is place centred, whilst a poltergeist is person centred. The former will have a visual apparition and little physical effect, whilst the latter is inviable and can wreak violent destruction. As we shall see, the two phenomena often occur concurrently and seem part of one whole. Hauntings can vary wildly and the sometimes manifest some of the strangest occurrences within Forteana.

WEIRD HAUNTINGS
The following selection of cases have been chosen simply for their high strangeness. I have ordered them roughly chronologically. As we proceed, the reader will notice how

many of these hauntings possess characteristics in common with monsters sightings and with the UFO encounters and fairy lore we will be looking at in later chapters.

THE DRUMMER OF TEDWORTH

The 1661 case of 'The Drummer of Tedworth' is notable for several reasons. It was perhaps the first haunting that was well documented as it unfolded. There were multiple witnesses including Mr Mompesson, his family, visitors and servants. It also included many very odd factors. The case was investigated by Joseph Glanvill, a member of the newly-created Royal Society.

The story unfurled in March. Mr John Mompesson was a magistrate living in the town of Tedworth in Wiltshire, England. Mompesson was at the neighbouring town of Ludgarshal, at the house of a friend who was the bailiff. Mompesson heard loud drumming from outside and. on asking about the racket. he was told that an itinerant drummer, William Drury had been at large in the town, demanding money for his performances with a licence that the bailiff thought was counterfeit. On confronting the drummer, Mompesson examined his pass. The drummer claimed that it was signed by none less than Sir William Cawly and Colonel Ayliff of Gretenham. Unknown to Dury, Mompesson knew both of these men and was familiar with the handwriting of both men. He had the drummer arrested by a constable and the drum taken from him and kept by Mompesson.

Mompesson had business in London and returned home on May 4th to have his wife tell him that there had been several attempted burglaries.

> ". . . *my wife told me that my house had like to have been broken up, and they had been much affrighted in the night with thieves, I rejoiced with her at the deliverance, and after I had been at home three nights, it was come again, so I arose and took some pistols in my hand, and went up and down the house, and heard a strange noise and hollow sound, but could not see any thing. . . So then it came oftener, five nights, and absent three. . .*
>
> *It would thump very hard all in the outside of my house, and then it came to the room where the Drum lay, being my Mothers Chamber, where he was thrown under a board, for my children did use to knock and play with it, and she delighting in their company caused it to be put there. There it would be four or five nights in seven, and make very great hollow sounds, that the windows would shake and the beds, and come constantly within half an hour after we were in bed, and stay almost two hours, and when it came we could hear a perfect hurling in the air over the house, and when it went away many times the Drum beat the same point of War that is usually beaten*

> *when guards break up as truly and sweetly as ever Drum*
> *beat in this world and so continued two months."*

Mompesson's wife gave birth and the strange disturbances seem to vanish for a period of three weeks. But then the strange noises returned.

> *"It would make Chairs, Tables, Trunks & all moveables walk*
> *up and down the Rooms. And often come tumble down the*
> *stairs, sometimes [making a noise] like a bowl & other times*
> *as if it drew a chain after it."*

Scratching noises began under the children's beds that sounded like they were made by 'iron talons'. The ghost focused on the baby, snatching candles away from its room and taking them up the chimney. At one point, a sound like a large bag of money being emptied on the floor was heard, though none was seen.

The servants found themselves victims too.

> *"It has taken our servants up in their beds, bed and all, and hath*
> *lifted them up a great height, and laid them down softly again,*
> *and lays often on their feet with great weight. Sometimes the*
> *candles will not burn in the room where it is, and though it*
> *come never so loud and on a sudden, yet no dog will bark: it*
> *hath often been so loud that it hath been heard into the fields*
> *and has wakened my neighbours in town.*

Soon the hauntings were preceded by light phenomena described as blue and flickering lights. One such light came into Mompesson's bed chamber and apparently it hurt his eyes to look at it. A strange sulphurous smell was noted in Mompesson's home as well. A disembodied voice was heard to cry, *"a witch, a witch."* During this time, it was common for such disturbances to be blamed on witches. It may seem strange, but are we any better now, blaming them on spirits of the dead?

By 1663 things were escalating. Items began to appear in unusual places, such as a long piked iron in Mompesson's bed, a knife in his mother's bed, and a Bible in the ashes of the fire. The entity would yank the sheets from the servant's beds and sometimes hold them down. One stout servant struck out with his sword at the ghost and it left him alone.

Another servant claimed to have seen the ghost at the foot of his bed. It had a great, dark, hulking body and red eyes. If such a figure was seen in a forest in modern times, it would surely be identified as a Bigfoot!

Sir Thomas Chamberlin, governor of the East India Company, visited the house. He asked if the Drummer had been sent by the Devil and to beat three times if he had.

Three beats of the drum were clearly heard.

In January 1663 Joseph Glanvill, the Vicar of Frome in Somerset, stayed with Mompesson. The description contained in his 1681 book *Saducismus Triumphatus* gives a detailed record of phenomena that he observed in the children's room.

> *"At this time it used to haunt the children, and that as soon as they were laid. They went to Bed that night I was there, about Eight of the Clock, when a Maid-Servant coming down from them, told us it was come. The neighbours that were there, and two Ministers who had seen and heard divers times went away, but Mr. Mompesson and I, and a Gentleman that came with me went up. I heard a strange scratching as I went up the Stairs, and when we came into the Room, I perceived it was just behind the Bolster of the Children's Bed, and seemed to be against the Tick. It was as loud a scratching, as one with long Nails could make upon a Bolster. There were two little modest girls in the Bed, between Seven and Eleven years old as I guessed. I saw their hands out over the Clothes, and they could not contribute to the noise that was behind their heads.*
>
> *I, standing at the Beds-head, thrust my hand behind the Bolster, directing it to the place whence the noise seemed to come. Whereupon the noise ceased there, and was heard in another part of the bed. But when I had taken out my Hand it returned, and was heard in the same place as before. I had been told that it would imitate noises, and made trial by scratching several times upon the Sheet, as 5, and 7, and 10, which it followed and still stopped at my number. I searched under and behind the Bed, turned up the Clothes to the Bed-cords, grasped the Bolster, sounded the Wall behind, and made all the search that possible I could to find if there were any trick, contrivance, or common cause of it; the like did my friend, but we could discover nothing.*
>
> *It went in to the midst of the Bed under the Children, and there seemed to pant like a Dog out of Breath very loudly. I put my hand upon the place, and felt the bed bearing up against it, as if something within had thrust it up. I grasped the Feathers to feel if any living thing were in it. I looked under and every where about, to see if there were any Dog or Cat, or any such creature in the room, and so we all did, but found nothing. The motion it caused by this panting was so strong, that it shook the Room and Windows very sensibly. It continued thus, more than half an hour, while my friend and I stayed in the Room, and as long after, as we were told.*
>
> *During the panting, I chanced to see as it had been something (which I thought was a Rat or a Mouse) moving in a Linen Bag, that hung up against another bed that was in the room. I stepped and caught it by the*

upper end with one Hand, with which I held it, and drew it through the other, but found nothing in it at all. There was no body near to shake the Bag, or if there had, no one could have made such a motion, which seemed to be from within, as if a Living Creature had moved in it."

The idea of some small creature crawling around is repeated in other hauntings, though why this should be, is a mystery.

The phenomena had a weird effect on horses stabled at the house too. Glanvill, found his horse sweating one morning, as if it had been ridden all night. This effect is known as being 'hag ridden' and is well known in folklore. A well-fed and rested horse is found exhausted in the morning. Tradition has it that the horse had been ridden by fairies or witches. Glanvill took his horse for a modest ride and found it was lame. Three days later it died. Mompesson himself found his own horse on its back with one of its hind legs in its mouth.

As for William Drury himself, whilst in Gloucester jail, he told a visiting man that he had plagued Mompesson for the taking of his drum. He also confessed that during his time as a solider under Cromwell, he had gotten some books off an old man he took to be a wizard.

Mompesson brought accusations against Drury based on the witchcraft act of 1604. He and three other men testified at the court that they had seen chairs, stools and bedstaves move when no one was near them, witnessed parts of the house visibly shake, and heard the beating of drums in the air over the house. A servant testified that he had heard Drury confess to having caused the phenomena: The Drummer said,

> *"...it was because he took my Drum from me; that trouble had ever befallen him, and he shall never have his quiet again, till I have my Drum, or satisfaction from him."*

Dury was acquitted of witchcraft, but found guilty of being a rogue and a vagabond. He was transported to the American colonies. Thus ended the strange case of the Drummer of Tedworth.

THE HAUNTING OF EPWORTH RECTORY

Epworth Rectory, in Lincolnshire is notable from two reasons. Firstly, John Wesley, founder of the Methodist movement lived here as a boy. Secondly, for several months in 1716-1717 it was the setting of a strange haunting. Once more, the records of this case are recorded. This time in a series of letters to and from Samuel Wesley Jr, John's elder brother, about the happenings, John Wesley kept these letters and later published them.

Young Sam was away in London and when the strange events began, some of them feared that it was a sign of Samuel Jr's death and his letter to them came as a relief.

The hauntings began with groans akin to those of a man in his death throes along with objects being moved and latches lifted. Soon after, a swishing noise was heard like a silk nightgown being dragged invisibly around.

The first letter from Samuel's mother details the events.

To Mr. Samuel Wesley, from his Mother, January 12, 1716

"Dear Sam, This evening we were agreeably surprised with your pacquet, which brought the welcome news of your being alive, after we had been in the greatest panic imaginable, almost a month, thinking either you was dead, or one of your brothers by some misfortune been killed. The reason of our fears is as follows; On the first of December, our maid heard, at the door of the dining-room, several dismal groans, like a person in extremes, at the point of death. We gave little heed to her relation, and endeavoured to laugh her out of her fears. Some nights (two or three) after, several of the family heard a strange knocking in divers places, usually three or four knocks at a time, and then stayed a little. This continued every night for a fortnight; sometimes it was in the garret, but most commonly in the nursery, or green chamber. We all heard it but your father, and I was not willing he should be informed of it, lest he should fancy it was against his own death, which, indeed, we all apprehended. But when it began to be so troublesome, both day and night, that few or none of the family durst be alone, I resolved to tell him of it, being minded he should speak to it. At first he would not believe but somebody did it to alarm us; but the night after, as soon as he was in bed, it knocked loudly nine times, just by his bedside. He rose, and went to see if he could find out what it was, but could see nothing. Afterwards he heard it at the rest. One night it made such a noise in the room over our heads, as if several people were walking, then run up and down stairs, and was so outrageous that we thought the children would be frighted, so your father and I rose, and went down in the dark to light a candle. Just as we came to the bottom of the broad stairs, having hold of each other, on my side there seemed as if somebody had emptied a bag of money at my feet; and on his, as if all the bottles under the stairs (which were many) had been dashed in a thousand pieces. We passed through the hall into the kitchen, and got a candle, and went to see the children, whom we found asleep. The next night your father would get Mr. Hoole to lie at our house, and we all sat together till one or two o'clock in the morning, and heard the knocking as usual. Sometimes it would make a noise like the winding up of a jack, at other times, as that night Mr. Hoole was with us, like a carpenter planing deals; but most commonly it knocked thrice and stopped, and then thrice again, and so many hours together. We persuaded your father to speak, and try if any

voice would be heard. One night about six o'clock he went into the nursery in the dark, and at first heard several deep groans, then knocking. He adjured it to speak if it had power, and tell him why it troubled his house, but no voice was heard, but it knocked thrice aloud. Then he questioned if it were Sammy, and bid it, if it were, and could not speak, knock again, but it knocked no more that night, which made us hope it was not against your death. Thus it continued till the 28th of December, when it loudly knocked (as your father used to do at the gate) in the nursery, and departed. We have various conjectures what this may mean. For my own part, I fear nothing now you are safe at London hitherto, and I hope God will still preserve you. Though sometimes I am inclined to think my brother is dead. Let me know your thoughts on it."

S. W. "

Once more we have the odd detail of the sound of a bag of money. Mr Hool, mentioned in the letter to Jack dated 16th September 1726, was the rector of the nearby village of Haxey. Mr Hool confirmed his experiences in a letter.

"As soon as I came to Epworth, Mr. Wesley telling me, he sent for me to conjure, I knew not what he meant, till some of your sisters told me what had happened, and that I was sent for to sit up. I expected every hour, it being then about noon, to hear something extraordinary, but to no purpose. At supper too, and at prayers, all was silent, contrary to custom, but soon after one of the maids, who went up to sheet a bed, brought the alarm, that Jeffery was come above stairs. We all went up, and as we were standing round the fire in the east chamber, something began knocking just on the other side of the wall, on the chimney-piece, as with a key. Presently the knocking was under our feet, Mr. Wesley and I went down, he with a great deal of hope, and I with fear. As soon as we were in the kitchen, the sound was above us, in the room we had left. We returned up the narrow stairs, and heard at the broad stairs' head, some one slaring with their feet (all the family being now in bed beside us) and then trailing, as it were, and rustling with a silk night-gown. Quickly it was in the nursery, at the bed's head, knocking as it had done at first, three by three. Mr. Wesley spoke to it, and said he believed it was the devil, and soon after it knocked at the window, and changed its sound into one like the planing of boards. From thence it went on the outward south side of the house, sounding fainter and fainter, till it was heard no more. I was at no other time than this during the noises at Epworth, and do not now remember any more circumstances than these."

Jeffery was the name given to the ghost by the family. Sam wrote back to his mother, asking if the new servants may have been playing tricks on the family. His mother wrote in answer.

From Mrs. Wesley to her Son Samuel – January 25th or 27th, 1716.

"Though I am not one of those that will believe nothing supernatural, but am rather inclined to think there would be frequent intercourse between good spirits and us, did not our deep lapse into sensuality prevent it; yet I was a great while e'er I could credit any thing of what the children and servants reported, concerning the noises they heard in several parts of our house. Nay, after I had heard them myself, I was willing to persuade myself and them, that it was only rats or weasels that disturbed us; and having been formerly troubled with rats, which were frightened away by sounding a horn, I caused a horn to be procured, and made them blow it all over the house. But from that night they began to blow, the noises were more loud, and distinct, both day and night, than before, and that night we rose, and went down, I was entirely convinced that it was beyond the power of any human creature to make such strange and various noises. As to your questions, I will answer them particularly; but withal I desire my answers may satisfy none but yourself; for I would not have the matter imparted to any. We had both man and maid new this last Martinmas, yet I do not believe either of them occasioned the disturbance, both for the reason above mentioned, and because they were more affrighted than any body else. Besides, we have often heard the noises when they were in the room by us; and the maid particularly was in such a panic, that she was almost incapable of all business, nor durst ever go from one room to another, or stay by herself a minute after it began to be dark. The man, Robert Brown, whom you well know, was most visited by it lying in the garret, and has been often frighted down bare-foot, and almost naked, not daring to stay alone to put on his clothes; nor do I think, if he had power, he would be guilty of such villainy. When the walking was heard in the garret, Robert was in bed in the next room, in a sleep so sound, that he never heard your father and me walk up and down, though we walked not softly I am sure. All the family heard it together, in the same room, at the same time; particularly at family prayers. It always seemed to all present in the same place at the same time, though often before any could say it is here, it would remove to another place.

All the family as well as Robin were asleep when your father and I went down stairs, nor did they wake in the nursery when we held the candle close by them; only we observed that Hetty trembled exceedingly in her sleep, as she always did, before the noise awaked her. It commonly was nearer her than the rest, which she took notice of; and was much frightened, because she thought it had a particular spite at her, I could multiply particular instances, but I forbear. I believe your father will write to you about it shortly. Whatever may be the design of Providence in permitting these things, I cannot say. Secret things belong to God: but I entirely agree with you, that it is our wisdom and duty to prepare seriously for all events."

S. Wesley.

Mrs Wesley and her children had kept the events from Samuel senior for the first couple of weeks. The fact that she was, at first, convinced that the noises may have been down to rats, shows she was pretty levelheaded. The idea of ghosts was not her first thought.

The trembling of Hetty in her sleep is of great interest. Many poltergeist cases seem focused on an adolescent, mostly, but not exclusively a girl. One theory is that the phenomena is unconscious telekinesis from the focus. Clearly, this cannot be possible in many of these instances, as they cannot account for all the phenomena such as apparitions, but it seems there is some kind of link in at least some cases.

Sam's sister also wrote to him, confirming the events.

From Miss Susannah Wesley to her Brother Samuel – Epworth, Jan 24. 1716.

"Dear Brother, About the first of December, a most terrible and astonishing noise was heard by a maid-servant, as at the dining room door, which caused the up-starting of her hair, and made her ears prick forth at an unusual rate. She said it was like the groans of one expiring. These so frighted her, that for a great while she durst not go out of one room into another, after it began to be dark, without company. But, to lay aside jesting, which should not be done in serious matters, I assure you that from the first to the last of a lunar month, the groans, squeaks, tinglings, and knockings, were frightful enough. Though it is needless for me to send you any account of what we all heard, my father himself having a larger account of the matter than I am able to give, which he designs to send you; yet, in compliance with your desire, I will tell you as briefly as I can , what I heard of it. The first night I ever heard it, my sister Nancy and I were set in the dining room. We heard something rush on the outside of the doors that opened into the garden; then three loud knocks, immediately after other three, and in half a minute the same number over our heads. We enquired whether any body had been in the garden, or in the room above us, but there was nobody. Soon after my sister Molly and I were up after all the family were a-bed, except my sister Nancy, about some business. We heard three bouncing thumps under our feet, which soon made us throw away our work, and tumble into bed. Afterwards the tingling of the latch and warming pan, and so it took its leave that night. Soon after the above mentioned, we heard a noise as if a great piece of sounding metal was thrown down on the outside of our chamber. We, lying in the quietest part of the house, heard less than the rest for a pretty while, but the latter end of the night

that Mr. Hoole sat up on, I lay in the nursery, where it was very violent. I then heard frequent knocks over and under the room where I lay, and at the children's bed head, which was made of boards. It seemed to rap against it very hard and loud, so that the bed shook under them. I heard something walk by my bedside, like a man in a long night-gown. The knocks were so loud, that Mr. Hoole came out of their chamber to us. It still continued. My father spoke, but nothing answered. It ended that night with my father's particular knock, very fierce. It is now pretty quiet, only at our repeating the prayers for the king and prince, when it usually begins, especially when my father says, "Our most gracious Sovereign Lord," etc. This my father is angry at, and designs to say three instead of two for the royal family. We all heard the same noise, and at the same time, and as coming from the same place. To conclude this, it now makes its personal appearance; but of this more hereafter. Do not say one word of this to our folks, nor give the least hint. I am, Your sincere friend and affectionate Sister, Susannah Wesley."

Another sister, Emily, wrote to Sam around the same time.

From Miss Emily Wesley to her Brother Samuel .

"Dear Brother,

I thank you for your last; and shall give you what satisfaction is in my power, concerning what has happened in our family. I am so far from being superstitious, that I was too much inclined to infidelity, so that I heartily rejoice at having such an opportunity of convincing myself, past doubt or scruple, of the existence of some beings besides those we see. A whole month was sufficient to convince anybody of the reality of the thing; and to try all ways of discovering any trick, had it been possible for any such to have been used. I shall only tell you what I myself heard, and leave the rest to others. My sisters in the paper chamber had heard noises, and told me of them, but I did not much believe, till one night, about a week after the first groans were heard, which was the beginning, just after the clock had struck ten, I went down stairs to lock the doors, which I always do. Scarce had I got up the best stairs, when I heard a noise, like a person throwing down a vast coal in the middle of the fore kitchen, and all the splinters seemed to fly about from it. I was not much frighted, but went to my sister Suky, and we together went all over the low rooms, but there was nothing out of order. Our dog was fast asleep, and our only cat in the other end of the house. No sooner was I got up stairs, and undressing for bed, but I heard a noise among many bottles that stand under the best stairs, just like the throwing of a great stone among them, which had broke them all to pieces. This made me hasten to bed: but my sister Hetty, who sits always to wait on my father going to bed, was still sitting on the lowest step on the garret stairs, the door being shut at her back, when soon after there came down the stairs behind her, something like a man, in a loose night-gown trailing after him, which made her fly rather than run to me in the nursery. All this time we never told father of it, but soon after we did. He smiled, and gave no answer, but was more careful than usual, from that time, to see us in bed, imagining it to be some of us young women, that sat up late, and made a noise. His incredulity, and especially his imputing it to us, or our lovers, made me, I own, desirous of its continuance till he was convinced. As for my mother, she firmly believed it to be rats, and sent for a horn to blow them away. I laughed to think how wisely they were employed, who were striving half a day to fright away Jeffery, for that name I gave it, with a horn. But whatever it was, I perceived it could be made angry. For from that time it was so outrageous, there was no quiet for us after ten at night. I heard frequently between ten and eleven something like the quick winding up of a jack, at the corner of the room by my bed's head, just like the running of the wheels and the creaking of the iron work. This was the common signal of its coming. Then it would knock on the floor three times, then at my sister's bed's head in the same room, almost always three together, and then stay. The sound was hollow, and loud, so none of us could ever imitate. It would answer to my mother, if she stamped on the floor, and bid it. It would knock when I was putting the children to bed, just under me where I sat. One time little Kesy, pretending to scare Patty, as I was undressing them, stamped with her foot on the floor, and immediately it answered with three knocks, just in the same place. It was more loud and fierce if any one said it was rats, or anything natural. I could tell you abundance more of it but the rest will write, and therefore it would be needless. I was not much frighted at first, and very little at last: but it was never near me, except two or three times: and never followed me, as it did my sister Hetty. I have been with her when it has knocked under her, and when she has removed has followed, and still kept just under her feet, which was enough to terrify a stouter person. If you would know my opinion of the reason of this, I shall briefly tell you. I believe it to be witchcraft, for these reasons. About a year since, there was a disturbance at a town near us, that was undoubtedly witches; and if so near, why may they not reach us? Then my father had for several Sundays before its coming preached warmly against consulting those that are called cunning men, which our people are given to; and it had a particular spite at my father. Besides, something was thrice seen. The first time by my mother, under my sister's bed, like a badger, only without any head that was discernible. The same creature was sat by the dining-room fire one evening; when our man went into the room, it run by him, through the hall under the stairs. He followed with a candle, and searched, but it was departed. The last time he saw it in the kitchen, like a white rabbit, which seems likely to be some witch; and I do so really believe it to be one, that I would venture to fire a pistol at it, if I saw it long enough. It has been heard by me and others since December. I have filled up all my room, and have only time to tell you, I am,

Your loving sister, Emilia Wesley."

Once more, we have seen a small, odd animal turning up in the midst of a haunting and seeming to be part of it. The 'badger without a head' sounds like some fevered description of a witch's familiar gained by Matthew Hopkins as he tortured some innocent. Would a headless badger be something a young girl just made up off hand? A small, hairy animal with a name akin to Jeffery would one day become a superstar of Forteana over 300 years later. We will be meeting him in the penultimate chapter. In the meantime, back to Epworth.

By March, the phenomena seemed to be winding down, there were just a few knockings and these faded away.

Another sister wrote to confirm the haunting.

Miss Susannah Wesley to her Brother Samuel – March 27th 1716.

> *"Dear Brother Wesley, I should farther satisfy you concerning the disturbances, but it is needless, because my sisters Emilia and Hetty write so particularly about it. One thing I believe you do not know, that is, last Sunday, to my father's no small amazement, his trencher danced upon the table a pretty while, without any body's stirring the table. When, lo! an adventurous wretch took it up, and spoiled the sport, for it remained still ever after. How glad should I be to talk with you about it. Send me some news, for we are secluded from the sight, or hearing, of any versal thing except Jeffery."*
>
> *Susannah Wesley.*

John recorded the account of the servant in 1726.

> *"The first time Robin Brown, my father's man, heard it, was when he was fetching down some corn from the garrets. Somewhat knocked on a door just by him, which made him run away down stairs. From that time it used frequently to visit him in bed, walking up the garret stairs, and in the garrets, like a man in jack-boots, with a nightgown trailing after him, then lifting up his latch and making it jar, and making presently a noise in his room like the gobbling of a turkey-cock, then stumbling over his shoes or boots by the bed side. He was resolved once to be too hard for it, and so took a large mastiff we had just got to bed with him, and left his shoes and boots below stairs; but he might as well have spared his labour, for it was exactly the same thing, whether any were there or no. The same sound was heard as if there had been forty pairs. The dog indeed was a great comfort to him, for as soon as the latch began to jar, he crept into bed, made such a howling and barking together, in spite of all the man could do, that he alarmed most of the family. Soon after,*

being grinding corn in the garrets, and happening to stop a little, the handle of the mill was turned round with great swiftness. He said nothing vexed him, but that the mill was empty. If corn had been in it, old Jeffery might have ground his heart out for him; he would never have disturbed him. One night, being ill, he was leaning his head upon the back kitchen chimney (the jam he called it) with the tongs in his hands, when from behind the oven-stop, which lay by the fire, somewhat came out like a white rabbit. It turned round before him several times, and then ran to the same place again. He was frighted, started up, and ran with the tongs into the parlour (dining room)."

The sound of machinery, such as the winding of a jack, or corn grinding, is frequently reported in all kinds of Fortean events. Recall the sounds of underground machinery in some Bigfoot cases?

THE HORROR OF GYB FARM
Frederick George Lee (1832-1902) was an Anglican priest who wrote several books on the supernatural. These were anthologies of stories mostly taken first hand from the witnesses themselves. His books have now been mostly forgotten by all save the most ardent Forteans. One of the strangest stories was a testimonial written for him on February the 10th 1877 of strange events occurring around 1837. The writer was one David Eustace 'the eldest surviving member of an old yeoman family of Buckinghamshire. His nephew Joseph Eustace also wrote a testimony on February 11th 1877. David's narrative runs thus...

"Some forty years ago my father resided at a small farm-house, the back part of which faced a large unenclosed common (since inclosed) and stood close to four cross roads, two of which lead to what thereabouts is called "Uphills," the Chiltern Ridge from Tring to Wycombe and Stokenchurch. The spot is very lonely even now, but was much more so then: for, at that time, there was not a single human habitation within a quarter of a mile of my father's abode.

Our house had always been called "The Gyb Farm"-why, we did not exactly know—but because, as we afterwards found out, there had been often erected near the site of it, a gibbet for the punishment of malefactors, and many a person who had taken his own life (let alone the murderers, highwaymen and sheepstealers) had been buried at the side of the road there: but the name of the farm, as a law-parchment states, seems to have been altered about the year 1788, when a much less disagreeable name was then adopted for it.

In the year, and about the time, that King William IV died (i.e. in 1837), my father and mother, two of my sisters, a younger brother and myself were all at home. One night, when we had all been in bed for some time, quite in the smaller hours, we were each suddenly startled and awakened by the most frightful, shrill and horrid shrieks and noises just outside on the roadway that ever man heard. Partly human and partly as if made by infuriated hogs, violently quarrelling, the roar and the screeching simply appalled us. I never heard the like of it in my life. It went through and through me.

For a little while we all endured it: but in about five minutes we gathered half-dressed at the top of the staircase—father, mother, my brother and I— and went to a long front window overlooking the road, in order to learn the cause.

The night was rather dark, and our tinder-box would not light, we were looking out, without any candle or lamp, towards the spot from which this horrible and hellish row came, when all of a sudden a white face-a face most awful in its pallid aspect and miserable imploring look—was pressed from outside against the glass of the window and stared at us wildly. We all saw it, and I could mark that even my father was deeply affrighted.

The indescribable and unearthly noises still continued, and even increased in their discordance and frightful yelling for at least four or five minutes. Then by that time a candle had been procured.

My father at once opened the lattice: and there by the light of the sky, such as it was, we saw a collection of the most hideous black animals, some of them like large swine, others horrid and indescribable in their appearance, grubbing up the ground and half buried in it, scattering the earth upwards where the graves were, fighting, screaming and roaring in a way that no mere words can properly tell or set forth. Some of them, judging by their motion, seemed to have no bones in them.

We were all very much terrified. My mother implored the Almighty to protect us, and I confess that, overwhelmed by fear, I prayed most heartily to God for his assistance. In a minute or two after this, with shrieks increased in intensity the frightful creatures (whatever they were) rushed screaming down one of the roads.

In the morning there was not a sign nor sound to be seen. The ground had not been in the least degree touched, scratched up nor disturbed. But the "Ghosts of the Gibbet," as we afterwards discovered, had been seen by others than us."

It sounds like a vista of hell painted by Hieronymus Bosch. I can recall no other report quite like it. It would be interesting to try and identify Gyb Farm, if it still stands today.

THE COALBAGGIE KOBOLD

On May 15th, 1894 the *Dubbo Dispatch* a newspaper based in Dubbo, New South Wales, Australia, ran an extraordinary story. It concerned a German-born man Peter Stein, who had brought a 2560-acre property at the isolated Coalbaggie Creek, some 21 miles north of Dubbo. He had brought with him his English wife, children, and a man named Daley, an adopted son. Stein had arrived in 1891. The property had nothing but an old hut on it and Stein had commenced to build a house on the site. No sooner had they settled in, however, the family began to be plagued by hideous screams and moving furniture. A weird, goblin-like creature was also sighted.

The following month, on June 1st, Peter Stine had ridden to Dubbo to seek help with the manifestations. He visited the offices of the *Dubbo Dispatch* and was interviewed there. The paper published a follow up article.

Strange sounds had been heard in the old hut when the family had first arrived, but the haunting did not really begin until they had built their new house and moved in. It threw around candlesticks and furniture, shook bedsteads, as well as hammered tin dishes so hard it put dents into them. The entity also spoke. It claimed that it's mother and sister had been fatally burned in the hut after it's mother's dress caught fire as she was cooking. The dying mother had given her sister sixty pounds and a gold watch, asking her to keep it for her son. But the boy's uncle had struck him upon the head with the handle of a stock whip and left him for dead. The implication being that the ghost was that of the murdered boy who had been robbed by his uncle. Ghosts, however, are notorious liars and it seems that something much stranger was at work in Coalbaggie. Stein says...

> *"One night mother and I were sitting by the fire, and clods were pelted at us as if by some person in the fireplace. We looked and saw a strange figure. It had the body of a child, about five years old, and a most peculiar face, with a whitish beard on it. I went to catch it and it dissipated. On another occasion I saw something like a hand coming over a box, and when I tried to grasp it there was nothing."*

The entity could cause chaos in four different places in the house at once. The thing hated religious paraphernalia, breaking a crucifix into pieces and hurling it into the faces of the family. Some blessed candles they had brought from Dubbo were destroyed.

Other people independently witnessed phenomena when visiting the Steins. A Mr M'leod and a Mr Dwyer and his son all witnessed mysterious fires break out in four different parts of the house.

Of the fires Stein said...

> *"On Friday and Saturday last it was very bad and it took us all we could*

do to prevent it burning down the place. We can smell like fire before we see it and the house will be on fire in four different places at once, and the bedclothes and articles of female apparel also burning."

The creature seemed to be able to shapeshift as well.

"On one occasion it came in the shape of a bear, got up on the wall of the kitchen, and when we went to chase it away, it disappeared in a white smoke. On another occasion, a big mouse about a foot long, came on the roof and mysteriously moved about and another day a wallaby was near the house and would not shift for my sisters. They tried to put the dogs on it but they came back with their tails between their legs. My brother and I put two kangaroo dogs on it, and it ran into the creek and disappeared as if into the ground. The dogs came out on the other side looking terribly frightened. That night it talked to us and said it was the wallaby that we were chasing. it said it could appear as a lizard or snake or any shape it liked.

It told us that it was no use to bring out the priests. It said it would haunt us not only while we were at Coalbaggie, but would follow us about wherever we went. It talks in two voices, and sometimes speaks sensibly enough, while at others it seems quite mad, and uses language that could not be beaten by the lowest Sydney larrikin."

When asked why it broke the crockery, it answered that when its mother was burned, it had been carrying crockery and dropped it in shock. Therefore, it amused it to brake crockery. Peter Stein claimed it had broken £100 worth of crockery, a huge amount for a poor family in the 1890s. The thing also sang.

"It seemed to know all sorts of songs. It seems fond of 'The Banks of the Clyde' and 'The Ship that Never Returned'. It sings the last one pretty fair but it is quite horrible to hear it singing 'The Banks of the Clyde' - it's quite sickening.

What finally happened at the house is not recorded, though Stein said that the manifestations had become quieter after he had spoken about it to two priests.

That Stein was of German heritage is very interesting. In Germanic folklore there is a type of fairy called a kobold. Some inhabit mines and others lurk in houses and farms. They are often described as looking like short, naked people with beards. They could also manifest as balls of fire and shapeshift into animals. Kobolds can be useful, doing chores around a house or farm if rewarded with food and drink, but can also cause utter chaos. The mine dwelling kobolds seemed to be more malevolent than the other kinds, often tricking miners into digging up poisonous deposits instead of ore. This case could have just as easily fitted into the section on fairies that appears later in this book. One wonders how many poltergeist outbreaks in past centuries were attributed to fairies. And who is to say that attribution was wrong?

THE CASE OF THE GIANT GHOST CRAB

Perhaps the weirdest phantom on record is a giant, floating, ghostly crab recorded in the late 1800s in South Africa. The case is mentioned in a book called *They Walk in The Night True South African Ghost Stories and Tales of the Supernormal* by Eric Rosenthal. This is the quote verbatim.

> *"Here is a tale set down by Mr C.H Basson, an eyewitness of what went on in the home of Mr J. van Jaarsveld of Haartebeest River in the district of Uniondale shortly after the Jameson Raid. It appeared in the daily paper "Dagblad", of Cape Town, as follows:*

> *"In consequence of what he heard, he went to the farm of a Mr van Jaasveld. Shortly after sunset, the spook commenced his pranks and certain noises were heard coming out of a chest. The spook seemed especially attracted to Mr van Jaasveld and his niece, Miss Mayer. Whenever the latter dared to take a seat on the chest, she was moved about and the chest moved also.*

> *But the most weird thing of all "proceeded Mr Basson' "happened at night time. Miss Meyer went to lie down. We blew out the candle, but no sooner had this been done than she called out to us to light it again. We did so and lo! The spook had, during the few seconds that the candle was extinguished, tied her hair firmly to the bedpost. We untied it and plaited her hair into one tress, tied it at the end with a firm knot and made her lie down again. We then ranger ourselves round her bed, each with a box of matches in his hand. The candle was blown out again. Immediately afterwards, she cried out that the ghost was tugging at her hair. We all struck a match and found that one strand of the plait had been twisted out and tied as firmly as ever to her bedpost. Three of those present were able to see the spook. They say it resembled a phosphorescent crab with two huge pincers. They saw it 'floating' about the room touching here and there. On a former occasion it assumed the form of a skeleton hand with two fingers."*

The Jameson Raid mentioned in the text was a botched British raid on the South African Republic that took place from 29th December 1895 to 2nd January 1896 and allows us to roughly date the case. You have to ask, who in their right mind would make up such an unlikely ghost if they were spinning a yarn? A giant, glowing, floating ghost crab? Such a weird detail makes me more inclined to believe the story.

THE THINGS AT ABBEY HOUSE

Abbey House was built around 1590 and stands near the centre of the university town of Cambridge, England. No strange events were reported at the house until October of 1903. Professor J.C Lawson, a classics scholar and his family moved in and seemed to trigger a weird haunting. On the very first night of their residence, the disturbances began. Two maids who were sharing a room were awoken by a pounding on their door that was so savage they thought the door would be ripped from its hinges. When the noises subsided, the two maids thought it may have been the nursemaid who was sleeping in a room with the family children across the hall. The pair hurried across the hall to the nursemaid's room. There, they found the nursemaid trying to console the two children who had wakened and who were sobbing with fear. When they discovered that none of them had caused the banging, the three women huddled together in the nursemaid's room for the rest of the night and vowed that they would not spend another night in Abbey House.

The next day Professor Lawson told the women that the family's Newfoundland dog had caused the noises as it searched for the children in the new home. The three maids agreed to stay. The professor, of course, had lied in order to keep the women in his service.

The professor and his wife themselves were awakened many times by some unseen thing walking noisily along the corridor outside their bedroom. Each time they looked, nothing was there. The noises also came from the room next to their bedroom. Mrs Lawson's brother slept there when visiting Abbey House. After the first night, her brother said that "it" walked around the room all night.

> *"Oh, I could've slept through footsteps, I'm sure,"* he said, *"but it--whatever it was--tripped over my boots and made a most unholy noise. Then I heard it walking round the room, but at least it avoided colliding with my boots after that!"*

The ghost of a nun would appear between midnight and 4 am. Mr. and Mrs. Lawson stated that they observed the Nun many times walking across the master bedroom from the door to the foot of the bed. After a few moments, the Nun would turn, walk to the window, and vanish there. Jane, three, had her own experience with the Nun. For a short time, she slept in her parent's bedroom. One night, Mrs. Lawson woke to hear her daughter crying in the crib beside her. She rocked her back to sleep and in the morning asked her daughter what had happened. The girl said...

> *"Something came and stared at me, Mummy, it came in at the door. Then it came to my crib and stared at me."*

During the summer of 1907, when Mrs. Lawson was ill for a few weeks and resting in bed, she grew impatient with the Nun's nightly visits. Mrs. Lawson needed her rest, but the Nun kept standing at the foot of the bed and sighing loudly. Finally, one night, Mrs. Lawson sat up and made the sign of the cross. Then she said, *"In the name of the Holy Trinity, poor soul, rest in peace."* The Nun walked away towards the curtain and vanished. According to Mrs. Lawson, she didn't see the Nun again after that.

But there was a stranger thing at large in Abbey House. A creature that the children christened 'Wolfie' was seen by both the adults and children. It resembled a huge brown hare or rabbit but with short ears. It ran around on its hind legs.

Professor Lawson described it thus...

> *"...a nondescript kind of animal resembling a large-sized hare, but with close clipped ears. It was always seen and heard running about on its hind legs, the patter of its footsteps being very distinct and characteristic. It is never seen standing still or moving slowly. Its haunts were the downstairs parts of the house...It was generally seen in twilight but it was also seen quite clearly by artificial light in the drawing room."*

Professor Lawson himself heard the thing coming down a passage, and watched it as it ran into view and ran past him. It looked perfectly solid.

John, the Lawson's son, was the first to see Wolfie in 1904. John, who was three at the time, was in the nursery when his mother came to get him.

"Where's it gone, Mummy?" he asked. He was standing in the middle of the room, looking around as if he had lost something.

"Where has what gone?" his mother asked.

"That little brown thing--it was standing there," he said, pointing to the door. *"It ran to the curtain and now it isn't there."*

Mrs. Lawson looked at the window and saw that it was closed. Then she searched the room with her son, but found nothing. In her diary, Mrs. Lawson wrote that her son *"looked strange--though not frightened--and I did not like to question him much"* for fear that she would frighten the boy.

Their daughter, Jane, saw it. One night, as she was being tucked into bed, Jane told her mother, *"Sometimes after you go downstairs again at night, Mummy, I see a lot of little brown things walking in at the door and round the room. Do you think they is wolfies?"*

"There aren't any wolves around here," Mrs. Lawson assured her. *"Maybe it's the cat."*

"No, it's not the cat," Jane said. *"There are lots of them and they're all brown, but I'm glad it's not the wolfies."*

Squire Butler. The Squire, who lived in Abbey House in the 1700s, supposedly had a pet dog that he trained to walk on its hind legs. Some think that Wolfie was the ghost of this dog, but that would not explain why groups of the creatures were seen together.

The Lawsons moved out in 1911, but ghost researcher Alan Gauld, contacted the Lawson children to see what they remembered. At the time, in 1972, John, who was over seventy-years-old, wrote Gauld that he had "no recollection of the events recorded by my parents....

> *"Years after we left Abbey House I was very surprised to hear from a third party that I had lived in a 'haunted' house, and asked my parents whether this was true. Even when they described events which I had reported to them it did not bring back any personal recollection."*

However John Lawson said...

> *"My father had considerable critical ability and would not have accepted his own experiences and recorded them without having carefully tested and cross-examined himself. My mother was a religious woman who would have been naturally unwilling to accept such events...."*

Later tenants also encountered the ghosts of Abbey House. In 1920, Mr G. Granville Sharp moved into Abbey House. Mrs Sharp later wrote to Professor F.J.M Stratton of the Society for Psychical Research.

> *"We had been in the old Abbey just a week when I heard, one morning, my little girl Charmain, crying in the hall. I went to see what was the matter and found her in the doorway of the dining room crying hard with tears running down her cheeks. When I asked her what was the matter, she pointed to the far corner of the dining room, by the cupboard, and said "Will it hurt me? "I answered her that there was nothing to hurt her, but she would not go into that room at any time.*
>
> *About a week later the same thing happened again-and this time I asked her what it was she saw, and she said it was an 'animal' as she called it.*
>
> *From this time she often spoke of the 'animal' and said it came in at the window.*

She was only two years and nine months old at this time, and had heard nothing whatever about the subject, nor would she have understood if it had been mentioned.

I myself had heard nothing more than that the house was haunted, until afterwards. And she was always with me, and couldn't have been told of it by anyone else."

The little girl also saw the nun that Mrs Lawson thought she had banished.

One night at bedtime, Charmain told her mother that she didn't like being alone in her bedroom because *"mummies"* came into her room at night. Her mother told her that nothing could hurt her.

"But I don't want to be alone" she said

"But why?" her mother asked.

"Mummies come in and lean over my crib. They look at me. And they only come in when you've gone downstairs."

Not long after, Charmain's mother discovered that Charmain slept in the room that the Lawson's had used for a bedroom, the same room that the Nun frequently visited.

The Abbey House animal was last seen in 1947. Mrs Celia Schofield, a friend of the tenant, said in a letter to a friend of how her son Christopher saw Wolfie. He was running ahead of his parents into the kitchen and called out, *"Oh, look! Tiny doggie!"* His parents knew the reports about the animal, but had taken great pains to protect their son from them.

The hauntings seemed to run down and nothing was reported between 1965 and 1980. The new owner of the house, Professor Danckwerts, reported that no one in his family had seen any ghosts in the house. He was convinced that Abbey House was not haunted.

In June 1980, however, an elderly woman who rented a room from the Professor experienced a strange encounter. She said...

"At approximately 3:30a.m. ...I was in bed but awake when suddenly the figure of a man appeared in the doorway. The door was open and the figure was framed in brilliant white light.... The man was conventionally dressed in light grey jacket, light coloured shirt, and dark tie. The complexion was ruddy, rather mottled, and the eyes were dark. The hair was grey and wavy. Every detail was quite distinct. He did not move but looked straight at me. I suppose it was a matter of seconds until the

figure and light vanished. Immediately afterwards a 'procession' of whitish nun-like figures passed quickly across the room in front of the door in an arc formation as if climbing up a step or two and down the other side. About 5 or 6 passed in this way and then all was normal."

Again we have a haunting in which a small, strange animal is seen alongside more conventional ghosts.

HOUSE OF BLACK ANTENNA

Prolific ghost hunter, Elliot O'Donnell, records a truly weird haunting that could have come straight from the pen of M. R. James. In his 1908 book *Some Haunted Houses of England and Wales* he reproduces the diary of gentleman 'since deceased' who once lived in a house in Portishead, Somerset. A professional man, he asked O'Donnell to uses pseudonyms. The author refers to the residence as 'Harley House' in his book. The diary runs thus...

"Before I commence my story, I think it expedient to state that both my parents are dead, my father having died many years ago and my mother quite recently. The latter had lived to the very ripe age of ninety, had possessed an unusually strong will, was a most devout Roman Catholic, and took the deepest interest in everything that concerned our welfare. She had two peculiarities: (1) A strange aversion to children; (2) a positive loathing and dread of black beetles. The house stands alone, some thirty yards or so from the road, and is well concealed from view by a high brick wall and numerous trees. There are four bedrooms upstairs, two on either side of the landing--which for clearness I will number--viz., No. 1 occupied by my wife and I; No. 2 my sister Mary's room; No. 3 my sister Joan's room; No. 4 the spare bedroom in which my mother died. The top storey consists of two attics inhabited by the servants.

"January 1, 1906, we first became aware of the disturbances—violent knockings being heard about midnight on the walls and floor of room No. 4. On hurriedly entering it, we could discover nothing.

But on leaving the room the noises were repeated and kept up till two or three in the morning.

"January 5. A recurrence of the disturbance--only much louder.

"January 6. Have in a carpenter who makes a thorough examination of the wainscoting and reports 'no traces of rats, mice nor any other animals.

"January 10. Tremendous knockings again in room No. 4, the door of

which is swinging to and fro violently. A loud clatter on landing as though half a dozen children were engaged in the roughest horse-play. The uproar terminates in a terrific crash on the panel of No. 3 door. Joan rushes out of her bedroom thinking the house is on fire and sees a strange, green light some six by two feet long moving across the landing. It disappears in room No. 4.

"January 15. We are all awakened by a loud crash and on reaching the landing find a big, black oak chest from the coach-house, lying there on its back. Every one much alarmed.

"February 1. My sister Mary awakened at midnight by feeling something tickle her cheeks. She puts out her hand to brush it away and encounters something cold and scaly. Her shrieks of terror bring us all into her bedroom--there is nothing there.

"February 3. My wife and I are aroused by feeling our bed gently lifted up and down, and on my getting out for a light, I tread on something indescribably disgusting. It feels like a monstrous insect!!

"February 4. The knocking very bad all night--particularly in room No. 4.

"February 5, 6, 7, ditto.

"February 10. The clothes mysteriously taken off Joan's bed and transported to room No. 2.

"February 15. Both servants undergo our experience of February 3.

"February 16. The knockings still continued and distant sounds heard as of someone coming upstairs and turning the handles of all the room doors.

"February 17. Scufflings on the landings, and in the passage as though caused by a troop of very noisy children.

"February 19. Knockings in room No. 2. The washstand and a heavy mahogany wardrobe moved some feet out of their places. Mary, who was awake at the time, saw the shunting of the furniture, but could detect no sign of any agent.

"March 1. About 8.30 A.M. after Martha had laid the breakfast things she went downstairs to finish a cup of tea. On her return to the breakfast room she found it in the wildest state of disorder; chairs overturned, ashpan and front of grate removed to furthest extremity of room, all the pictures taken down from the walls and laid face upwards on the floor, and the cups, saucers, plates, knives and forks piled in one heap in centre of table; all this had been done without either breakage or noise.

Terrified out of her wits Martha rushed upstairs to our door, and nothing would induce her to enter the breakfast room again alone.

"March 3. On returning home about 10 P.M. from a neighbouring town, we found the servants sitting huddled together, half dead with fright in the kitchen. They had heard knockings and the most appalling thuds ever since we had gone out; and on entering our room (No. 1) we found it in an absolute turmoil: the bed-clothes in a promiscuous pile on the floor, the duchess table turned round with its face to the wall, the pictures ditto--but--nothing broken.

"March 15. Awakened in middle of night by three loud crashes in room No. 3, after which we distinctly heard our door open and someone crawl stealthily under our bed. We at once lit a candle--no one was there.

"March 18. Knockings in both the attics. The servants badly scared.

"March 21. As Joan was running downstairs about mid-day, she received a violent bang on her back as if someone had hit her with the palm of their hand. She came to my study in a very exhausted condition, and it took her some minutes to recover.

"March 24. Found my mother's shoes, which we were certain had been locked up in a bureau, placed where she had always placed them in her lifetime--_i.e._, on the hearth-rug before the dining-room fire.

"March 31. My mother's favourite arm-chair found upside down in front of the fire-place in room No. 4.

"April 2, 11 P.M. As Mary was stooping to look under the bed for fear of burglars, she was suddenly pushed down and the mattresses and bedclothes were thrown on the top of her. Her frantic struggles and muffled screams being, fortunately, overheard by my wife (I was in London at the time), she was immediately extricated. No injury, only bad shock.

"April 3, midnight. The contents of a large chest of drawers in room No. 3 suddenly emptied on to the floor. Loud crashes in all parts of the *house.*

"April 10, 11 P.M. On going up to bed, we find room No. 4 aglow with a pale green light and filled with a faint sickly odour, which we at once recognised as identical with that smelt there at the time of my mother's decease and which we considered was peculiar to her disease.

"I must mention that after her death, the room had been thoroughly renovated, the old flooring replaced by new, the walls repapered and everywhere well disinfected with the strongest carbolic. My mother had died at 11 P.M.

"April 12, 13, 14, 15; 11 P.M. The same light and smell.

"April 20. Joan fell over some large obstacle in the hall, hurting herself badly. She could see nothing, but was half suffocated with a stench similar to the one already described.

"April 30, 2.20 A.M. Both my wife and I distinctly felt something brush across our faces. We lit a candle and perceived to our horror two long black antenna (like the antenna of a monstrous beetle) waving to and fro on our pillow. We spent the rest of the night on the drawing-room chairs and sofa.

"May 1. Shut up the house."

This bizarre case has many classic features; strange lights, vile smells, weird noises, objects being moved about, but it also has a giant ghost beetle! It is strange that after the mother's death, the two things she loathed most in life, children and beetles, manifest in the house, but writ large. Sounds like gangs of urchins making a racket and giant, face fondling beetles crawling around at night. Strangeness seldom gets higher than this. If only we knew exactly which house in Portishead was the house of the black antennae.

THE SPIKE ISLAND HORROR

Dermot MacManus was told of a particularly unpleasant phantom, that haunted the grounds of a doctor's house, on Spike Island in Cork Harbour. The story was related to him by Mrs Eileen Ganly. It occurred in 1914, when her family were living on the island in some barracks, Eileen's father being a solider. Eileen was sent each day to fetch her father's paper from Cobh, on the mainland. On her walk down to the launch she would pass the house of a local doctor. The house stood in a grove of what Eileen described as 'Arthur Rackham Trees', gnarled old growths. The garden was surrounded by a five foot wall. It was pleasant summer's day when Eileen saw something horrific in the grounds.

"It must have been a very tall thing that was looking over the wall, because I could almost see to its waist-and the wall was at least five feet high. It was the rough shape of human being-that is, it had a head, shoulders and arms-though I did not see the hands, which were behind the wall.

Except for two dark caverns which represented its eyes, the whole thing was one colour, a sort of glistening yellow. The only thing like it is the glistening luminosity of rancid butter when it has been left in the sun.

As the wall was parallel to the road and on my left, the thing was looking past me-across the little road and straight across to Cobh.

I cannot tell how long I stood in motionless fear, gazing at this thing but eventually it began to turn its head very, very slowly towards me.

Petrified as I was I heard a voice in my ear- 'if it looks straight at you Eileen, you will die'.

Eileen managed to break the thing's paralysing hold of fear and ran to the cottage of a woman called Mrs Reilly, who did the washing for the family. On telling what she had just seen, Mrs Reilly told her that she wasn't the first to see that thing and she wouldn't be the last.

THE GROWING GHOST AND THE HEADLESS HORROR.
In 1966, the *Sunday Mirror* asked Dr A.R.G Owen, geneticist, biologist and mathematician and reporter Victor Sim to investigate and report upon a number of alleged hauntings. These were later published in a 1971 book called *Science and the Spook*. One of the strangest cases took place in an ordinary looking council house, near Gravesend, London. Previous tenants had reported nothing amiss with the property but that changed in 1962, when the Maxted family moved into 16 Waterdales. The family consisted of Mr Sidney Maxted and Mrs Joyce Maxted, a young couple and three children, Kevin 6, Linda 4 and baby Claire. The children would come downstairs complaining of scratching noises from under the bed and having their bedcovers pulled away.

Something invisible slapped and pulled at them. Mr Maxted himself heard the scratching. Mrs Maxted grew anxious about heavy footsteps she would hear upstairs and small objects would vanish. Other minor occurrences would be lights in unoccupied rooms switching themselves on and the clock gaining two hours. The events waxed and waned during the years that the family lived at number 16. Joyce recalled that the events were often accompanied by a smell akin to petrol. Gas Board officials could find no leak.

Matters came to a head in August of 1965. It was 2.00am and Sidney was awoken by

Joyce calling out *"Linda."* He saw her face frozen in terror. She began to scream and then collapsed. Later she explained that after having attended to baby Claire, she had gotten back into bed, but was surprised to see the figure of a little girl in the doorway. The child was aged about six and was the same height as her daughter Linda. She thought it was Linda until she noticed the child had fair hair, whereas Linda had red. The apparition advanced towards her and seemed to grow in height as it came. The thing's face was so awful that it caused Mrs Maxted to have a breakdown. She flatly refused to describe it further and the investigators were kind enough not to press her as she was still recovering when they spoke with the family. The Maxteds left for good soon after.

The house was then occupied by Mr Eric Essex and his wife Margaret. Apparently the Maxteds had tried to dissuade the Essex family from moving in, but they had needed a home. Soon the sounds of footsteps from empty rooms and a noise like furniture being moved about was heard. Mrs Essex said there was also a mould smell in the air and a low frequency ringing in the ears. These disturbances were tolerated until August of 1966. Eric awoke to a whistling sound and found that the bed was vibrating. Then the bed seemed to be heaved up at one end. Looking around he saw, standing next to the bed, a pinkish-orange, glowing form that seemed transparent around the edges. It seemed to have on a long dress but it was completely headless. The family left next morning to stay with Mr Essex's mother and father. When interviewed, he stated categorically that the ghost terrified him and he would never stay at the house again.

Owen and Sims also spoke to Mrs Margaret Harrison, a young housewife who lived next door at number 14. One of the bedrooms of number 14 was over the hall and staircase of number 16. The bedroom Margaret Harrison slept in. Her husband had been away shortly after the Essex family had left and she had been sleeping alone in the room. She was startled to hear a loud booming and thudding from below, like a big object was bouncing down the stairs, then against the ceiling. The noise was right under her bed. Then the bouncing transformed into a scratching noise that seemed to cover all of the underside of her floor. She found the noise very frightening. The case was never resolved but one can easily see the same patterns of a classic haunting occurring. It is as if each haunting has base phenomena, onto which its own weird characteristics are then placed, be that headless badger, headless women, or monster beetles.

The ghost that changes size, growing or shrinking as it is observed is found in other cultures too. In Japan, the Nyūdō-bōzu is a ghost that appears as about three feet tall when first observed. As it draws closer it becomes larger, the size of a person and as the observer looks up at it, the ghost continues to grow. Neil Geddes Ward is an author, artist and researcher into ghost and fairy law. Once, whilst working in the broadcasting industry he struck up a conversation with a security guard who had moved to England from the West Indies. The man told him that once, he and a friend were walking home one night along a way they regularly took. The path lead them past a church. Both men saw a gigantic woman, some fifty feet tall sitting astride the church roof with her legs on each side. The man described her as being as white as that fridge, whilst pointing to a refrigerator. As his friend hid behind a grave stone in terror, the man picked up a rock and rashly threw it at the giant woman. The huge white figure stood up and got off the church. She began to walk towards the pair, but as she came she shrank down with each step until she was of a normal size before vanishing completely.

FOOTPRINTS ON THE CEILING

Ian Deakin wrote into *Fortean Times* in 2000, about a strange occurrence in his youth. In the late 60s his family were living in a terraced house in East Park, Wolverhampton in the English Midlands. One night his older sister and his mother, who had been up late talking, saw a man in a sailor's uniform on the landing. The sailor walked towards the bathroom and vanished. After this, adult and child size footprints began appearing on the walls and ceilings of the house. The family tried painting over the prints but they re-appeared through the paint. The prints could only be covered up by wallpaper. When the family sold the house in the mid-1980s, they stripped the wallpaper and found the prints still there. On another occasion Ian's aunt felt what seemed to be an invisible child brush past her in the hall.

The family connected the hauntings with the tragic case of Harry Park Temple, who in the 1920s drowned whilst trying to rescue two young boys from a pond. The boys also died. Ian never contacted the family who bought the house to see if the footprints ever returned.

CRIPPLED DWARFS AND GIANT CATS

Atop Montpelier Hill, County Dublin, Ireland stands the ruins of the Irish Hellfire Club. It was originally built as a hunting lodge, by William Conolly in 1725. In 1737 Richard Parsons, the 1st Earl of Rosse and James Worsedale, formed the Hell Fire club, modelled on those existing in England. Members drank "scaltheen," a mixture of whiskey and hot butter and left a chair empty for The Devil. They supposedly sacrificed black cats by dousing them in whiskey and setting them alight. One account tells of a priest who came to the house one night, and found the members engaged in the sacrifice of a black cat. The priest grabbed the cat and uttered an exorcism upon which a demon was released from the corpse of the cat. Another time a crippled dwarf was said to have been ritually killed in a black mass there. But one night a footman supposedly spilt a drink over member Thomas Whaley's coat. The infuriated Whaley poured brandy over the man and set him alight. The fire spread to the building and killed several members. After this the club would meet in the Stewart's House.

The Stewart's House was a building that stood across the road from Killakee House and its estates. After Killakee House was demolished people began calling the Steward's House, Killakee House, rather confusingly. For the sake of clarity I will refer to it as the Steward's House.

In 1968 the O'Brien family bought the house with plans to turn it into an art centre and tea rooms. Men who were working on the house refused to stay there, leaving without even picking up their wages. They complained of a cold atmosphere and a door that would constantly unbolt itself and open. One carpenter saw the door open and a panther-like cat, the size of an Alsatian dog, walk through. The animal stared at the man before vanishing. The O'Brien's continued to work on the house themselves, with the help of friends. Several weeks after the workmen had left, artist Tom McAssey, who was helping to renovate the house, had a strange encounter. He and two other artists were renovating the hall. The heavy, 18th century door swinging open, revealing swirling mists. Standing in the doorway, was a small man, no more than three feet tall, who appeared to be crippled. The tiny figure addressed them in a guttural voice saying that the door must be not be closed. Then, the witnesses said that the dwarf transformed into a huge black cat with glittering eyes.

Later, McAssey painted a portrait of "The Black Cat of Killakee" which is still in the house today.

In the 1970s, a series of séances and Ouija board sessions were carried out, in an attempt to lay the ghosts to rest. However, the disturbances only worsened. Lights turned on and off independently, furniture was broken into pieces during the night and bells were heard ringing. The dwarf was seen walking with a bloodstained man. Patches of a sticky, glue-like substance were found throughout the house. Furniture and ornaments were smashed to bits and paintings slashed to ribbons. Weirdest of all, the house was invaded by all kinds of hats. The hats seemed to materialise all over the house and included babies caps, sun caps, woolly hats and women's hats. Some contained coins from France and Poland. The cat was seen again, after a bunch of drunken actors held a séance there, the ghosts of two nuns began to appear as well. A medium visited soon after, and claimed that the spirits were those of women who had assisted in satanic rituals, done by the notorious Hellfire Club.

A priest finally exorcised the house in 1977 and the hauntings stopped. The Stewart House is now a private residence.

FACES IN THE FLOOR

In August of 1971, Maria Gomez Pereira noticed a stain forming on her kitchen floor in the Spanish town of Belmez. Gradually, the stain seemed to form a face. Terrified, Mrs Pereira tried to remove it by scrubbing at it several times, to no effect. Her husband Juan and son Miguel upped the ante by taking a pickaxe to the floor, then laying a new one with cement. All was fine for a while, until another face seemed to form on the new floor. The story spread around the town and the Mayor took steps to preserve the face. The face was carefully removed and preserved.

Surveyors came to the house, and the floor was once more dug up. The workmen found a number of skeletons buried ten feet beneath the kitchen floor. Some were missing their skulls. The bodies were dated to some 700 years old. They were re-interred in a local graveyard and the floor was cemented over for a second time. Within a fortnight the face came back and this time it wasn't alone!

Different faces appeared and then vanished, one making way for another. The excavation seemed to have made the phenomenon worse. The story had now gone international and gained the attention of German paranormal investigator Dr Hans Bender. A full-scale investigation was launched. Samples of the concrete used were sent to the Instituto de Ceramica y Vidrio (The Institute of Ceramics and Glass) for study. The ICV could find no evidence of any pigment, dyes or paint used, which ruled out a hoax. The remainder of the floor was photographed in sections and covered with a jacket that was sealed at the edges. Finally, the door and windows were all sealed with wax. A German film company filmed the process as local dignitaries looked on. Satisfied that nobody could tamper with the scene without detection, they left the kitchen alone for three months. When they officially unsealed the kitchen, the faces had moved and changed.

The faces continued to appear on the kitchen floor, even after Mrs Pereira passed away in 2004.

In 2014 the investigative journalism show *Cuarto milenio* investigated the case. Dr José Javier Gracenea, a chemical engineer and Luis Alamancos, a forensic chemist carried out tests on the faces. Dr Grancena concluded *"They weren't made with paint and according to scientific knowledge and techniques employed in the analysis, there is no external manipulation or elements in the faces."* Alamancos then attempted to reproduce similar images, through the variety of methods considered valid in previous investigations, including but not limited to concrete solvents. He failed to reproduce the faces and confessed utter bewilderment.

COILS IN THE DARK, MY OWN EXPERIENCE.
The Centre for Fortean Zoology, the cryptozoological organization for which I work, used to hold an annual conference in the village of Woolfardisworthy, North Devon. Each year, experts from various fields of Forteana would converge to lecture in the village hall. The conference ran for years and was very popular with several hundred attendees. I lectured there each year, usually speaking on the cryptozoological expedition I had taken in that particular year. In 2008 I was there with my then girlfriend, and stopping at the now defunct Hilltop Holidays near Bucks Cross. The sight consisted of a farm house, with cottages attached, and a number of caravans as well as a large, stand-alone cottage. The site was run by a family who were friends with Jon Downes, director of the Centre for Fortean Zoology. The family had been known to me for many years. My former girlfriend and I were stopping in the standalone cottage.

One afternoon the proprietor, Mrs Kaye Braund-Phillips, asked us if we could make sure that the lights and the TV were turned off when we left the cottage for the day. I was surprised as I'm usually careful about doing this. We made sure all the lights and the TV were off before we left for the day. On our return that evening, we found that they were switched on again. It was odd but I thought little of it.

That night, as we lay in bed, we both heard the sound of heavy footsteps, like boots on flagstones downstairs. Thinking there was a break in, I ran down stairs, ready to tackle a burglar, only to find there was nobody there. At this point I realised that the house was carpeted. There were no bare flagstones anywhere in the house. Puzzled I returned to bed.

When we awoke next morning, my girlfriend found she had a series of bruises along her right arm. They were in the shape of a huge hand. The imprints of fingers and thumb were clearly visible. The hand prints were far larger than mine and besides, in order to have grasped her arm like that, I would have had to lean right over her in my sleep.

We mentioned these happenings to Kaye, who became very upset, repeating the phrase *"don't say that, oh don't say that."* She sounded frightened and wouldn't talk about it.

But the story was to have a much stranger sequel. The following year, during the Weird Weekend, my friend, the author and excellent Fortean researcher Neil Arnold, and his wife Jemma, stayed in the same cottage and the same room. Neil told me that he awoke in the night to feel a great pressure around his legs. Looking down, he was horrified to see, in the moonlight, a vast snake like a python, coiled about his legs and tightening upon them. Neil described the reptile as huge, about thirty feet long. It was yellowish with brown, mottled markings. The description resembles the Burmese python (*Python bivittatus*). His cries of alarm woke Jemma, who also saw the huge snake, that was now hauling Neil out of the bed. She grabbed her husband and tried to pull him back into bed, but the snake dragged him onto the floor. Jemma hit the light switch to find that the huge serpent had vanished. Despite the monster having dissipated, large bruises were visible around Neil's legs.

Hill Top Holidays has now closed and no explanation has ever been forthcoming for the strange events that happened there. A phantom python haunting a house in rural Devon; if that's not high strangeness I don't know what is.

THE SNAKE SCHOOL
Weird as it may seem, my story of the giant phantom snake, is not the only case of an eerie ophidian. In 1925, Annie Whitney and Caroline Bullock, were collecting stories and folklore from the Pennsylvania and Maryland areas. They spoke to William Johnson about a bizarre occurrence that happened when he was 16, that would have been around 1886.

The haunting was triggered by the construction of a new school, at a crossroads near Jenner Township, Somerset County, Pennsylvania. An old man warned the townsfolk that the area was haunted, but he was ignored. As it turns out, the place *was* haunted, weirdly haunted.

A vast snake would appear, coiled around the school. The thing would only manifest during moonless nights. The serpent's head and tail were never visible, just the coils that were covered in sharp scales. Anybody trying to attend night classes when the creature was coiled about the school, would be thrown to the ground if they tried to step over the massive coils. Weirdly, not everybody could actually see the monster, but even those who could not were still thrown to the ground if they tried to enter the school.

The monster then began to coil around other properties in the area. The mega-serpent coiled around the house of a man called Frame, who was so scared of its appearances, that he sold his house and moved. The property was brought by Joe Leversos, who moved in with his family. The family lost their fear of the apparition as they became used to it.

Sometimes local men would pluck up courage to attack the entity with fence posts. The serpent was not in the least affected by these attacks.

The monster was still manifesting when Johnson left the area at the age of thirty, around 1900.

The church snake reminds me of old Scandinavian stories of lindorms, huge serpentine dragons that would coil around churches and crush them.

One wonders if the old schoolhouse is still standing today.

THE DRAGON OF QIU MANSION
Often Fortean phenomena will merge and overlap. In this case we have the most ancient and iconic monster of them all, turning up in a haunting case.

The Qui Brothers lived in a small village, but migrated to Shanghai in search of a better life. The brothers made a living by working for a German. They worked for him until World War II broke out and the German fled China, leaving behind his business assets.

The Qui brothers, who looked after their bosses' business assets, were left in charge after their employer fled. Because of World War II the docks were closed. with exports grinding to a halt. The Qui brothers opened one of the warehouses and found a stockpile of paint cans. Owing to the war. the price of paint skyrocketed and the Qui brothers were at the right place. at the right time. to cash in. They turned into millionaires overnight.

They built two identical mansions next to each other. with a sprawling garden. They kept animals like tigers in the gardens, an artificial lake full of crocodiles and peacocks roaming the lawns. However. both brothers seemed to vanish mysteriously. Their animal collection was sold off to other zoos. One of the Qui mansions was razed due to its dilapidated condition. and it was decided that the other mansion would be moved to make way for skyscrapers.

Paranormal activity at the Qiu Mansion does not seem to have begun until 2009. Workers at the site spotted ghostly animals, and many were attacked by these animals. Workers working the night shift had to be rushed to the hospital with animal bite marks, but nobody ever found animals in Qui mansion. A woman, Mrs. Ye, who lived near the construction site in the alley off Wujiang Lu, once spotted a monstrous dragon coiling the boom of a construction crane. A mason once tried to kill his employer with a hammer, and when he came back to his senses, said that a lizard like creature made him do it.

MONSTROUS MIASMAS
One class of apparition seems to take the form of a shapeless cloud or miasma. In some areas these things are called 'old boneless.' They are, for the most part, black or white.

In February of 1695, a spate of strange happenings began in the farmhouse of Andrew Mackie of Rincroft, Scotland. Cows were found untethered and moved on two nights, then a quantity of burning peat was brought into the house, and would have set fire to it, if someone had not alerted the family first. After dousing the fire, they could find no agency for it's being there. In March, the classic poltergeist trick of stone-hurling beset the house, and soon after they were joined by knocking sounds. The antics became more violent as people were dragged out of bed and struck with staves. Andrew Mackie himself said that something pulled his hair and scratched his scalp with invisible nails. The children were pulled from their beds and slapped by unseen hands. During prayers, the entity cried 'hush, hush', and made groaning sounds. Balls of fire appeared in the house but vanished. Many visitors to the house were also attacked by the unseen assailant. Human bones were uncovered just outside the house. On April 8th, the local magistrate ordered that the Laird of Coiline should examine every person still alive, that had ever lived in the house, that was built twenty eight years before. All of them were made to touch the bone, an old test mentioned often in witchcraft trials but nothing was established.

Five ministers tried to exorcise the house, but failed when they were pelted with stones, and some were lifted clean off the ground by some agency.

Through April, the entity threw stones, shook the house and flung mud in the faces of those at prayer. Then the thing began to speak, calling the farm's inhabitants witches and rooks and that it would take them all to hell. It began to frequently set fires in the house as well.

On April 30th. Charles Macklellen of Coiline, and several neighbours, were praying in one of the barns when they saw a black shape in the corner. The thing grew bigger, swelling into a black, shapeless cloud of huge size. The black mass hurled mud and barley-chaff at them, and grasped several of them by the arms with such force that they were in pain for five days afterwards.

The next day the entity vanished for good.

Ruth L Tongue, was told of an encounter with Old Boneless that occurred early in the 20th century. The witness was a policeman, whose beat encompassed the Minehead to Bridgewater road in Somerset. The witnesses experience was related by his sister-in-law.

> *"He told her it was darksome above Putsham Rise and the tide was in far below-he could hear it plain down two hill fields, and then his lamp lit up a white Summat across the road. It weren't fog. It were alive-kind of woolly, like a cloud or a wet sheep-and it slid up and all over him on his bike, and was gone rolling and blowing and stretching out and in up the Perry Farm Road. It was so sudden he didn't fall off his bike-but he says it was like a wet, heavy blanket and so terrible cold and smelled stale."*

The experience so scared the policeman, that he was switched to another beat.

On the 23rd of November 1904, Godfrey H. Anderson was strolling down a street in Edinburgh, Scotland. He saw an odd, black shape in a gutter. Anderson described it as black, four feet long and two and a half feet tall. It was shaped like an hourglass, and moved in a crawling motion like a caterpillar. Suddenly, the thing leapt at the throat of a passing horse that reared up in terror. As it did so, the apparition vanished.

While travelling on his bicycle at Cappagh White, County Tipperary, Ireland, Thomas Fahey stated that a strange black blob landed on his handlebars. The weird entity slowed the bicycle down considerably, before moving off and disappearing along a path. This occurred in 1910 along the road to Ironmills Bridge.

In 1915, on the island of Cuba, two armed guards stumbled across something odd, whilst performing a routine night patrol, near a sugar cane field near Jicotea, Las Villas. The two men noticed that their horses were snorting and seemed afraid of something. Then the men noticed a white, sack-like shape on the ground near the trail. The white bundle seemed to be a living thing that was crawling closer towards them and causing the horses to panic. One of the guards shot the white, slithering thing with his pistol, but each time the bullet struck, the shape seemed to grow larger and larger until it was as tall as the horses. The mounts, wild with fear, threw their riders and fled into the night. The men themselves ran in terror from the shapeless white mass that was still growing and creeping towards them.

Next morning, with the bravery brought by sunlight, they returned and found their horses grazing, unharmed, and no trace of the weird white mass that had confronted them the night before.

Christian Jensen Romer or CJ for short, is a parapsychologist, who works with the Association for the Scientific Study of Anomalous Phenomena (ASSAP). He is also a good friend of mine. In an online talk he delivered in 2021, he related a strange tale told to him by his late mother, June. She had been out on a bike ride with her friends, Winnie and Joyce, in the countryside lanes beyond Borley, north Essex. Reaching the village of Belchamp Walter, they stopped to look back over the fields towards Borley. Suddenly, they see a brown, cloud-like thing the size of a house. The shape of the thing was akin to a loaf of bread. It came rolling across the fields and brought with it a sickly -sweet smell, she described as being like mimosa (a genus of neo-tropical herb). Suddenly, it changed direction and came straight for them. Jumping on their bikes, they rode away terrified. Looking back, they saw the thing roll out onto the road. She never did find out what it was.

One moonlit September night, during the 1950s, railwayman John Davies, was riding his motorbike back home, to his cottage in Derbyshire's Longdendale Valley, when he saw a black mass crossing the road not far ahead. Moments earlier, he had felt an uncanny, seemingly reasonless compulsion to brake, and as he did so, he saw what looked like a huge black slug sliding across the road and up the moor, making a scraping noise as its massive but near-shapeless form moved along. Up closer, it looked a little like a massive whale, and even possessed an eye-like structure, that seemed to swivel around as it slithered along, Davies later learnt that it had been seen by others. One such observer was a friend of Davies, who had seen it sliding across the valley below Ogden Clough, where it was also observed on a separate occasion by another of his friends. Both of them were convinced that whatever it was, it was definitely evil, and both had fled in panic after seeing it.

The town of Runcorn lies on the south bank, on the River Mersey, to the south east of Liverpool. Sixty eight year old Sam Jones had lived there, in a little house at number 1 Byron Street, for thirty three years with no odd occurrences. At the time that the strange events began, on Sunday 17th of August of 1952, the small home was a crowded place. As well as Sam, there was his 17 year old grandson John Glynn, his widowed daughter-in-law Lucy Jones, and Miss Ellen Whittle, an elderly lodger. Sam and his grandson had to share a double bed in order for the lodger to have her own room.

On the night of the 17th, a strange scratching noise emerged from a dresser. It was dismissed as the work of a mouse, but over the coming days the scratching was joined by knockings and movement of furniture, in the formula that should be familiar now. The police were called in and two of them were thrown from a chest of draws they sat on. But there was little the police could do. A local medium Philip France, of the Runcorn National Spiritual Church, was brought in and held a séance during which objects were thrown around. France claimed to have communicated with a spirit and passed some kind of message on to the family.

For three weeks all was quiet, but then the thing came back with a bang. It was rocking and scratching the dresser again. Sam had begun sleeping down stairs and a frightened John convinced his friend Johnny Berry to share the room with him. The story had got around after a story was written in the *Witness Weekly News,* who dubbed the ghost 'the Runcorn Thing'. A hundred or more people would converge on the little house each night, trying to get a look at the ghost they had named 'Jooker'. Sam invited many of them in and they saw objects flying about.

The Reverend W.H Stevens, the Superintendent Methodist Minister of Widnes, heard the dresser rocking in the dark and swiftly turned the light on, to find both boys had their hands under the bedclothes. When the Stevens asked the entity to knock three times if it understood him, the dresser rocked three times. Soon after, books and an alarm clock flew off the dresser and hit Stevens on the head.

Many witnesses claimed to see little points of light during these outbursts. Most of the phenomena happened in the dark, but as the days went on the poltergeist began to perform when the lights were on. Local businessman J.C Davies, saw a heavy table beating itself against a wall. He also saw objects floating as if held by invisible hands.

On October 22nd the lodger, Miss Whittle, was out walking when she fell into a quarry called Frog's Mouth, and died.

The very next day a film crew from the BBC arrived, but the spectre failed to perform for the cameras. Indeed the phenomena at Byron Street seemed to be winding down. On December 13th, it performed its last trick, folding up the bedroom carpet.

Where was the miasma in all of this you may ask? Well, amazingly, a parallel haunting was occurring at Pool Farm on Heath Road, owned by Sam's employer Harold Crowther. On August 10th, Crowther said he saw the ghost of his farther-in-law wandering the farm yard. Soon after, he found three of his pigs dead. A vet examined the bodies and found that their windpipes had been crushed. To crush the trachea of a pig would take immense strength. Within two weeks all fifty three of Mr Crowther's pigs were dead. He said he saw a 'thing' floating around the pigsties. It was a tall, black cloud around seven feet high. It was shapeless, except for two horns or prongs that seemed to stick out of it. Three days later, his wife saw the same entity. Three weeks after that Harold saw the thing in the kitchen of the farmhouse. He rushed past it heading for the light switch and felt the horns, like blunt sticks, at his throat. When he switched on the light, the thing vanished. Crowther also said he saw the horror manifest at number one Byron Street, hovering over John Glynn's bed.

The final time he saw it was on December the 13th. The shapeless apparition tried to strangle him but his two dogs attacked it and it retreated and vanished, never to return.

Over in the US, High School science teacher, Tom D'Ercole, had a bizarre encounter in 1975. D'Ercole lived in Oyster Bay, Long Island, and had just stepped out of his house that morning when he noticed an object like a small, dark cloud hovering above his house. The thing was about the size of a football, and was much lower than any normal cloud. The cloud seemed to be moving against the wind and as he watched, it grew in size. Floating back and forth across the roof, it changed into an ovoid shape and finally a long shape stretched out to six feet high and one and a half feet across. The thing seemed to grow vaporous lips, inhale and then spit at the teacher, soaking him in liquid. It then seemed to dissolve before his eyes.

Ever the man of science, D'Ercole bagged up his soaking shirt, and took it to his lab at Garden City Junior High. Running a pH test he found that the liquid was simply water.

The woods around Kinderhook (Children's Corner), a town in New York State, has seen several encounters with phantom, cloud like things.

New York State Library genealogist, Bruce Hallenbeck, was ten years old when he and his cousin encountered the thing whilst playing behind his house in 1962. Hearing a whistling noise, they both looked up to see a white thing peering at them from behind a tree. Perhaps peering is the wrong word, as the white blob had no visible eyes or features. Both kids ran away.

Two years later, a man hiking through the woods, saw a big white blob gliding down the hill towards him. The man was so scared that he leapt six feet over a pond to escape it.

He later plucked up courage to return with a friend. The pair had armed themselves with pitchforks and shovels, but when they saw the white blob floating amongst the tree tops, they dropped their weapons and fled.

Finally in 1964, two 14 year old campers, Barry Scott and Russell Lee, heard a tramping sound outside of their tent. Investigating, they both saw what looked like an archetypal, white 'bed-sheet ghost', bell-shaped and rippling, floating towards them. The thing then seemed to vanish into the thicket and was gone.

In the mid-1980s, parapsychologist Robin Furman, was informed of an exceptional weird haunting at the abandoned St Botolph's Church in Skidbrooke, Lincolnshire. The church, dating from early 13th century, had been empty and redundant since 1973. Robin was contacted by a student of land surveying, called Mark, who had been visiting old churches with five other students. He said the area had an oppressive atmosphere and the three girls in the group wanted to leave, but the three boys talked them into staying. They found the remains of small fires that seemed to encircle the church, and the melted stubs of black and red candles inside the church. They found a number of tombs had been disturbed, and the lids pulled aside. One of the girls was so badly affected by the place, that she began to cry. Things got stranger as the group saw blackish green clouds of mist seeming to rise up out of the opened tombs and two, cloaked hooded figures watching them from the bushes. Then points of twinkling golden light started to appear in the mist as the clouds seemed creep closer towards them. Filled with terror, the students fled and Mark admitted to crying with fear on the drive home.

With Mark acting as a guide, Robin took a team of investigators to St Botolph's Church. These included Robin's son Andy, Janice Paterson and Rodney Mitchell as well as Robin's dog Ben. Andy had brought with him an invention of his own, a 'tracton beam'. This was a hand held, boosted strobe light, contained in a tube of metal that could be pointed like a gun. It was the team's belief that the light could dissipate certain entities. As Mark has said, there were indeed the remains of a circle of fires, burnt-out candles and disturbed tombs. They also found the remains of many birds that seemed to have been drained of blood. Some seemingly newly-killed birds would become dry and desiccated when examined again only a short time later.

Suddenly, the team saw the clouds of green mist rising up from the tombs, just as Mark had described. Robin states that they had an air of decay about them. A feeling of oppression and nausea swept over the team and they retreated inside the church. The clouds seemed to coalesce, into one great mass, that followed them into the church. The points of light could clearly be seen in them and Robin likened them in size and colour to the heads of brass tacks. Then the cloud went back out of the door and a loud booming sound could be heard from inside the church. Robin says he felt as if they were being told to leave at once, but the team stood their ground. Gazing out into the churchyard, they saw that one of the tomb lids had been utterly shattered. The green fog returned, this time making a deep growling sound and oozing back into the church. Andy shot his concentrated strobe at the thing, and the pulses of light seemed to bounce back off the brass coloured particles within the cloud. The entity roared as the reflected light temporarily blinded the group. Then the cloud seemed to explode and vanish.

What are we to make of this adventure? It seems so strange that it could have dripped from the pen of H.P Lovecraft. I've met Robin twice, and he seemed not only sane, but exceptionally clever. He thought that the cloud was something called up by cultists and that it was feeding by draining the blood from the birds.

Occult activity seems to be continuing at St Botolph's Church. As recently as 2020 the *Lincolnshire Live* website ran an article stating that ...

> *"The centuries-old Grade I listed church, has suffered extensive damage due to vandals over the past few years, and has most recently been the site of a number of 'satanic' rituals.*
>
> *Local residents have reported seeing chicken carcasses with their throats slit, pentagrams sketched into stonework, while rings of candles and salt have been left inside the building. Churchyard gravestones have also been cracked. One local farmer named Martin Chapman, who visited the church to clear away what had been left from Friday night rituals,*

said that he had been left feeling sick at what had been carried out."

Many cases of violent ghosts involve black shapeless masses. We have seen a couple of supposedly fatal encounters with these 'things'. There are a number of others in her 1989 book *'Ghosts on the Range: Eerie True Tales of Wyoming'*. Debra Munn recounts the experience of a family that encountered a shadowy thing. In front of the whole family and a guest, the story goes, it picked up one of the sons and hurled him across the back porch and into the house, through a closed door. The mother was furious and chased the thing into the garage, where it turned on her and attacked, wrapping its arm like tendrils around her. She said it felt cold, and she felt as if the life were going out of her until a friend pulled her away.

The user 'Quercus' on the Forteana Forums website related a strange personal case that they experienced first-hand, in Ireland in 2001.

> *"One evening around October 2001, I'd stayed late visiting some friends in the Donegall Road area of Belfast, Northern Ireland. I'd let time get away from me, and had half-forgotten that I'd promised to phone my girlfriend, who was at university in England. It was well after midnight - and probably nearer one o'clock in the morning - when I finished my coffee, left my friends' house, and got into my little blue Ford Fiesta to drive the twelve miles or so home to Bangor.*
>
> *As I made to start the car, my mobile phone beeped, and I realised a number of unread text messages from my girlfriend had just come through. Back in the early 2000s, mobile networks weren't great and Belfast had a number of areas where phone reception could be patchy. This part of South Belfast was one of them. The messages began cheerfully enough by asking when I was going to call, and became more annoyed and finally despondent as the hours had passed without any reply from me.*
>
> *Angry with myself at being so slack, I quickly rang my girlfriend's number – and, having now woken her up, my garbled apologies and feeble excuses didn't go down too well. She was having a tough time adjusting to university life, and I'd really hurt her by breaking my promise to call. We ended up getting into an argument, and by the time I hung up, I felt wretched and consumed with self-loathing. I fired up my car, alone on the dark street, and began the drive home.*
>
> *I was a mess of emotions, and as I hammered along the empty dual carriageway towards Bangor, I realised I didn't want to go straight home. On an impulse, I pulled off the main road and headed towards Helen's Bay, a small coastal village where the two of us had often gone together to walk along the beach. I had happy feelings about the place; right then*

I felt like I needed to recapture some positive emotions.

Because it was late and there was no-one around, I parked up on double yellow lines on the Grey Point Road, right beside a cut-through path that led down to the beach. It was a calm night, overcast but dry, and not too cold for mid-autumn. The path down to the beach was lit by a sodium arc street light, so it wasn't that hard to see. I kicked moodily across the dark sand, stood for a while by the water's edge glaring at the lights on the lough shore opposite, and ended up slumped on a wooden bench about halfway around the bay, still eaten up with guilt and remorse.

I'd probably sat there for around half an hour, reading through the texts from earlier, my thoughts becoming darker and darker. Suddenly, I became aware of a peculiar buzzing noise coming from further along the beach path, towards the woods. It was soft, and the pitch oscillated slightly – more like an insect in flight than something mechanical like a motor. But it was growing louder, like it was coming closer.

I peered along the path, into the darkness. It was long past two o'clock in the morning, very quiet, and I'd seen no-one at all since parking the car. The odd noise jolted me out of my thoughts, making me alert but not yet alarmed.

Then, as the buzzing noise intensified, I made out what appeared to be a small, white cloud, maybe two feet across, coming out of the darkness. It was drifting slowly at head height along the path towards me, and appeared to be the source of the buzzing noise.

I stood up, becoming rather more alarmed at this point. I had never seen anything like this before, and although I didn't exactly feel in any immediate danger, I wasn't sure what might happen next. I began to walk quickly back towards the car, looking over my shoulder as I did so. The white cloud was still visible, coming closer to the bench where I'd just been sitting, while the buzzing, pulsing noise seemed to be getting louder.

I broke into a run, feeling ridiculous as I did so, but the rising panic in my chest spurred me on. I glanced backwards, now unable to see the cloud but still hearing the buzzing noise, which now seemed to be coming from all around me. A weird phrase kept repeating through my mind, over and over again – 'it's not human and it never was'.

At the top of the path, just beside my car, I stopped at a pair of bollards and wheeled round, to face the direction I'd just come. The buzzing noise had faded into the distance as I ran, but now was building in intensity again.

I'd never done this before, but as I stood there at the top of the path, I made the sign of the cross with my two index fingers and aimed it down the path towards the buzzing noise. At the time, I was attending a Pentecostal-style church; while the existence of things like demons and spirits were not emphasized, they were however acknowledged. Tonight, I felt that calling on the Divine might somehow help to stop whatever-this-was.

There was no visible cloud to be seen, under the streetlight's glow, but the weird oscillating buzz remained. I felt that the entity was still there, and close by. I was muttering under my breath at it, saying stuff along the lines of "You just stay there," and "Don't come any closer, in Jesus' name."

I still couldn't see anything, but I had the feeling that the entity had stopped about ten yards away from me. I could still hear the buzzing, but it seemed to be getting softer – not further away, just softer.

My nerve broke, and I rushed round to the driver's side of my car and leapt in. I fired up the little Ford, and tore off at speed back along the road towards Helen's Bay Village, trying to make sense of what had just happened.

I passed through the quiet of the village, under a railway bridge, and past a country park before turning left into Crawfordsburn village, another settlement outside Bangor. As I drove past a darkened filling station, I suddenly heard the same low buzz start up again, quieter now – but this time, inside the car with me.

Without taking time to think, I pulled over in the main street and jumped out, leaving the door wide open. I lifted the tailgate and pulled open the passenger door, then commanded whatever-it-was to get out of the car and leave me alone. After a few moments, the buzzing noise gradually faded out. I stood there in the empty village street, feeling pretty stupid but unsure of whether or not it had actually gone. Eventually, as a couple more minutes passed, I got a 'lifting' sense that the entity had now fully departed. Still feeling a bit shocked, I closed up the car doors, and drove on home without any further incident.

Although this wasn't the only experience I've ever had, it was the only one quite like this. I have no explanation for what I heard and saw; my rational mind says that it couldn't have happened like that, but I recall the whole incident very clearly. Maybe it was all in my head, but if so it involved a number of phases over a span of about ten minutes.

If I had to throw out a theory, ignoring the fact it goes against all rational thought, I'd say that somehow my very negative feelings as I sat there in that lonely spot seemed to attract something from the woods that wanted... well, I don't know what it wanted. To feed on these negative emotions? To attach itself on to me? Worse? I don't know.

In recent years, I've heard a few stories about strange occurrences reported in the adjoining Crawfordsburn Country Park, but nothing like what I experienced. Despite everything, I didn't feel that the entity was actively evil, just out and about doing its thing. But I was in the wrong place at the wrong time, and it was enough to frighten me quite a lot.

Regardless, after that night I stopped hanging around isolated spots by myself late on. I have no particular desire for such an experience again."

BLACK DOGS, THE HOUNDS OF HELL
The domestication of the dog began over 40,000 years ago. That formidable predator and competitor, the wolf, soon became man's best friend. This may have happened surprisingly quickly. Breeding experiments with foxes showed that aggression could be bred out of the animals very quickly. Dogs helped our ancestors to hunt and protected their human owners. In fact, there is a theory that the domestication of the dog may have given us the advantage over the more powerfully built, and better adapted Neanderthals. The human race may owe its existence to dogs! Dogs feature highly in the mythology of most cultures.

Small wonder there are many stories of ghostly dogs. However many of them are not simply the shades of departed, much loved pets. Phantom black dogs come in two types, firstly, standard sized dogs that seem protective or at least harmless. Then there are gigantic, shaggy phantom hounds, often said to have red eyes that shine like burning coals and bring with them a sense of deep dread.

These hell hounds range in size, from that of a Labrador to the size of a donkey. Think Hound of the Baskervilles rather than Lassie. Most, as the name suggests are coal black but other phantom dogs are white or even reddish. In the UK, there are legends of such demonic dogs in every county and they have different names in different parts of the country. Black Shuck in Norfolk, The Barghest in Yorkshire, Padfoot in Staffordshire, Galley Trot in East Anglia and so on. The black dog is also known throughout Europe and also in Latin America where they take the place of werewolves in folklore. But like werewolves, black dogs seem not to be satisfied with just remaining a legend. Sightings of these ghostly creatures continue to this day.

The black dog is infamous for instilling The Nameless Dread. In his 1813 book *A Relation of the Apparitions and Spirits in the County of Monmouth and the Principality of Wales,* The Reverend Edmund Jones related the story of a young woman, who intended to walk from Laugharne, Carmarthenshire one evening. Her mother warned her of a phantom black dog, seen near a water filled pit, by the side of the road leading to Laugharne, but she took little heed.

> *"On coming back before night, though it was rather dark, she passed by the place; but not without thinking of the apparition: but being a little beyond this pit, in a field where there was a little rill of water and just going to pass it having one foot stretched over it and looking before her, she saw something like a great dog, one of the Dogs of Hell, coming towards her; being within four or five yards of her, it stopped, sat down, and set up such a scream so horrible, so loud and so strong, that she thought the earth moved under her, with which she fainted."*

Sidney Benton of Horncastle, Lincolnshire, told veteran Fortean researcher Nigel Watson of an encounter he had in 1923. He was employed by a dairy to deliver milk. His round took him to a very old part of the town, past an old foundry, a vicarage and over a small bridge with a water wheel. One evening, a massive black dog appeared in the bath in front of him. Though talking in 1974 he clearly recalled the glowing eyes appearing in front of him, several feet from the ground. He leapt back over a fence and the creature seemed to transfix him with its glowing eyes. He was paralysed by fear. Finally the apparition vanished. Benton's employer and his family noticed how white and frightened he looked. He never visited the area at night ever again.

Such an encounter can stay with a person for all of their lives. The following account is from a letter sent to Fortean researchers and authors, Janet and Colin Board, by a witness who saw it as a girl during WWII.

> *"At the time, because of the war, my mother and I usually stayed with an elderly gentleman, who had kindly taken us in as refugees from London. We only went back to the capital when the bombing ceased. The cottage we lived in, is still in existence in Bredon, Worcestershire. My encounter took place in late afternoon in summer, when I had been sent to bed but was far from sleepy. I was sitting at the end of the big brass bedstead, playing with the ornamental knobs, and looking out of the window, when I was aware of a scratching noise, and an enormous black dog had walked from the direction of the fireplace to my left. It passed round the end of the bed, towards the door. As the dog passed between me and the window, it swung its head round to stare at me-it had very large, very red eyes that glowed from the inside as if lit up, and as it looked at me I was quite terrified, and very much aware of the creature's breath, which was as warm and strong as a gust of wind. The creature must have been very tall as I was sitting on the old fashioned bedstead, which was quite high, and our eyes were level.. Funnily enough, by the time it reached*

the door, it had vanished. I assure you that I was quite awake at the time, and sat on for quite some long while wondering about what I had seen, and to be truthful, too scared to get into bed, under the clothes and go to sleep. I clearly remember my mother and our host, sitting in the garden in the late sun, talking, and hearing the ringing of the bell on the weekly fried-fish van from Birmingham, as it went through the village! I am sure I was not dreaming, and have never forgotten the experience, remembering to the last detail how I felt, what the dog looked like etc."

Dermot MacManus records an even more dramatic encounter with a black dog in his book *The Middle Kingdom*. It was Easter and the witness was fishing near a river.

"As he was standing on the dry, gravelly edge of the bed, casting into a small pool, he suddenly felt constrained to look to his right along the river. He could not see far as there was a bend less than a hundred yards away, and there the hedge from the next field ran down to the bank. But as he looked he saw a huge black animal come into sight, padding along in the shallow water. He could not at first make out what it was, whether dog, panther or what, but he felt it to be intensely menacing, so without wasting a moment he dropped his rod and jumped for the nearest tree on the bank, a youngish ash, and climbed it till it bent dangerously under his weight.

Meanwhile the animal continued padding steadily on, and as it passed it looked up at him with almost human intelligence and bared its teeth with a mixture of a snarl and a jeering grin. His flesh crept as he stared back into its fearsome, blazing red eyes which seemed like live coals inside the monstrous head. Even so he could only think of it as a wild, savage animal which had, presumably, escaped from some travelling circus.

It passed on and was soon lost to view around the next bend, and once he felt it was well on its way, he slid down from his precarious perch, grabbed his rod and raced back to his house. His father was out, but he got his shotgun, loaded it with the heaviest shot he could find, and went out in search of the animal, feeling that no one in the neighbourhood would be safe while it was still at large. However, he drew a blank. Everyone he met, including those who must have been in its path, denied all knowledge of the animal."

In some cases, it seems that black dogs can kill with fear. In her book, *English Folklore*, Christina Hole recounts a story she heard herself whilst staying with a friend in Norfolk as a child. Her friend's mother described how, thirty years before, her brother had been sent on an errand at dusk. On reaching his destination, he found the house shut up and turned to leave. Just then, a massive black dog seemed to rise up out of the ground and placed its paws on his shoulders. Breaking free he ran home in a state of terror and told his parents what had happened. Gradually he calmed down and went away to his room to study. But at midnight, he fell down stone dead, apparently from the delayed effects of shock.

Pierre Van Paassen (1895-1968) was a Dutch born Canadian journalist and foreign correspondent who wrote for papers such as the *New York Evening World* and the *Toronto Star*. During the course of his career he got himself thrown into Dachau concentration camp – and later out of Germany, for criticising Adolf Hitler back in 1933. Then went on to cover the Italian invasion of Abyssinia and the Spanish Civil War, before giving it all up to become a Unitarian minister. In the spring of 1929, he had taken lodgings in a private house, in Bourg-en-Foret, France. One night he was alarmed to see a huge black dog pass him on the stairs. The creature seemed to vanish as it reached the landing. A search of the house turned up no dog.

He didn't mention the odd event to anybody in the house, and had to leave on business for a few days. Upon his return he found other members of the household visibly upset, and on enquiry found that they too had seen the black dog. Van Paassen decided to wait up late in the hope of encountering the "animal" again, and he invited a neighbour, a Monsieur Grevecoeur, and his young son to join him as corroborating witnesses. The black dog appeared at the head of the stairs again that night. Grevecoeur whistled to it, and the dog wagged its tail in friendly fashion. As the three of them approached it, the animal simply faded away.

A few days later he waited for the beast again, this time accompanied by two police dogs. The black dog materialized and came part of the way down the stairs, before vanishing. A moment later it seemed that the police dogs were involved in a savage fight with an invisible foe.

> *"The dogs pricked up their ears at the first noise on the floor above and leaped for the door. The sound of pattering feet was coming downstairs as usual, but I saw nothing. What my dogs saw I do not know, but their hair stood on end and they retreated growling back into my room, baring their fangs and snarling. Presently they howled as if they were in excruciating pain and were snapping and biting in all directions, as if they were fighting some fierce enemy. I had never seen them in such mortal panic. I could not come to their aid, for I saw nothing to strike with the cudgel I held in my hand. Then one of my dogs yelled as if he were in his death-throes, fell on the floor and died."*

An examination of the police dog's body failed to reveal any external signs of injury.

This terrified Van Paassen's landlord, who summoned a priest. This man, Abbé de la Roudaire, arrived and stood watch with the journalist the next night. The black dog appeared and snarled at the cleric, then vanished as he approached it. Summoning the landlord, Roudaire asked him if he had any young girls in his employment, explaining that there was often a link between girls of a certain age and apparitions. The landlord said that he did indeed employ a girl. After she was dismissed (on what grounds we do not know) the black dog was never seen again.

This is a fascinating case on many levels. Firstly, it is a multiple witness case. Secondly, the black dog had the power to kill one of its mortal counterparts. And finally, the link with a teenage girl. Poltergeists are often associated with adolescents, mainly girls. One theory postulates that the poltergeist is nothing more than externalised psychic energy, subconsciously emanated by the teenager. Others feel that external, non-human entities can use this energy to manifest. I take the latter view.

The Reverend Dr Donald Omand, whom we met earlier, had his own run in with a hell hound. Before joining the church he had visited Kettleness, a desolate promontory on the Yorkshire coast. He was there with his family just prior to WWI. He felt that the area 'seemed laden with corruption'. Years later, as a reporter he interviewed a man who claimed to have seen a monstrous black dog in the very same area. Later still after being ordained and becoming an exorcist, the Reverend was on holiday in Whitby with two friends. Looking over to Kettleness, all three saw a huge black hound that vanished leaving behind an overwhelming feeling of evil.

One of his friends, a schoolmaster, asked Reverend Omand to exorcise the monster. After making preparation they returned. A black hound that Omand says was bigger than any mortal dog, came loping towards them across the desolate sands. The schoolmaster fled to his car but the exorcist held firm. Casting holy water at the brute he cried out...

> *"Be gone in the name of the Lord Jesus Christ. Be gone to the place appointed for you, there to remain forever. Be gone in the name of Christ."*

The monster melted away and the atmosphere of corruption and menace vanished with it.

Another, even weirder dog-like apparition, was encountered by the grandfather of Dermot MacManus who wrote about it in his book *The Middle Kingdom*. In the 1860s he was walking through an orchard one summer evening on the family estate in Ireland. He was going to join a party that was going on in the summerhouse. Upon hearing a rustling sound, he looked round to see a fox-like creature the size of a wolf approaching him. When it reared up on its hind legs, he saw to his horror that it had no head. Despite this, he knew it was looking at him and pouring out feelings of 'hate, bestiality and evil'. He made the sign of the cross at the thing and it fell back on all fours and ran away. In 1901, at the age of 91, he made a signed statement that the event was still fresh in his memory.

Black dogs can have electrical effects too. One morning in 1972, Tom Morgan (alias), a farmer living on Dartmoor, Devon, England and his wife were awoken by the sound of scratching at their bedroom door. Arming himself with a poker, Mr Morgan peered out of the door but saw nothing but a gloomy corridor. But as he moved towards the stairs he saw a huge black dog with glowing red eyes. The beast began to walk towards him and filled with fear, he lashed out at the spectre. The hound vanished in a blinding flash of light. Every window in the house broke, there was a total electrical failure and tiles were scattered from the roof. It was as if there had been a huge electrical discharge, though Morgan himself was unharmed.

A similar story was told in the weirdly named book *Nummits & Crummits* written by folklorist Sarah Hewett. A man was walking from Princetown to Plymouth in Devon one evening. He was on the Plymouth side of the reservoir. He saw a black dog the size of a Newfoundland. When trying to pat the animal, his hand passed right through it. Frightened, the man walked on quickly but the dog kept pace with him until they reached a crossroads, At this point the hound vanished in a flash of light that threw the witness to the ground, senseless.

We have seen the effects monsters can have on electrics. Ghosts too seem to have this power, as do UFOs, as we shall see later. There is also a connection with weird smells and light phenomena. In 1893, two men driving a cart along a road at Rockland, Norfolk, found their way barred by a huge black dog with glowing red eyes. The driver, despite his companion's warnings, drove the cart straight at the entity. The black dog vanished in a ball of fire, accompanied by a sulphurous smell. Black dogs also seem to be associated with water, much like the worms and serpents of legend. It has been noted that phantom dogs seem to strictly stick to territories, manifesting along certain stretches of road. One such is Dog Lane that runs along the boundary of Devon and Dorset at Uplyme. It was the scene of a very strange encounter reported by a local woman in 1857.

> *"As I was returning to Lyme one night with my husband down Dog Lane, as we reached the middle of it I saw an animal about the size of dog meeting us. "What's that?" I said to my husband. "What?" he said "I see nothing." I was so frightened I could say no more then, for the animal was within two or three yards of us, and had become as large as a young calf, but had the appearance of a black, shaggy dog with fiery eyes. Just like the description I had heard of the 'black dog'. He passed close to me, and made the air cold and dank as he passed along. Though I was afraid to speak, I could not help turning round to look after him, and I saw him growing bigger and bigger as he went along, till he was as high as the trees by the roadside, and then seeming to swell into a large cloud, he vanished into the air. As soon as I could speak I asked my husband to look at his watch, and it was then five minutes past twelve. My husband said he saw nothing but a vapour or fog coming up from the sea."*

This is a fascinating report for several reasons. Firstly, only the woman is able to see the black dog, just like the case of the canal Bigfoot from the last chapter, when only two of the three present could see it. Secondly, the atmospheric effect, the lowering of the temperature when the black dog is near. We see that in other ghost cases. Finally, the fact that the apparition seemed to swell in size, to as large as the trees then became cloud-like before vanishing. This suggests that the entity is ethereal in nature, and can change shape, an ability we have seen before, and will see again.

In another case from Templeleogue, a suburb of Dublin, two people were walking home at about 9.30pm in the 1890s. One saw a huge black dog accompanied by a sound, like dragging chains crossing their path. The other did not see the hound, but heard the clanking of the chains.

Another black dog haunted the area around Torrington, North Devon, and was investigated by Mrs Barbara Carbonell, who travelled in the region in the 1920s. At the village of Copplestone, villagers told her that the black dog had been seen near a fourteen foot tall, granite pillar, dating from before the sixth century and carved with Celtic symbols. It stands at the centre of a junction where three roads meet. From Copplestone the road is joined by an ancient lane, that runs into the crossroads at the hill top village of Down St Mary, where a Saxon church stands. The villagers said that the dog would run up the lane on dark nights, past the blacksmith, and between the church and the village school, the edge of which juts out into the road. The black dog is said to strike the side of the school, causing a sound like falling masonry but no actual damage is ever done. One old man told Mrs Carbonell *"I've a-heard the stones fall as he goes roarin' by and I've heard the same from my father and grandfather time and time again. It's true right enough."*

Another local man described an encounter he had as a boy. He and several friends were making their way home after a choir supper at Christmas. As they neared the blacksmith's they heard the black dog coming. They dived into a field and saw it lope by. They saw it was as big as a calf, with glowing eyes and making a baying noise. As it passed the school they heard a noise like stones falling in a quarry. The boys waited for the baying to die away before running home. Unexplained sounds like ticking, crashing, cranking or chains dragging are common in all forms of Fortean phenomena.

Black dogs have an adverse effect on animals too, showing that they cannot just be figments of the human witnesses imagination. Mrs Grace Pearse wrote to the folklorist Theo Brown about an encounter she had with a black dog in Devon in the 1890s. She was returning from a tea party, with her young daughter, who was riding on their pet donkey, Romeo. As they were on the road that passed Okehampton Castle, a huge black dog, as big as a pony, leapt out from the castle grounds and glared at them. Romeo refused go further. Even after the dog vanished, the donkey would not walk along the lane and they had to take the long way home.

Robert Chambers in his *Book of Days* tells us the following story. In 1751, a chimney sweep called Thomas Colly had drowned an old couple, John and Ruth Osbourne, whom he had believed to be witches. Colly was hung for his crime. His body was hung in a gibbet, at the spot of the crime, in the parish of Tring, Hertfordshire. Since that time, a tradition that the place was haunted by a black dog grew up. A schoolmaster related how late one night he and another man were driving home in a horse-drawn gig. As they drew near to the place in question, they saw a flame 'as large as a man's hat' in a ditch near the road.

> *"What's that? I exclaimed. Hush! Said my companion all in a tremble: and suddenly pulling in his horse, made a dead stop. Then I saw an immense black dog lying on the road, in front of our horse, which also appeared trembling with fright. The dog was the strangest looking creature I ever beheld. He was as big as a Newfoundland, but very gaunt, shaggy, with long ears and tail, eyes like balls of fire, and large, long teeth, for he opened his mouth and seemed to grin at us. He looked more like a fiend than a dog, and I trembled as much as my companion. In a few minutes the dog disappeared, seeming to vanish like a shadow or sink into the earth and we drove over the spot where he had lain. The same canine apparition is occasionally witnessed at the same place or near it."*

Another folklorist, Ruth L Tongue, wrote to Theo Brown with a story she collected from the Quantocks, in Somerset. This case also involves an animal being effected by the appearance of a black dog.

> *"A rider returning in late autumn dusk over the Quantock Hills after stag -hunting found his weary horse in a sweat of fear and saw a large black dog pacing alongside, about ten feet away. The horse was too terrified to swerve or gallop and had to be urged to walk on over a distance of about a mile. The rider said he was sweating in cold terror too. The dog seemed to come from one cairn and follow along to disappear to another. At once the horse broke into a gallop and arrived home exhausted. As my hunting friend does not want to be laughed at, I cannot give the locality, but judging on the impression made on him and the nerve of his horse, they certainly saw the Black Dog this year (1957)."*

Black dogs are still seen today. Most witnesses encounter the phantoms whilst driving at night. A fine example was the sighting by a Mr Lee, who was driving to his home in Molland, Exmoor, North Devon. At 8.30pm on Halloween of 1984, he saw a huge black dog, like a Great Dane with a huge head, running down the road towards him.

> *"I had on my anchors and in the headlights it slowed down and walked right up to the car. I could see its eyes plain as day, green and glassy there where and he looked right over the bonnet at me, he was that tall*

and then he just went!

When asked if the animal had run off, Mr Lee clarified.

"Oh no, it just went like a light going out, I just couldn't see it any more. It isn't real like an ordinary dog. I could feel the hairs on my neck standing up. I just started the car and drove off, wasn't hanging about with that thing out there!"

Victoria Rice-Heaps was driving home from her boyfriend's house on the morning of May 11th 1991. The time was 2.14am. She was on the Blythe road heading for her house in Worksop, Nottinghamshire.

"I travelled about a mile, almost up to Hodsock Priory when I saw in my headlights two red dots in the distance about 150 yards in front of me. I slowed right down to a crawl and saw a huge black dog. It looked like something from hell. It had very shiny fur and a short coat; the nearest thing I've seen to it in size was a Great Dane, but it had a good 18 inches over that breed.

Its ears were erect and it seemed to be dragging something quite large across the road. I've lived with dogs all my life but I've never seen anything like this creature. While I was waiting for the dog to get out of my path, headlights approached from the road ahead. The car slowed right down as it pulled up and I could see it that was a red Montego estate.

The male driver stopped and wound down the window, as I did as well. He asked me if there was something wrong and I asked him if he could see the dog in front of my Fiat Panda. At this moment he shouted "Oh Jesus" and sped off into the night. I looked in front of me again and to my joy and amusement the creature had vanished. I drove home as fast as I could. I did a little research later and found a tumulus, a river and an old boundary, as well as the priory."

How much more terrifying would it be to encounter one of these hell hounds whilst on foot, without the safety of a car around you like a shark cage? Mr E.G Thompson wrote to the magazine *Fortean Times,* to tell of one such encounter he and his family had in 1996. Mr Thompson's 19-year-old daughter had been working at a holiday camp in Norfolk. One night, around midnight, Mr Thompson and his wife met their daughter after her late shift in order to walk her back home. Suddenly, they heard a rustling from some bushes and a deep growling noise. A shape like a huge dog appeared.

"It stood there snarling but unlike any dog we had ever heard—and we have always had dogs. To say we were nervous was an understatement.

To cap it all, it had glowing red eyes. There was no way it could be a trick of the light as the house lights were behind it. It stood there as we walked fearfully past. After a few yards I looked back, as the snarling had stopped. The creature was nowhere to be seen, though we had heard no sound of it moving."

Though the UK is the spiritual home of the black dog, they are also reported from overseas. A user calling themselves 'Tamyu', on the venerable Forteana Forums message board, told a story of a black dog encounter in the USA, though they do not specify what part.

"I had a personal encounter with a black dog, or at least I think that's what it was. I didn't realize that black dogs were such a common phenomenon.

It happened about 6 years ago. I was visiting family in the US around the winter holidays. I would say it happened in January, it was cold, and there was snow on the ground.

The relative I was staying with had received a little dog from friends because they had moved to an apartment - we rigged up a chain attached to the hand rail at the side of the steps going down from the back door. There was no fence around the place so we would always clip the chain to the dog's collar and let it out.

Anyway, I was there alone and had let the dog out. It was mid afternoon. The dog started desperately barking and yipping. I assumed that there was something out there upsetting it (cats, or squirrels).

It seemed to be really freaking out so I hurried and went out to unhook the chain and let it in. I looked around when I first went out and saw nothing. After I unhooked the chain and picked the dog up, I turned to go up the steps and saw a HUGE black dog walking slowly down the road, about 30 meters from me. It was staring directly at us, and stopped. Then it started a mad dash toward us. I was scared to death of being attacked and leapt up the steps and inside, slamming the door behind me. A few seconds later the big dog crashed into the door and it felt like it was going to literally break it down. I was still holding the little dog who lost bladder control and started whining.

I dropped it and it ran away somewhere.

The door had a little ventilation window at the top, and I cracked it open to look at the dog banging into our door. It smelled awful, like rotten meat and eggs. When I cracked the window it suddenly stopped, and when I looked out the crack I couldn't see anything odd.

Of course I got scared and ran around the house checking all the windows and was in fear that it would come crashing through the glass at any time.

I didn't see it again, but didn't risk going outside for a few hours.

Up until this point, I hadn't thought it supernatural at all. Stray dogs (Or just really poorly kept ones) were not all that rare. It was huge, but seemed like just a really big and stinky dog. When I went out later to try and persuade the little dog to actually come out and do it`s business outside, I looked to see which way the dog had gone and if it had been back (Because there was at least 10 cm of snow on the ground).

There were no tracks. NOTHING had crossed the area between the road and the back door. The little dogs prints were very clear, as were mine from going out to get it, but there was no sign of any other animal coming close.

Then when I thought back about it, I realized that I hadn't heard it bark or growl the entire time. In fact, the only sound was the banging on the door.

Having read this thread, I really think it may have been a "black dog."

The vile smell and lack of tracks resemble other Fortean entities, and seem to mark this out as something other than simply a big, aggressive dog.

The strangest black dog report of all occurred in the USA. On September 9th, 1973, (surprise, surprise) a group of youths in Savannah, Georgia, saw a landed UFO in Laurel Grove Cemetery. Doors opened in it and ten huge black dogs emerged and ran off into the night. The craft then turned off its lights. The terrified youths fled the graveyard. This is not the only time black dogs have been associated with UFOs. In a case from 1980, in Todmorden, Yorkshire, a policeman claimed to have been kidnapped by aliens. Under hypnosis he recalled seeing a huge black dog on the ship with him. Yorkshire is famous for huge, ghostly black dogs. We will look at this case more closely in the next chapter.

Another case linking UFOs and black dogs comes from the UK. My friend and fellow researcher Nick Redfern was told of a black dog encounter near Castle Ring in Cannock Chase, England. In December of 1991, the witness noticed a small area of dense fog and investigated. When he got to within twenty feet of it, he felt an effect like static electricity, that made his hair stand on end. There was also a smell like burning metal. Then a huge black dog, the size of a small horse, emerged from the fog. He began backing away from the beast. When he was about 150 feet from the dog, a ball of light hovered over the fog and shot down a beam of blue light. In an instant the fog, ball and dog vanished.

Now, we have seen bigfoots, lake monsters and black dogs in association with UFOs . Makes you think doesn't it?

GHOST TREES
Trees dominate most landscapes. It was said that at one time the forests of Britain were so vast, that a squirrel could travel from Land's End to John O' Groats, without putting a paw on the ground! Today, there are vast swathes of rainforest in Africa, Asia and South America that are still unexplored. In Australia there were once giant mountain ash trees that grew to 435 feet tall, larger than the biggest American redwoods, until they fell before the white man's onslaught.

Trees loom large in the world's mythology. The Norse Yggdrasil or World Ash was a tree that had its roots in the underworld, where they are gnawed by Níðhöggr (dread biter) a great dragon. Its branches support Asgard, realm of the gods.

In Arabic law certain trees are haunted by jinn. If sacrifices are made to them they will help the sick who sleep beneath them, by giving them prescriptions for cures in their dreams.

The Egyptian Book of the Dead mentions sycamores as part of the scenery of the afterlife. Is it any wonder then, that there are ghost trees?

Trust Japan to have the weirdest ghost tree of all, a vampire tree! A Jubokko is created when a tree grows amidst the carnage of a battlefield. Its roots suck up the spilt blood, and nourished by death, they develop an unnatural thirst. A Jubokko might seem like a normal tree, until an unsuspecting human draws within its reach. Then the yokai tree's branches whip out to seize the poor victim, hoisting them up and draining every drop of blood from their body.

There are a number of monster trees in Japanese lore. Some believed that once a tree became more than 1000 years old, it becomes a spirit tree with sentience.

One story tells of a woodcutter named Musabi no Gen, who went up to a mountain forest to fell a tree. He found a huge, gnarled old tree, and was about to hew it with his axe, when he heard a voice warning him that the tree was about to fall. Startled he looked around but saw no one. He was about to swing his axe again, when he heard the sound of a tree crashing down. He leapt back from the trunk, but no tree fell. Each time he attempted to cut the tree the loud crashing noise stopped him. Finally at nightfall the tree underwent a horrific transformation. A twisted mouth formed in the trunk and evil looking eyes opened in the knots in the bark. It uprooted itself and began to glow a lurid blue. Its eerily luminescent branches snatched up the hapless woodcutter and enveloped him.

Furusoma, meaning 'old timber,' is the ghost of a person killed by a falling tree. It haunts the mountains of Nagaoka District, in Tosa Province. It cries out "*ikuzō ikuzō*" ("its going, its going"), and then there is an almighty crash, like a huge tree falling. Upon investigation no fallen tree is found.

In Gifu Prefecture, central Japan, there is an old persimmon tree that is believed to grow human hair. The tree grows in a graveyard, on the grounds of the Fukugenji temple, in the town of Yoro. The temple is now abandoned, and with good reason. At night the tree is said to glow a weird blue. Its branches sprout hairs that when burned smell like human hair. Removing this hair brings bad luck. Several decades ago, two youths from the town cut some of the hair from the tree. One fell into a fever and died and the other had a 'fatal accident'. Both were dead within a month of defiling the tree. In 1978, a group of hikers picked persimmons from the tree. On their way back home they met with a bad accident.

It is said that in 1681, the father of a man named Ishii Mitsuno-jo, was murdered by Akabori Gengoemon in Osaka. Ishii Mitsuno-jo, a 26-year-old samurai pursued his father's killer in order to take revenge. However, Akabori killed the samurai by strangling him and hid the body under a persimmon tree, before fleeing to safety. Some 29 years later, two brothers of Ishii Mitsuno-jo, named Genzo and Hanzo caught up with Akabori and killed him. The tree that Ishii Mitsuno-jo was buried under began to grow hair. The tree finally died and fell over, but a new one grew in its place, and this too sprouted hair. The tree glows blue at night.

In 1971, a professor at Tokyo Agricultural University examined the hair and said that it was actually the fibers of a plant that resembled human hair. What came first, the odd plant with a legend growing up to explain it, or the legend supported by the growth of a hair like plant?

That controversial 'king of ghost hunters' Elliot O'Donnell, was told of a ghost tree that haunted a garden of a family called Andrews in Warwickshire. The phantom ash was said to appear just before a family member died. Mr. Andrews was disturbed by his dog barking one night and decided to investigate. On the lawn in front of the dog's kennel was a shadowy ash tree, where none had stood before. It swayed violently as if buffeted by a storm, though the night was still. As Mrs. Andrews joined her husband in the garden, the tree vanished. Soon after they found out that one of their children had met with a fatal accident whilst on holiday abroad.

In his 1959 book, *The Middle Kingdom,* Dermot MacManus writes of demon inhabited trees in Ireland.

"I know of one tree that has a particularly malevolent surrounding. In reality there are three trees, two thorns and a boutree, but they look like one--for the tree straight stems are almost touching, certainly not more than two inches apart, and their branches are inextricably mingled into one bushy crown...This tree is set in a low lying field of poorish land and some thirty feet from a narrow country road so one looks down on it as one passes.

It is guarded by three malevolent demons who, after dark, haunt that stretch of road. There are eerie and disturbing tales roundabout, of passers-by late at night, who have had their arms fiercely gripped so that the marks could be seen next day, or who have heard bone-chilling, inhuman laughs or angry spittings, as if from some enormous cat; and sometimes even dim and horribly misshapen figures have been made out moving in the darkness."

Jon Pertwee, the actor famed for playing the Third (and in my opinion the best) incarnation of The Doctor in *Doctor Who,* had a childhood encounter with something stranger than anything he met whilst playing his best known roll. His experience was recorded by Richard Davis in his 1979 book, *I've Seen a Ghost.*

As a small boy Jon used to go and stay with a school friend in an Elizabethan manor House in Sussex. The family lived in one wing and the rest of the house was not widely used, except when they were having parties. There was also a dining area with a minstrels' gallery running around it. Leading off the gallery was another room used as a bed room. On this particular occasion, Jon was asked if he minded sleeping in this room, as all the other rooms were full during the holiday season. He distinctly recalled his friend's father saying *"Do you think that's wise?"* His mother replied *"Oh yes, that's all right, he's a sound sleeper."*

On the first night he awoke feeling an awful nausea and proceeded to vomit on the bedclothes. Dreadfully embarrassed he cleaned the sheets with water and hung them up to dry. In the morning he told his hosts he had slept well.

The following night he found out just what had made him so ill.

"The next night I went to bed again and again I woke up and this time I was able to realize what had made me sick. In the room there was the most overpowering smell of putrefying flesh, it was exactly like a dead sheep, and it permeated the room.

I shot up out of bed and again felt violently sick. I looked up and about four feet from the end of my bed was a thing I can only describe as a sort of tree trunk. It was a light greenish colour and it undulated, and as far as I could see it bubbled: it seemed to have bubbles that blew up at the side of it and didn't burst exactly but disappeared. This thing was moving very, very slowly towards me."

The thing frightened Jon so badly, he wet the bed and went running down the gallery to the wing where the other people stopped. Whilst being comforted by his friend's mother, he heard her husband say *"You see, we should never have put him in there."*
On asking his friend about it, he was told that other people had seen the thing and the family never put a guest in the room. They had thought that Jon, as a child would be a deeper sleeper and would not be awoken by the thing. He never did find out what the crawling, glowing, bubbling, stinking tree stump was.

The idea of a ghost tree may sound odd (why not a ghost cactus of a ghost sunflower?) but with the amount of trees destroyed every day by humans around the world, the question might not be 'why ghost trees? But why aren't there more ghost trees?

ROAD GHOSTS AND PHANTOM HITCH-HIKERS
Richard Studholme, was a guitarist who played with the band *Chicory Tip*. Driving home one night in 1974, he picked up a young girl hitch-hiking near Blue Bell Hill. She asked to be dropped at West Kingsdown. She also requested that he drove to a Swanley address to inform her parents of her whereabouts. Studholme agreed and after dropping her off, he drove to her parent's home, only to be greeted by the bereaved father who explained his daughter had been killed while hitching a ride home two years ago. Studholme thought he had been the victim of some sort of cruel hoax until he read of others who experienced exactly the same thing around Bluebell Hill. He says he touched the girl, helped her into the car and held her handbag.

Road Ghosts are a distinct group of phantasms with a flavour all of their own. Some of Britain's roads, such as Watling Street that links Dover in the southeast to Wroxeter in the northwest, date back to Roman times. Driving alone, especially at night in a rural, lonely area can be a spooky experience. Some have even suggested that the driver can even fall into a kind of trance, moving along in their isolated metal box.

Bluebell Hill in Kent, is home to the most famous phantom hitch-hiker, a pretty young woman. Some stories may have had their origin in a real accident, in which Susan Brown and her two friends were killed. On November 19th, 1965, the 22-year-old was returning from her hen night, when her Ford Cortina spun out of control on the A229, colliding with a Jaguar heading the other way. Susan was due to marry RAF technician Brian Wetton the following day. One of the group, Patricia Ferguson, was killed at the scene, but Miss Brown and Judith Lingham died a few days later in a Maidstone hospital.

However strange sightings seemed to have occurred before the terrible events of that night. For example, in 1962, Bob Vandepeer said he gave a lift to a girl on the hill only to later turn around and discover the hitch-hiker had vanished in the back of the car. If phantom hitch-hikers were the ghosts of people killed in car crashes, then the roads of the world would be awash with ghosts.

The story of the phantom hitch-hiker, with some variations, runs like this. A person, usually but not exclusively a man, is driving along a lonesome road at night. He sees a young woman hitch-hiking and stops to pick her up. The girl is usually very pretty and seems normal. She asks to be dropped off at a certain address and the man takes her there and waves goodbye. Later he finds that she has left some item in his car, usually a piece of clothing such as a sweater. The man takes it back to the address and is met at the door by one of the girl's parents who tell the man that she had been killed years ago in a car crash along that particular stretch of road.

Other variants include a driver knocking over a figure in the road. On 13 July, 1974, local bricklayer, Maurice Goodenough, ran into Rochester Police Station and reported he had run over a girl around 10pm on Blue Bell Hill, who suffered several cuts, and had left her at the roadside wrapped in a blanket. When the police returned to the scene, the girl had disappeared, and despite an extensive 10-mile radius search of the area, the girl was never found.

In the 1980s a group of four women driving up Blue Bell Hill claimed to have driven right through a white figure that appeared on the road.

Perhaps the strangest twist occurred also in Kent, but close to the village of Hutton near Maidstone. On September 20th 1985, at 11.30pm, Peter Russell, a lorry driver was motoring up Hutton Hill when he ran over the body of a woman lying in the road. A doctor and policeman were already in attendance when the incident happened. The policeman told him he had not killed the woman and he should go home and not worry about it. In shock, Mr Russell did just that. Later he phoned the police station. The duty officer told him that there had been no report of such an incident and there had been no doctor or police officer at Hutton Hill.

The phenomena occurs in many countries. The N9 in South Africa has a well-known tradition of a phantom hitch-hiker, seen between the towns of Willowmore and Uniondale. In 1973, a man from Cradock, Mr. Leonard Fraser and his wife, Catharina, were driving on that road around midnight when they spotted a girl in a long white gown. *"She was standing on a storm-water drain. She didn't look human. I could see through her"* said Catharina. Apparently, when they stopped to pick her up, she disappeared.

On the cold, rainy evening of May 1st, 1976, when Anton La Grange was driving his Mercedes on the road near of Uniondale.

> '*I saw a girl standing at the side of the road. I stopped. She opened the door and got in. I asked where she wanted to go and she said 'Porter street 2, de Lange.' She was dark-haired with a pale face, dressed in a*

dark coat and slacks. "After a few miles, I turned my head and she was gone!"

Puzzled, La Grange drove into Uniondale, reported the incident to the sergeant on duty, Cornelius Potgieter, then resumed his journey.

"Just outside Uniondale, I heard the most chilling sound I've ever heard in my life," he remembered. "To this day I can't tell you if it was a laugh or a scream. It was right in the car, really loud... although I was absolutely alone!"

Terrified, La Grange sped back to Uniondale, where he and Potgieter searched the car, finding no trace of the girl. La Grange again set off, the sergeant following him in his police van.

"Just outside Uniondale, while I was driving at a steady 43 mph, the right rear door of my car slowly opened and closed. It was a controlled gentle opening and closing, exactly as if someone got out and then shut the door behind them."

Sergeant Potgieter remembered:

"La Grange drove through a patch of mist and about 200 yards beyond the mist, I saw the right rear door slowly open and close as if somebody got out."

Rattled from the experience, Sergeant Potgieter returned to the police station, locked up for the night and went home.

At about 8.00am the next morning, Sergeant Potgieter confronted police sergeant Pat McDonald, and asked him if he remembered the girl that was killed in the car accident a few years before. Pat McDonald was the first officer to the scene of the accident that day. McDonald explained that he had found a Volkswagen Beetle off the road, and a girl was lying on her back with her head against the embankment, she had died of head injuries. He later ascertained that her name was Maria Roux, Ria for short. Her fiancée the driver, had survived. He explained to sergeant Potgieter what the girl actually looked like, which was exactly the same description that was given by La Grange. Later, when Anton La Grange was shown the photo of Maria, he said that there was no doubt it was the same girl. La Grange admitted that the indecent had scared him so much he had never returned to the area.

Maria Roux had been killed in a crash during a storm on April 12th, 1968. The car was being driven by Michiel Oberholzer, who had asked Maria to marry him the day before. They were travelling to Riversdale from Pretoria to visit Maria's parents and discuss their wedding plans. Michael lost control of the car in the terrible rain. He survived with bad injuries, but his bride to be had been killed.

On Easter weekend 1978, Army Corporal Dawie van Jaarsveld, was doing his national service at Oudtshoorn army base 100 miles from Uniondale, he was on his way there to spend the holidays with his girlfriend. His journey took him along the N9.

> *"When I reached the intersection, I saw somebody standing on the side of the road. Just as I turned to the right, she kind of lifted her arms up – like 'oh no, aren't you going to stop'."*

Dawie stopped his motorcycle and offered a lift to the young woman, with dark hair and dark clothing. She climbed on, putting his spare helmet on her head and an earpiece in for some music.

> *"I asked her to please hold tight around my waist so that I can feel if something goes wrong. After a kilometre or two, the bike had a twitch, I thought she fell off. A lot of scenes went through my mind. I turned around, I wanted to see if I still had someone with me, there was nobody. I turned around, went back with the motorcycle to see if there was anybody lying along the road. I saw the spare helmet was back on my luggage rack, then I just had to move off because I realized then that I didn't actually pick somebody up."*

Dawie van Jaarsveld also said she was a shortish girl, short dark hair, and she had slacks on and a jersey.

> *"I didn't think there was something wrong at the time. I didn't notice that it was a spirit or something. But I did feel strange."*

Strangely the earpiece was melted when Dawie examined it. Like other Fortean phenomena, it seems electrical effects are part of the road ghost's tool kit.

Some two years later, on Good Friday 1980, motorcyclist Andre Coetzee, had a terrifying experience while riding along the N9 past Uniondale looking for a friend who he thought might have run out of petrol in the desert.

> *"All of a sudden I felt hands around my waist, I could actually feel the pressure, and as I looked down, I saw the hands. I was very very scared and I felt I must get away from that place. That is when I accelerated and about 150km she gave me three wacks against the head and then she just disappeared. I never saw nobody standing next to the road or nothing."*

Andre drove to a local cafe in Uniondale for help.
> *"He could hardly speak when we asked him what had happened. But*

gradually it dawned on us that the woman ghost had appeared once more."

...said Jeanetta Meyer, the cafe owner.

Whilst on his honeymoon, in 1983, Danny De Kock, had a personal encounter with the phantom hitch-hiker.

"My wife and I stopped next to the road to take a rest. My wife was asleep in the passenger seat. It was late at night and there wasn't the slightest wind. I got out and walked around the car (VW Beetle) for a wee, when a terrible sound like a rushing wind came rolling down the road, and went dead quiet as it reached the car. I got the fright of my life and ran back to the driver's side, to be met by someone sitting in my seat. I ran back to the other side again to pull my wife from the car as I thought we were being hijacked, but upon wakening my wife I discovered the other person was gone. We left the spot in a hurry and about 500 meters down the road, we saw this woman in what seemed to be dressed in a gown, waving at us to slow down. I reduced speed in case it was someone in need of help but before we could reach her she disappeared. I stopped at a garage in Uniondale and told a stranger about my experience, and that was the first time I learned about this ghost. You can imagine the chills that went down my spine. This is the honest truth."

But according to Maria's brother, Frans Roux, the ghost is not that of his sister. He felt she was far too conservative to have hitch-hiked, and would never have gotten on a motorbike in her life.

"We are more than tired of it. It's a bunch of nonsense, but it just doesn't end. The organizers of the Klein Karoo National Arts Festival even use it as an advertisement for their festival, Anton Goosen sings about it, and then there are the people who go camping at the accident scene on Easter weekend, year after year in the hope of seeing a ghost. Even Uniondale is benefiting from the tourism.

We want to say to the people: release Ria. There is no 'ghost' of her and there has never been one. People need to know that Ria was a happy girl and she was a compassionate person. She loved jokes, loved music, loved animals, and she had a bubbly personality. If Ria, where she is in heaven right now, was looking at all these monkeys and their stupid ghost stories here on earth, I'm sure she would have a good laugh. She was that type of person."

If the phantom is not Maria, could some kind of phenomenon have built up in the area after the accident. Because people believed the road was haunted, it actually became haunted by something that just used Maria's image?

Åsa Ljungström notes two traditions from Sweden. There is a story of a girl trying to get a ride, usually standing by the roadside, not far from Mora in Dalama County. Drivers pick her up now and then. After a short while, the girl starts screaming: *'Let me out', let me get out!* So they stop to let her get out. In doing so she turns away from them and then they can see that there is a hole in her back, and it is all eaten by worms. It is interesting to note here that several supernatural female figures in Scandinavian folklore, such as the huldra, are said to have 'hollow backs' (like a tree eaten away by woodlice). That girl was murdered, she is a haunting ghost. She did in fact stand there, trying to get a ride. Somebody picked her up, but soon he turned into the forest on a narrow road and she was raped and murdered. There is another version of the story where car drivers notice her hitch-hiking but pass on without stopping. Then all of a sudden, she will be sitting in the car right beside the driver. Then she screams and wants to get out.

Another variation is an old man who appears in the back of a moving car. Again Åsa Ljungström has noted an example from Sweden. In the northern inland of Sweden, in the forests of Lapland, there is a small village called *Old Man's Lake*. Here there is talk of a ghost, a shepherd, appearing at night. There are two roads, one right through the village and the other goes around it in the outskirts, some five kilometres longer. You have to choose either the shorter, straight and broad way through the village where nothing exciting will happen to you, or the longer one around. If you pass a certain spot alone in the car, there may suddenly be a man sitting beside you in the car. He complains severely, but you are not supposed to talk to him, just drive on until he gets off.

Similar cases are reported in the UK. Rob Gandy has collected and investigated more cases of road ghosts than anybody else in Britain and this is just one of many from his files.

> *"It was a Wednesday night in early 2010, around 9.40pm, when Thomas was driving home with his friend Keith, after working at a youth club. Thomas told Keith about a phantom hitch-hiker type experience that his ex-girlfriend's father had had on White Gate Hill, a road leading out of the village of Hunmanby, Yorkshire. After dropping Keith off at his house, Thomas continued his journey alone towards White Gate Hill, listening to music on the car radio. It was dark and the weather was dry as he came to a point on the road, before reaching the actual hill. Just then he looked to his left and, to his shock, he saw a man sitting next to him in the passenger seat! He was completely 'real', solid and in full colour Thomas genuinely thought that somehow someone had got into his car, which he naturally found to be very scary. The man did not turn to look at him; he just looked forward expressionless. He was wearing a*

flat cap, possibly grey in colour, and a green coat, like one of the wax jackets people wear in the countryside. Thomas jumped out of his skin. He was travelling at around 60 miles per hour but he slowed down and pulled over to the side of the road so that he could get out of the car. However when he stopped he saw that the man had gone; he had simply disappeared, as Thomas would have been fully aware if the man had tried to open the car door, or similar. The whole event must have lasted around 10 seconds. Thomas got out of the car to calm himself down before carrying on home."

THE BREATHER

A woman called 'Donna' (not her real name), had an encounter with an unseen entity that apparently entered her car. But just because it was not visible didn't make it any the less frightening. One night, in September of 2009, Donna was driving east along highway 70 heading for Rudioso, New Mexico. She had her three dogs with her. As they did not get along they were seated in separate areas. A large Doberman on the back seat. A middle sized dog on the passenger seat and a small dog on her lap. As she drove past Round Mountain, a landmark topped by a big white cross, her car jerked to one side as if she had a puncture. Pulling over, she got out to look at the damage but found there was none. Getting back into the car she found all three dogs on the passenger seat, hackles raised, teeth bared and growling. They all seemed to be staring into the back seat. She tried to get her Doberman into the back seat but he refused to go. She set off with all three of them huddled on the same seat.

Soon she noticed that her radio had died. Dead air was all she could pick up on any channel. Suddenly she was overcome by the feeling that there was some unseen presence on the back seat of the car. Then a slow, ragged breathing began behind her. The back window began to fog up as if someone was breathing in the back the car. The dogs began whimpering again. The horrid breathing continued for around twenty miles until she reached the edge of the Mescalaro Reservation. Then the air pressure in the car seemed to alter. The radio came on again and the dogs returned to their places.

Once more we see radio interference and an effect on animals.

A SUMATRAN SPOOK

I have visited the Indonesian island of Sumatra six times on the trail of a new species of ape called the orang-pendek. But outside the deep jungles, other, less physical things are encountered.

In 2004, Chandra Widodo was driving between Jambi and Palambang at night. Up ahead he noticed a woman in white waving him down. He slowed down intending to give her a lift. She was very pale and very beautiful. He wound down his window and asked if she needed help. The woman remained silent but a strange grin appeared on her face. Then to his horror Chandra saw her features change from a beautiful young woman to an ugly old hag with white bulging eyes and a wide, toothy mouth. Chanda put his foot down hard on the accelerator and began to speed away. but his lights flickered and the car almost stalled. The car seemed to lack power. and looking in the rear view mirror. he saw the woman slowly waving at him as she stood some twenty feet away. Then there was a rumbling noise and the car began to shake. The further he got away from the figure. the less influence she seem to have on the car. and eventually he got away. The events still haunt him nearly 20 years later. He thinks what he encountered that night was a kuntilanak, a kind of vampiric ghost from Indonesian folklore.

THE PHANTOM DOCTOR

Some road ghosts seemed to be dressed in apparel from times passed. A personal friend of mine hand an encounter with one such phantom. Jon Pertwee (no not the *Doctor Who* actor but a friend of mine and, like me a fan of the show, who changed his name by deed poll) and his friend. were driving across Woodbury Common, Devon, in the early hours of 25th February, 1989. Sometime between 3.00am and 4.00am they were travelling north on the B3180. The weather was clear, but the night was dark as they approached the main Exeter and Exmouth Road (A3052), with the isolated Halfway Inn just about visible on the far side of the crossroads. They headed across the trunk road aiming to continue up the B-road to the right hand side of the pub. In the car's headlights, they caught sight of a man standing at the pub entrance, just to the left of the corner, and assumed that it must be the landlord, checking something now the pub was closed. But as they crossed the road, the man stepped out from the doorway and walked into the middle of the road in front of them.

He was clear in the headlights, and both Jon and his friend could see that he was wearing a black top hat and black frock coat, and he was carrying a Gladstone bag. Jon said the man, in his Victorian garb, reminded him of the character Dr Mopp from the kids TV show Camberwick Green. They could both see everything in detail, even down to the stitching on the bag. The man turned his head to look straight at them, although his normal-looking face seemed blank and unaware of the presence of the car.

Jon's friend slammed on the brakes, but the car simply went through the man! Both Jon and his friend were convinced that they had killed someone and got out of the car as soon as it came to a halt. Jon's friend burst into tears; he was that shocked. Tentatively, using torches, they looked behind the car, and then under the car, but there was nothing to be found. They searched for several minutes, but there was no sign of the man. This completely freaked Jon's friend out, and he was in tears and shaking. They decided not to call the police before setting off back on their journey because, quite reasonably, they thought that they would not be believed. A week later, Jon decided to pay a visit to the Halfway Inn to talk to the landlord about the ghost; his friend point blank refused to go with him. The landlord told him that several other people had had similar experiences over the years. He believed that what had been seen was the ghost of a doctor who had attended a birth at the inn, and upon leaving had stepped out into the road, only to be run over and killed by a horse and carriage.

THE PETER B'S ROAD GHOST

Road ghosts seem to have an affinity with musicians. The *Peter B's* were a pop group from the early 60s. Peter Bardens who led the band would go on to be the keyboard player in the 70s prog rock band *Camel*. Drummer Mick Fleetwood would later co-found the rock band *Fleetwood Mac*. The *Peter B's* had been performing at the Bird Cage Club in Portsmouth, and were returning to London via the A3. There were four of them in a van, with Peter driving, when they encountered a ghost. It was 2.30am and they had just left Cobham.

Peter's father Dennis Bardens recorded a tape interview with them on 27th of November, 1965. Mr Bardens provided a transcript to Dr A.R.G Owens and Victor Sims, who included it in their book *Science and the Spook*.

There were patches of mist but the night was generally clear. As they went round a bend in the road a figure appeared. Peter said...

> *"We all saw it simultaneously and, on the realisation that it wasn't an ordinary guy, Mick sort of screamed. He was walking towards us along the pavement, but he wasn't looking at us. I think that Mick got a longer look at him than we did, because I was sitting on the inside. I only caught a flash of it, but I realised that it wasn't an ordinary bloke because he was abnormally big and radiating a kind of pale light; and there were no sorts of lights around, except ordinary street lamps, which are pretty dim and which could not have produced that effect for an ordinary bloke. He was a big figure, going on for seven feet. The hands were down at his sides, pointing slightly outwards, as if he were almost marching, except that his arms were parallel and it seemed to be not so much walking as gliding. It wasn't above the level of the pavement, and was staring fixedly in front, just looking ahead, not at us or the van. The whole thing seemed to be a greyish-whitish-yellowish colour.*
>
> *All four of us realised that it wasn't an ordinary bloke, you know--that it was something horrible. I shuddered and gripped hold of the steering wheel, I temporarily lost my mind. Mick screamed and was sort of semi-hysterical and the other two were a bit shocked. Mick started shouting for me to drive on faster and get away from the thing, while Pete shouted at me to turn round so we could get another look at it. But my instinct was to get the hell out of it. I didn't get such a good look at it; I just knew that it was not very nice. It resembled a man in his fifties. The face made the most impression; there was a big body; it looked just like a great, long overcoat going down to the feet almost. Or it could have been a kind of shroud. It had some hair but I have no detailed recollection of it."*

It is interesting in the effect it had on the witnesses, causing screaming and fear.

Drummer Mick Fleetwood gave his own account.

"It appeared to be at about fifty to one hundred yards, judging by the time it took to actually see it, to register and then pass it. I was sitting on the left-hand side of the van farthest away from Peter, who was driving. And suddenly...we were going along normally, we weren't talking about anything which suggested this in any way; and just suddenly I looked up and saw a very tall figure. It could have been about 6 foot 9 inches, nearly 7 feet high, and it had what I would assume to be a very long coat nearly down to the ground and it was light grey, slightly florescent; it seemed to have a light of its own against everything. I was the only one in the back seat of the van. Thank goodness Peter didn't see its face, you know. I really screamed; I was absolutely petrified.

The face was so vacant looking; that was so terrifying. It was just a face and quite an old face; it could have been about sixty. It was walking straight ahead on the right hand side, coming towards us on the pavement. It was walking very stiffly in quite a military sort of way. Its arms were moving very deliberately, in a sort of straight fashion. It was going straight along, and it's face—I really saw its face and it was really expressionless—looking straight ahead, forlorn, quite drawn. I didn't see the eyes, but I saw where the eyes were, there was a sort of darkness. The whole figure, including the head and the hair was this light colour, this light florescent grey. But I don't recollect that I could see through it. His overcoat or robe was the same length as a field coat, almost down to the ankles."

Peter Green, the guitarist also later of Fleetwood Mac confirmed that the group had seen a weird figure.

"It didn't dawn on me straight away but the only thing that made me think it was way out of this world was that it was staring straight ahead and it was so late at night. We passed him right by and he was just staring straight ahead all the time. It wasn't really his appearance but just the way he was walking, very slow and heavy, and like I said it didn't really dawn on me till afterwards.

It was a very blank old thing, very light. A light, long, macintosh he was wearing seemed to be catching all the street light. It seemed to have an odd complexion about the face, definitely an odd complexion.

I wouldn't say definitely it was a ghost, because there are a lot of weirdos out, it could have been one. The walk was very unusual like something out of a Frankenstein film—great strides, arms moving very steadily. The macintosh was buttoned up. I couldn't actually see the eyes.

Mike was upset but only gave a weak scream. Peter only just saw it as it got to the side of the window—it just sort of sent a shudder through him."

David Ambrose, another guitarist, agreed in all respects with the descriptions of his band mates. Describing the figure as having a blank, staring, expressionless face, being of a glowing grey colour, very tall and seeming to glide along.

THE STOCKSBRIDGE HORROR

Construction began on Stocksbridge Bypass, Yorkshire, in 1987. Linking Sheffield to Manchester, this small section of the A616 was built to ease congestion in the former city. Almost as soon as construction had begun, strange stories began to emerge. Workers reported seeing children in medieval clothing dancing in a circle and singing *Ring-a ring-a roses.* Security guards, Steven Brooke and Peter Goldthorpe, also saw the children at night dancing in a field. This was odd due to the late hour and remoteness of the site. When the men tried to approach them, the children, who were dancing close to a pylon, vanished. The puzzled men were then overcome with a sense of being watched. Looking towards the newly constructed Pearoyd Bridge the men saw a tall figure in what seemed to be a hooded robe. Thinking he was a trespasser they drove up to the bridge only to find that their car's headlights seemed to pass right through him. They felt this monk-like apparition had an evil and threatening presence. The hooded figure vanished and the guards drove off in fright, reporting what had happened to their supervisor, Peter Owens, who could tell the pair were terrified.

Next morning the men went, in person, to the local police station. PC Dick Ellis could see they were scared but said it was not a police matter, and suggested they contacted a priest. Goldthorpe and Brooke when straight to the nearest church. PC Ellis received a phone call from the rector of Stocksbridge church, asking to sort the matter out as the security guards were so frightened they wouldn't leave the church. Both security guards left their job. One quit within three days, the other within a few weeks. Neither would return to the bypass.

Subsequently PC Ellis and PC John Beet drove to the bypass to investigate and of course they went up there at night.

The coppers parked up and after a while thought they could see a shape moving on the bridge. PC Ellis investigated and found it was a sheet of polythene flapping about in a breeze. The two policemen were confident that they had solved the case of the phantom monk. But as they were about to leave, PC Ellis got a feeling of someone walking over his grave. A figure appeared at his right hand side, pressed up against the window. Ellis could clearly see the torso that looked to be clad in a robe tied at the waist with rope and had a kind of 'v-neck'. The entity suddenly vanished and almost instantaneously re-appeared at the passenger window next to PC Beet. Again, it vanished. Despite his fear

PC Ellis got out of the car and looked around. There were high banks on both sides of the car and nowhere for anybody to have run. He even looked under the car, to find nothing.

As they tried to drive away, the car refused to start. Once it had finally got going there were a series of loud bangs from the rear of the car, as if somebody was hitting it with a baseball bat. Both officers felt a great sense of dread.

Others too have experienced the Stockbridge horror. Graham Brooke (no relation to the security guard) and his son Nigel, were out running at dusk when they both saw a monk, who appeared to be walking with its legs sunken into the ground. Its face was blank with empty eye sockets. The ghost brought a musty smell with it.

In July of 1990, David and Judy Simpson, were driving to work when they saw what they thought was a jogger. But they realized that the grey figure was three feet above the ground. The thing, which seemed to have no face, bounded off an embankment by the road and landed in front of the car, vanishing upon the moment of impact,

THE BARKING MAN
In July of 2014, truck driver Paul Drew was parked up on a stretch of US Highway 95 about fifty miles north of Las Vegas. He was due to meet another driver, swap trailers and head home. Paul often did this and most of the time arrived a couple of hours before the other driver. He used this spare time to catch up on some sleep. On this particular run he had arrived at the swapping point, an empty parking lot in the Nevada Desert at about 2am. He settled into his sleeping compartment for a nap.

As Paul was dozing off, he was awoken by the sound of a dog barking wildly. This was strange as the nearest town was over 20 miles away. The dog sounded very big and it seemed to be right outside his cab. Looking out of the driver's window he saw a tall man, who he assumed was the dog's owner. He wound down the window and shone a torch at the man hoping to speak to him. To his bafflement he saw that it was the man barking. His face was contorted, and had a snarling expression and his eyes were beady and looked like the eyes of an angry dog. The noise he was making did not sound like a human imitating a dog, but was exactly like a dog.

Paul turned away to grab his phone and record the strange figure, but when he turned back again the barking had ceased and the tall man had vanished. The area was open and there was nowhere the man could have run to. He stayed awake for the next hour until the other driver arrived.

THE NEWPORT PAGNELL THING.
The Nameless Dread was experienced by author and ghost researcher Ruth Roper Wylde. Ruth has written several guides to haunted areas and investigated countless ghosts but this was the only one that truly scared her.

One night in early December 2019 Ruth was driving along a road out of Newport Pagnell towards North Crawley. It was 10.20pm with strong moonlight but a mist and fog starting to form. There was a frost and the moonlight lit everything up in shades of grey. Her headlights lit up a heavy set man facing away from her. He was wearing an old fashioned donkey jacket, like a work man. He seemed to be composed of shades of grey. He seemed to appear from nowhere. Ruth tried to convince herself to turn the car around for another look, but something told her not to. The figure had a horrible feeling to it. She says there was something very, very wrong about it. She still cannot drive along that road at night, and takes a long detour to avoid the road.

One of her friends experienced weird feelings as a child on the exact same road.

"One of my friends in the village here laughingly contacted me just a few days ago. She'd been listening to an interview with me online, where I had mentioned this ghost, and was asking "OK - what road is it exactly that I need to avoid from now on!" I explained which bit I was referring to, and she responded with, "Oh! You mean the creepy road!" When I asked why she called it that, she explained that as a child, when being driven home to Cranfield after dark, if they came down that bit of road she would shut her eyes and wait until they got "past" the creepy bit, because she was always afraid that "something" would come into the car, or that if the car broke down "something" would get them...."

Another account that Ruth was told came from a couple who were driving along the road at 3am. They were going on holiday and had made an early start. At exactly the same place as Ruth's encounter, they saw a looming figure, this time face on, this time in the middle of the road. He appeared to be wearing a long donkey jacket and the couple had to swerve to avoid him. They argued about what they had seen and if it was a real person or a ghost. They were going to drive back and check if the person needed help, but something told the both of them not to turn back again.

THE SLITHERING COLD, A PERSONAL EXPERIENCE.
After hearing Ruth's story on Paul Bestall's excellent *Mysteries & Monsters* podcast and corresponding with Ruth herself, fellow Fortean Jackie Tonks and I decided to investigate.

On Friday 4th of December 2020, Jackie Tonks and I were driving along the road that links Newport Pagnell to the village of North Crawley. It was around ten thirty at night. The North Crawley road becomes Brook End then becomes High Street and finally Crawley Street. We were about a mile beyond North Crawley, on the Crawley Street area of the road. We had turned around and started to drive back towards North Crawley.

Suddenly a strange sensation passed through the car. The sensation seemed to 'enter' the car from the right, effecting Jackie, who was driving. It was a cold, prickling feeling like standing next to a big open meat freezer, crossed with pins and needles. Jackie felt it in her legs and side, then it hit the right side of my face, then crawling around my scalp, neck and finally to my left cheek. Imagine a damp blanket with ice cold pins and needles draped across your head and neck, but with a dragging sensation as if something was moving over us. It lasted about twenty seconds. It was as if something moved through the car and us.

Jackie switched the interior light on and we both looked on the back seat. There was nothing there.

We turned around and drove back through the area but the sensation did not repeat itself.

Jackie later told me that she was badly frightened and that she felt that something was in the car with us for a while. She confessed she nearly screamed and burst into tears. She said it was one of the most frightening things she had experienced and she had survived the Nepalese earthquake of 2015.

I felt no sensation of fear, just the weird physical feeling of the ice cold prickle.

But human-like phantoms are far from the only things seen creeping around the world's roads. Some apparitions seem less than human, much less.

THE BELL GHOST
Sometime after midnight on November 13th, 1967, Philip Freeman (no relation) and his friend Angela Carter, were driving from Cranliegh to Shere in the county of Surrey. There was a dark cloud covering as they drove through the wooded, hilly countryside. Freeman stopped the car to wipe the windscreen that was misting over, as the car's engine had not been on long enough for the heater to kick in. As he got out of the car he noted a strange smell like a stink bomb. Upon getting back into the car a face or rather a lack of one pressed itself up against the passenger window. The oblong no-face was bright white, eight inches wide by ten inches long and had no features what-so-ever. The thing was glowing faintly. The thing was about four and a half feet tall, bell shaped and had a dark body. He saw one white arm resting on the car's bonnet.

Freeman jumped back into the car as the vile smell and a sense of terror overcame both him and Angela. As he started the car he saw the apparition move to the back of the car and press its lack of face onto the rear window. As he drove away he glanced in the mirror and saw that the thing was bell shaped in its outline.

THE RUSKINGTON HORROR
An even weirder thing was reported to Rob Gandy. Writing in the magazine *Fortean Times* he told of the thing that had become known as The Ruskington Horror.

> *"Kevin Whelan was driving down the A15 from Lincoln towards his*

home in Sleaford at about 60mph. On the horizon, just before the Ruskington turn-off, he saw something floating that looked like a large white bin-bag; he wasn't concerned, thinking it was probably just car headlights or something similar. However, when he got to where it was, a face suddenly appeared around the car's front pillar at the top-right of the windscreen. It was Greek looking, with dark hair and olive-green skin and was holding its left hand up. It had a pitted face and Kevin could see its teeth. Below the neck was a kind of white fluorescence, as with a photographic image when the camera flash is too bright. Kevin was shocked and scared but carried on driving for (he estimated) another 40-50 seconds. until he came to a dip in the road where whatever it was faded into obscurity down the side of the car. He then "just bombed it back home", where he ran into the house and woke his wife. He looked in the mirror; his face was white and his skin was covered in goosebumps. His wife later confirmed that he was panicking and crying when he told her about what had happened. In all their time together, she had never seen him so distressed."

The sighting occurred on Sunday morning, at about 2am, in December of 1997. Sarah Martin saw something at the same place as Kevin Whelan in 1997. She and her partner had been to the cinema in Lincoln and were heading down the A15 towards Cranwell.

"On the corner near the Brauncewell and Leadenham turn-off a black silhouette of a man ran out from the ditch and went straight in front of the car. Logically, they should have hit it – but there was no impact and seemingly nothing there. The figure had no face and Sarah cried out – but her partner who was driving didn't see anything. She said that "it was horrid... really frightening."

Another witness was Jenny Sellers.

"Jenny Sellars told how in the late 1980s she was driving from Sleaford to Lincoln, not particularly fast. She was on a rather dark stretch of road when she saw something like a sheet or bit of plastic, which came down in front of her windscreen. She slammed on the brakes, thinking she had hit something, but she knew she hadn't because there was no impact or anything on the screen. She stopped the car, and whatever it was then went around the side of the window. She opened the car door, but didn't get out. There was nothing on the ground."

THE RUNNING GHOST

In the 1980s, Stephan Penkridge, then 15, was walking the mile home from his friend, Paul Patton's rural house. It was 8pm and looked like rain. Stephan's father had needed the family car for work, so he was walking instead of getting a lift. His home was in the small town of Newport in Shropshire. He was walking along Chester road and had just turned the bend leading away from St Michael's Church. To his right was a high wall that was built on an embankment and ran for almost all of the highway. About fifty feet away something white moved at the top of the wall. He at first thought it was a white cat or a plastic shopping bag that had become snagged on the old brickwork.

Then the shape stood up. He was now within fifteen feet of it. He felt that it had been waiting for him, crouched, and stood up at his approach. It was a translucent apparition, dull white, human shaped but utterly featureless. Despite not having a face or eyes he got the feeling that it was watching him. Deciding it was too far back to run to his friend's house, he hunched down and walked swiftly passed the watching entity.
He glanced back several times and saw the white thing still atop the wall, watching him. When he was about sixty feet away he stopped and turn back again. To his horror he saw the thing leap from the wall and run towards him. He turned and fled in a blind panic. Luckily Stephan was a cross country runner and he was less than half a mile from his home. The thing seemed to be slowly gaining on him. When he reached home he banged madly on the door. When his mother opened it, both she and Stephan saw the white entity run past their house and along the street towards Chetwynd End.

There is a local tradition of the ghost of a woman called Madam Piggott, who is said to haunt the road. She lived at Chetwynd Park Estate in the 17th century and died in childbirth. She is said to leap out and scare motorists and even materialize inside their cars.

THE AUSTRALIAN HORROR
Emily Dentith was driving with her boyfriend Lewis, from Melbourne to Broome for a two week holiday in 2014. The journey covered 2,500 miles through the outback and would take four days, so the pair took turns at the wheel for six hour stints. One would sleep as the other drove. On the second day of the journey, whilst travelling along the Stuart Highway, just before Alice Springs, Emily awoke to the sound of Lewis crying. He was clutching the driving wheel and was bent over it. The car had pulled over but the engine was still running. The clock on the dashboard said it was 11.14pm.

On asking Lewis what was wrong he could only say *"It was a deer, it must have been a deer."*

She asked him to elaborate but he just kept repeating the phrase. He seemed to be in shock. Emily drove the last 40 miles or so to Alice Springs. Deer are not native to Australia but a few feral, introduced populations exist. They do not venture far into the desert though.

As the pair arrived at their hotel, Lewis seemed to calm down, so Emily asked him again just what had happened. Reluctantly, he said that he had seen something standing in the road. At first he thought it was a person illuminated in the headlights but as he drew closer, he saw it was extremely thin and walked with a weird gait, holding its arms out in front of it. Then he saw the ghastly face. It had an elongated snout and its large eyes were totally black. The thing looked at him as he drove past, and its gaze filled him with utter dread.

Lewis was never the same after the encounter. He developed insomnia, depression and finally killed himself the following February.

THE HAUNTED CAR OF BRAZIL

Legendary parapsychologist Guy Lyon Playfair recorded a case of haunting around a moving car in Brazil. On the night of 18th/19th of September 1960, six people were driving from Brasilia to Belo Horizonte. The group consisted of a newly-wed couple who had been married that very day. The bridegroom's parents, the driver and Dr Olavo Trindade, a distinguished doctor, who developed new ways of treating meningitis. They were driving along a road in the moonless night when the car, a Willys station wagon, began to overheat. They stopped, and opening the bonnet found nothing amiss with the car.

Suddenly they were bombarded with stones from the darkness. One, the size of a saucer, hit the bride and groom before falling into Dr Trindade's lap. The driver shot his gun into the darkness but the barrage of rocks continued. Knowing that there was a police station some two miles away they drove there for help. Dr Trindade showed the police some of the rocks. One of the cops accompanied three of the party back to the area, whilst the other three stayed behind at the station. Back at the spot Dr Trindade stayed in the car whilst the others explored outside. Suddenly the rocks began to fly again. The driver tried to shoot into the darkness again but his gun jammed. Stones hit him on the arm and scalp, drawing blood. The car's headlights revealed nobody in the scrubby wilderness.

The group returned to the police station and had coffee. Whilst it was brewing, a rain of gravel fell on the stones Dr Trindade brought to the station. Back on the road, gusts of sand blew into the interior of the car despite the doors being closed. Rocks continued to bang into the car as they drove.

Then something tried to open one of the doors from the outside as the car was moving! Dr Trindade had to use all his strength to keep it closed. The driver locked the door but the door unlocked itself again. The driver saw a vague figure outside of the window and tried to shoot it but the gun jammed again. The bridegroom's father took over holding the door shut as Dr Trindade's fingers were becoming sore.

They all began to pray (they were Catholics) and the pressure on the door ceased. The bridegroom's father noticed that his watch was broken, the glass smashed and the wristband stretched. Finally they reached their destination in Cidade Livre at 2am. The following morning the bridegroom's father's watch was miraculously repaired and the car did not have a scratch on it. Dr Trindade wrote up a report on the matter once they were home again.

LOOKING FOR THE TOWN CENTAUR

The strangest road ghost of all was reported from County Louth, Ireland in the spring of 1966. John Farrell and Margaret Johnson were driving past Lord Dillon's estate on a country road not far from Drogheda, when John broke hard as a large animal appeared in the road. Both witnesses were thrown back in their seats by the violent stop. Margaret described what stood in the road.

> *"...a huge horse with a man's face and horrible bulging eyes. I could see by John's face that he saw it too. I think I screamed, but both of us were so frightened that we were paralysed. The thing had a horse's body. But it was the face, leering and hairy and huge, that shocked.*
>
> *The animal stretched right across the road and completely blocked the car. It stayed there for nearly two minutes. We were petrified. Then it vanished. John quickly swung the car around and drove to my home, about a mile away.*
>
> *We were so frightened that we drove through the gates and knocked it off its hinges. I woke up my father and both of us told him the same story. We don't drink and two of us couldn't have imagined the same thing."*

Margaret spoke to the press about the sighting.

> *"They must have seen something awful. Both were still terrified when they came in here and each of their descriptions of the ghost tallied. Margaret was very sick for days after the incident."*

Amazingly, an identical beast was reported in England almost 30 years later. One night in 1994, the husband of Nicky Knott was driving home through King's Lynn, Norfolk, down a lonely rural back road, when he saw a large creature in a field to his right. As it moved closer, it seemed to be a horse, but the witness was horrified to see that it had the face of a man! Terrified, Knott slammed his foot on the accelerator and sped away, and even though he was sure that the 'thing' was pursuing him he never once looked back till he reached home. Once more we see the fear of pursuit in the mind of a witness. Mrs Knott wrote to cryptozoologist Dr Karl Shuker with the strange story.

THE HAIRY HANDS OF DARTMOOR

We cannot leave our round up of road ghosts without looking at the most infamous case of all. The hairy hands of Dartmoor. Dartmoor is 363 square miles of upland moor in Devon, England. It can be a spooky place, even in summer. It is studded with ancient standing stones and stunted woods. When the fog rolls in and the mist rises, the place can be treacherous. There are many legends associated with it, from feral big cats to pixies from black dogs to dragons but the best known Dartmoor tale is the hairy hands.

Sometime around the early 1900's, a series of unexplained accidents were reported along the stretch of the B3212 road which runs from above Postbridge to Two Bridges. Cyclists have said that suddenly the handlebars of their bikes were wrenched out of their hands, forcing the bike into the ditch. Pony and traps were also forced off the road and onto the verge. Drivers of cars and motor coaches were experiencing the same thing.

In 1921, an army officer was injured when his motorcycle was driven off the road, he survived and said that huge muscular, hairy hands materialized, clamped over his and forced the bike into the verge.

Later that year, a charabanc was carrying some tourists along the B3212. The vehicle suddenly swerved, shot across to the other side of the road and mounted a bank. At this point of time seat belts were not compulsory, and so several of the passengers were thrown out of the charabanc, one of which was seriously injured. Afterwards, the driver reported that a pair of ghastly hairy hands had grabbed the wheel and forced him off the road.

In 1924, the mother of folklorist Theo Brown was sleeping with her husband in their caravan in the ruins of Powder Mills. She awoke in the night and was horrified to see a huge, hairy, disembodied hand crawling up the outside of the caravan window. She felt that the hand was exuding an intent to do her and her husband harm. Frightened she made the sign of the cross and the hand retreated into the night.

In his book *Supernatural Dartmoor,* Michael Williams recounts a story told by journalist and author Rufus Endle. He claimed that, while driving near Postbridge on an unstated date, *"a pair of hands gripped the driving wheel and I had to fight for control."* He managed to avoid a crash and the hands disappeared. He requested that Williams not publish the story until after his death, for fear of ridicule.

In 1962 a visitor to Dartmoor, Florence Warwick, was driving along the road when, wanting to check her location, she stopped the car to look at a map. In her words:

> *"I looked up and saw a pair of huge, hairy hands pressed against the windscreen. I tried to scream, but couldn't. I was frozen with fear."*

Investigations on the B3212, showed that there was an adverse camber. When this was altered accidents on the road decreased dramatically.

The last known sighting of the hairy hands, that I am aware of, happened in 2008 when a motorist said a pair of huge hairy hands tried to grab his steering wheel. However the details are scant.

Next time you are driving at night, down some dark, remote country road, just think about what may be around the next corner, waiting for you in the shadows.

SHADOW MEN

A whole class of distinct ghost, shadow people, resemble three dimensional shadows or sometimes two dimensional holes in reality. Some have glowing eyes, all of them seem to be a negative presence. Sightings of these strange entities seem to be on the rise.

The Official Shadow People Archive is an online collection of such sightings. In 2004 a witness called Susanne recounted what had happened to her.

"Hi!

I too only realised today, as I came upon your site, how common these "shadows" really are. Here's my story:

A few years back at University, we attended a weekend field trip and were staying at a manor house near Boxhill, in England. A friend and I were sharing a room and at about 11.30pm lots of very strange things started to happen.

It began with what we thought at the time was smoke curling and twisting from a particular point in the room, at first only my friend could see it but after a few seconds I too could see it - all the more strange as there was no light. Yet it was thick, grey and coiling in a patch between our respective beds.

After watching the curls of what we thought was smoke, we started to notice pitch black 'shadows' gathering in a top corner of the room. The shadows were so black and opaque we could not see through it at all, but it got bigger and thicker, we could see what appeared to be more black blobby shadows joining the rest of the mass.

After a period of time the black shadows started to dart across the room, leaving my friend and I in sheer terror. I was too scared even to sit up and turn on the light at the foot of my bed. I remember at one point pulling my quilt up over my face and when I brought it down again a blob of this black shadow was literally nose-to-nose with me, I let out a scream at which point it darted across the room. My friend saw it dart across the room too. I can't remember at which stage it happened, but my bed began to rock (from top to bottom, not side to side) - not strongly but definite movement which scared the hell out of me! I can't remember how long this lasted, but it definitely lasted a while.

The next thing we saw was orbs of flashing lights - One which was shaped like an egg (but much, much larger) of a light blue colour. At the time my friend and I wanted to try and verify what we were seeing and we agreed to memorise the orb and the next day draw an annotated picture of it, in colour of the orb and the shapes of the black blobs. The pictures were identical as was the colour of the orb. There was some more orbs - one white one which didn't do too much. But the blue one was there pretty much all night, flashing regularly.

At the same time as the orb appeared we also noticed some lights on the wall where there shouldn't have been any. You know when light leaks through a closed door - well it was a bit like that but it didn't fit with the existing door. It was totally out of sync and there was no light on in the corridor outside. We also saw what appeared to be a fire in the unlit fireplace. This was quite faint but in colour.

All the while the orb continued to flash and the black shadows were darting across the room. This continued until dawn - a good 6 hours or so. Needless to say we both had a sleepless night that night, but it has stayed with me since.

So glad to hear other people have experienced similar - there have been lots of sceptics!

With best wishes Susanne"

Again we should note, that the ubiquitous strange lights are accompanying the phenomena once more. It also seems that, as with other Fortean phenomena, animals can sense shadow people, as the next case illustrates. It occurred in Eastern Canada in 2009. The user Paris wrote into the same archive.

"Two months ago, I got a great Dane pup who barks at everything. He sleeps in bed with me every night and frequently wakes me up growling at something. I always pet him, reassure him its fine and go back to sleep.

But two nights ago, he started barking and growling really loud. This woke me up, I was going to pat him on the head but as I came out of my sleepy daze, I felt this awful pressure. Kind of like the feeling you get when you're in an argument; you just feel the other person's anger. I felt like I was going to be sick. I saw something move out of the corner of my eye.

When I looked up, there was a tall shadow standing in the corner of my room, I'd say it was about six, six and a half feet tall and looked very much like the shadow of a man. My room is nearly pitch black except for the light on my alarm clock which casts a bit of a blue glow around the room. The shadow was as dark as could be, I couldn't make out any features, just its body shape.

When I saw it, I gasped and flipped on the light that's beside my bed. It just disappeared as I turned on the light. I'm 100% sure it wasn't just a shadow in the room, there's no tall people shaped things in my room that could possibly cast something like that.

When it disappeared, my dog stopped barking and laid back down.

I tried to go back to sleep with the light off but every time I did, I got the eerie feeling I was being watched. I ended up having to go back to sleep with the lights on, and even then, I still slept awfully.

Later that day I went right out and bought some sage which I burned all around the house and in all the corners (my mother used to do this every time our animals began to become uneasy for no reason at all. She was a firm believer they could see things that we cannot). Since then, my dog has not barked at anything in the middle of the night and I have yet to see another shadow in the corners."

YouTuber, Danielle Hallett, posted a fascinating account of her experiences in a haunted house in Swansea, Wales in the early 1990s. The phenomena she encountered included a shadow man that seemed to change shape, much as we have seen other phenomena morph from one form to another.

Danielle and her friend Deborah had decided to get a house together. They found a large house on a hill with a big garden, overlooking Swansea bay. As the house needed renovation the landlord let the girls stay there for free as they did it up.

From the beginning things were odd. The house was cold and had an oppressive atmosphere. Danielle was up a ladder fitting a lightbulb, when she was tickled from behind. Thinking it was Deborah she turned around, to find nobody there. A rocking chair, moved to make way for trestle tables, would always be back in its original position, no matter how many times it was shifted. Footsteps were heard upstairs when nobody was there and the old classic haunted house noise of a ball bouncing down stairs. As the days went on, shadowy figures were glimpsed fleetingly in mirrors and in the corners of rooms, mainly upstairs. Friends would claim to have seen these 'shadows' in upstairs windows when the girls were out and mistook them for one of the pair.

The next door neighbours mentioned that students had been doing 'devil worship' in the house in the past.

Deborah stopped staying in the house, leaving Danielle there alone. One night she was confronted by a huge, man shaped shadow thing on the landing. As it became aware of her, the man sized shadow, fell down on all fours like a dog and moved towards her.

Danielle fled down stairs in a panic. The shadow thing loped after her. In tears of fear, she ran into the living room, then realized that the perusing shadow had stopped in the hall barring her way to the door. The terrified girl opened the living room window and scrambled out.

Later, when she returned with Deborah, the kitchen window frame burst into flames. The girls doused the fire but were amazed to find no marks on the wood.

The neighbours arranged for some priests to exorcise the house.

Back in the house, the girls heard a deep growl like that of a big cat. Then an almighty crash came from the kitchen. Investigating, they found the garden tools they were keeping in a cupboard there. The tools had been thrown across the room with such force, that they were embedded in the skirting board on the other side of the room. Then a knock came at the front door. It was the two priests come to exorcise the house.

The clerics moved through the house room by room, carrying out their ritual, in one room the priest came across a cabinet and advised the girls to burn it after the exorcism was done. At the end of the ritual, the front door of the house opened, then slammed so violently, that it broke several panes of glass. The cabinet the priests pointed out toppled over onto its front with the doors wide open. The cabinet was thrown on a bonfire and according to Danielle the cabinet burned with blue flames for an hour until it fell to ashes. There was no further hauntings at the house.

The message board of the *Fortean Times* magazine and its successor *Forteana Forums* are also a mine of such stories.

In 2004, a user called 'Malatrice' wrote into the *Fortean Times* message board with the following story.

> *"I, too, have seen shadow people all my life. Most of them were associated with my great-grandmother's house. She was an Irish clairvoyant who saw bean-sidhe and often foretold the deaths of family members. Needless to say that sort of thing doesn't win you any popularity contests, but I still consider her one of the greatest women I've ever had the honour of knowing. Her house was always full of weird and wonderful stuff, and this included a lot of shadow people, who were a bit creepy, but never were particularly frightening. I honestly can't think back on the rooms of that house without a shadow person lurking in my memory of them.*
>
> *My other grandmother (on my father's side) had one shadow person that inhabited the back bedroom of her house. It was seen by several of my relatives. That bedroom was used as a storage room, predominantly, probably due to the fact that no-one in the family would ever sleep in*

there. The room was chilly, unfriendly, and difficult to be in even in broad daylight. I myself was often drawn to it with the same kind of morbid fascination that accompanies disasters - that sickly inability to look away. I used to sneak in there a lot and hide in the wardrobe or behind the bed. There was always a definite feeling of someone else in the room, and sometimes a shadowy figure could be glimpsed.

In my adult years, I've had three encounters with shadow people that were most unpleasant. All three occurred during a period of upheaval in my life that resulted in an absolute torrent of what may be termed by some as 'poltergeist' activity. The first of these encounters involved waking one night to see a shadowy figure hovering above me in mid-air, its face mere inches from mine. The second occurred on a sleepless night, when I opened my eyes to see a huge, hulking figure standing at the foot of my bed. My impression was that it was covered in eyes, though that seemed to be more of an impression than an actual visual. And the third was the same hulking figure, standing next to the head of my bed. It reached out and touched the right side of my face, and when I awoke the next morning, my right eye had swollen shut, my face was red and puffy on that side, and it felt as though that side of my face was suffering from a severe sunburn, or maybe a bad abrasion. The effects lasted about three days."

The bean-sidhe mentioned is the correct spelling of the anglicized banshee, the wailing, female spirit that foretells death in Irish folklore. The swelling and soreness of the face is reminiscent of the 'radiation burns' UFO witnesses have reported.

Another user 'Martingidlow' had the following to recount.

"I used to live on Chilkwell Street in Glastonbury several years ago in a very old house, it was a normal day and I was just walking upstairs to the bedroom. The bedroom door was open as I reached the top of the stairs when from the corner of the room appeared, as if from nowhere, a person shaped shadow. It looked exactly like a human being but it had no features as it was as black as the ace of spades. It seemed to be aware it had been seen and disappeared in front of my eyes. I saw it again two weeks later but this time where its eyes should be there was a red tint to them."

Shadow men are seen out of doors as well. Andy Hinkinson-Hodnett wrote to *Fortean Times* in 2005 with one such encounter.

"Some years ago, after a night of general chat, a friend and I were walking from another friend's house to my own in the leafy suburb of Brockley in South-East London. I had drunk tea or soft drinks all night

and consumed no drugs whatsoever. It was about 4am and dawn was breaking when we saw a figure walking up the street towards us. The word we coined later to describe its movement was 'lolloping'-a kind of up and down bouncy walk. It took a few seconds for us both to realise this was no human being. "See that man?" I asked. "Yes" "It's not a man, though is it? "I found myself saying. "No" said my friend, sounding scared, "It isn't."

The creature was entirely black and like a cardboard cut-out, flat and one dimensional. It had no features at all and arms that hung down to its knees. It seemed to be ignoring us, then seemed to realise we could see it and began to 'lollop' faster towards us. We ran to my front door and hid in the hallway as quiet and unmoving as possible when we saw the thing-we felt it was male-approach the front door and appear to look through the glass from the way it moved up and down and around. It then turned away.

We didn't sleep for some time after that, discussing what we saw. It was shaped like many descriptions of 'greys'. , but both of us came away with the impression that what we saw was not of this world but from a parallel dimension. We instinctively felt it was not a creature to try and communicate with, and not something that it was good to be near. We felt that if the creature had somehow got a hold of us, we would not be around today to tell the tale."

BBC Radio Sounds ran a great radio series called *Uncanny*, fronted by Danny Robins. In episode one, witness 'Ken', now a highly respected human geneticist who specializes in studying genetic diseases, spoke about events that befell him as a student in Belfast, Ireland, and have haunted him ever since.

In 1981, Ken was studying microbiology at Queens University, Belfast. His student digs were on the 6th floor of a 1960s tower block on the campus, called Alanbrooke Hall. The room was number 611 and he shared it with a friend called James. One night, in the dead of winter, Ken saw a silhouette of what appeared to be a man, standing by a desk in his room. The figure was the blackest black Ken had ever seen and seemed to bring with it a white noise in his ears. Ken said...

"...there was a very, very strong sense of pure, distilled evil coming from this figure. It felt like a force of nature, a feeling like everything good, everything called hope had disappeared so everything was despair and I felt it coming through me in waves."

Looking down, Ken could see ripples in the blankets of his bed, that seem to correspond with the waves of evil pulsing out of the being. He continued.

*"I also thought this evil was specifically against me. This thing wanted
to do absolute harm against me. It started to drift towards me."*

As the thing reached his shoulder, Ken panicked and threw himself at it, whereupon the figure and its cloaking evil seemed to vanish.

Speaking to James in the morning, he was amazed to hear what had happened to him at the same time. James had been trying to sleep when, despite having his eyes closed, saw a point of light. The point grew and grew, until it was like a window he was looking through. He could see the back of a man's head. The man, who appeared to be of student age, turned round and gave an evil glare as if he wanted to do James harm. James was deeply disturbed by his vision.

That night, Ken was alone in the apartment. He heard the lift stop at floor six and somebody approach the door. Then the same feeling of distilled evil hit him again. Suddenly the door was pounded so hard that it shook. Ken thought it would be knocked off its hinges. Bravely swinging the door open he found...nothing. The noise and the feeling of evil were gone but the lift was at floor six. All the lights were off in the other rooms on his level.

Both Ken and James heard rustling, like papers being crumpled, in cupboards high on the walls. Investigations found no mice. James saw a clip folder's papers turning and flicking on their own.

Later that week, Ken asked two other students, who had lived in the room the year before, if they had seen anything odd. They told him that they had seen a heavy book fly off a bookcase and hit the wall opposite. James, himself a scientist as well, contacted Danny and confirmed all aspects of the story.

Thankfully, Ken moved out of room 611 at the end of the year, but two new students moved in. Ken who was still on floor six, and the other students, decided to say nothing to the newbies. About six months into their stay they approached Ken in the communal kitchen and asked him if there was anything wrong with room 611. The newcomers told Ken that they had been plagued by very bad nightmares. Eventually the pair began having the same nightmare. Then they would have nightmares that seemed to leap from one to the other. Comparing notes they found their fractured dreams made a whole story. Then cutlery would start to fly about, and knifes would turn up beside their beds, where they would put their feet in the morning.

Ken and the others asked the cleaning lady about room 611. She said that for years and years she had felt deeply agitated when cleaning the room, as if somebody was watching her. She knew of three deaths of former students who had lived in the room. One had committed suicide by jumping, one had been shot in the early 1970s as they were going to, or coming from mass, and the third was found dead in the room for reasons unknown.

To this day Ken cannot rationalize what he saw.

Professor Gary Foster, a scientist specializing in viruses, was a student warden in Allanbrooke Hall some years after Ken's experiences. He lived on the top floor in a flat. During student holidays he was alone in the building. Gary would hear the lifts coming up despite being alone in the building. When he checked there was never anybody there. This was always at night. When he spoke to the lift engineers they told him that it was impossible for the lifts to move by themselves. The lights would switch themselves on when Gary was away, despite the tower block being locked. Gary put tape across the doors one night and locked Alanbrooke. When he returned later, the lights had been switched on but the tape was unbroken. The weirdness continued. Each floor had its own internal telephone, that could not take calls from outside. They were used for communication within Alanbrooke Hall. The phones would regularly ring when nobody but Gary was in the building, and always stopped before he could answer them.

One evening in his flat, all the lights began to shine with incredible brightness,

> *"I'm not joking, brighter than the sun. The main lights on the ceiling blazed to the point where I couldn't look at them. We all used to have these hi-fi stacks with a radio and turntable that was on like a little trolley on wheels. The cooling fans started to whir so fast that the actual hi-fi system started to move out across the room. Then everything popped. The microwave went 'ping', the desk lamps and light bulbs blew, broken glass flying everywhere. The hi-fi system completely boomed. We got the electricians out and there was no evidence of a power surge anywhere else in the building. Whatever had happened, had only hit my flat."*

Electricians could find no evidence of a power surge. Another time, as he was walking across his flat, he tripped over something and fell through a large open window, and found himself hanging on for his life. Later, looking at the floor there was nothing he could have tripped over.

As with other witnesses, Gary heard one night, a violent banging on his door like several people trying to kick it in. As soon as he touched the door the noise stopped. Nobody was outside. Security guards whom he had called, saw nothing.

Another former student, Billy, stayed in room 611 in 1982, the year after Ken and suffered from very similar visitations. A former RAF pilot who now flies commercial aircraft back in 1982, he was an engineering student sharing room 611 with a law student named Bill. Billy recalls cold spots where you could see your own breath. Sometimes Billy would wake in the night to see the shape of a man who stood out as being darker than the night around it, and that brought a feeling of despair. Billy saw it around six times.

Books, mugs and cutlery would be moved about. Billy saw, reflected in a window, a book fly across the room. Both Billy and Bill heard a violent hammering at the door and saw it bending inwards with the force. The two youths flung the door open and found nobody there.

On the day they left, Billy told Bill about the shadow man. Bill confessed that he too had seen the entity and prayed for it to go away.
Billy said that the cleaner, Ann, told him that the third death in the room was a suicide by hanging. One day they saw Ann run from the room screaming and would not tell them what she saw. She refused to clean the room without them being there in future.

Alanbrooke Hall was demolished in 2001, few could have mourned its passing.

The shadow people can also come in a dwarf form. 'Nova' wrote to *Fortean Times* with an encounter with these spooky midgets. Nova saw them at the age of 18 when visiting an abandoned drive in cinema, in Texas, with some friends. The friends would visit the old drive in and climb on the roof to watch the city lights and drink beer. To reach the dilapidated theatre they had to cross a field of knee high grass.

Nova saw a black figure off to the right.

> *"It was about waist high, very stout, no features-simply smooth, black and shiny. It appeared to be all shoulders. There was no discernible neck, and its head rested upon its body, showing no seams or creases. The only thing I could really compare it to, was if some sort of a small gorilla wore a tight fitting latex suit, with no breath holes or, or any other sort of gap in the smooth shiny blackness."*

It sounds almost like a young Bigfoot in a full body fetish suit!

The witness noticed another of these things as they climbed up the side of the theatre, peering around the side of the building. After a period of star gazing the three friends retraced their steps to the car, that was parked on the road beyond the field. Then another of the weird things appeared. Illuminated by street-lights this time, the pygmy shadow person was only fifteen feet away. This time Nova's friends saw it too. One was freaked out by the sight of the thing, but the other was interested and tried to get closer. The other two grabbed him and walked him swiftly back to the car.

All the while the thing stood still, and Nova related that, although it had no visible eyes it felt like it was staring at them.

Back in 1961, a Mrs Quinn and her daughters Brenda and Judy, had a horrifying encounter with a dwarf shadow man. The case was investigated by author and ufologist, Leonard Stringfield — who served as director of CRIFO (Civilian Research, Interplanetary Flying Objects) and public relations adviser for the UFO group NICAP (National Investigations Committee on Aerial Phenomena), who classified it as an extra-terrestrial, but as you will see, the thing had more in common with ghosts than aliens. The odd tale began at approximately 1:30am, sometime in August of 1961. Mrs Quinn awoke to find a ghastly creature hovering beside her bed. It was a fat, small, jet black creature with a round head and no neck. Its eyes were glowing wrinkles. Mrs Quinn looked on in horror as the shadow baby floated over to where her daughters slept.

14-year-old Judy awoke to see the thing hovering above her head, and described it as *"...a huge tar-baby with enormous eyes."* Judy said she could feel the creature tugging on her psychically, in an attempt to make her move against her will, and she slipped out of bed. She felt as if it wanted to take her somewhere. Suddenly the thing released its grip and vanished.

Philip Annetta was visiting his grandmother's house in Melbourne, Australia, in 1991. He was sixteen at the time. After a night at a restaurant with his family, Philip was sleeping on the sofa. He noticed a movement near a large window, diagonally opposite the sofa. As he watched, he saw a silhouette outside the window, just three feet from it. It was thin and had an 'elliptical' head that reminded him of the typical grey alien. It glided up to the window with a movement that was 'too smooth and too silent'.

Philip thought to himself *"If it doesn't make any noise I can imagine it's not there."*

Almost instantly, the shadow thing raised an arm and tapped three times, very deliberately on the glass. Terrified, he hid under the blankets. When he dared to look again, ten minutes later, it had gone. Next day he gauged the things height using the windowsill, and estimated it was 4.5 feet tall. Philip had experienced sleep paralysis before, and was insistent that this was different altogether.

Lynsey Morton and her boyfriend Ian Jenkins, moved into a house together in Cardiff, Wales, in 2014. A month into their tenancy, strange things began to happen. A pattern the reader will recognize began to unfold. Small objects were moved about, cans of food vanished from cupboards and were found stacked in weird formations and a sulphurous smell began to waft around the house. The couple's dog, Mitch, would seem to bark at something that they could not see.

One evening, whilst sitting on the sofa watching TV, they saw a short, black, featureless figure peering at them around the door out into the hall. All three saw the thing several times and Ian checked the hall but found nothing there. They described it's movements as jerky as if it were 'moving at a lower frame rate'.

This small figure was seen by them a number of times over the coming weeks. In relation to the door frame, the thing was estimated to be no more than four feet tall.

The events continued, with bumps and bangs from upstairs and the opening and slamming of doors. One time they returned home to find plates taken out of the cupboard and smashed on the other side of the kitchen.

On the 11th of August, 2014, Ian was working away from home and Lynsey was on her own in the house. At 4am she was awoken in bed by something shaking her foot. Turning over, expecting to see her boyfriend, she was confronted by a hulking figure seven feet tall. It was utterly black with no features. Hiding beneath the covers in fear, she heard the thing walk around the bed and out of the door. When the heat beneath the duvet became unbearable, she looked out to see that the shadow man had gone. She checked the house and found it empty, with no sign of entry. Lynsey phoned Ian in tears. She then went to stay with a friend.

On Ian's return, despite the fact that neither was religious, they went to visit a spiritualist church. A medium visited the house and said...

> *"There is more than one presence in this house and they are so far back behind 'the shroud' that I cannot see who they are. They speak in a language that I don't understand."*

The medium instructed Lynsey to move from room to room, holding a candle and reciting a prayer.

After the ritual the haunting ceased.

Whatever shadow men are, they bring the nameless dread with them in spades. They seem to exist to instil fear in the witness. Trying to explain them as simply spirits of the dead seems inadequate.

AND FINALY, THE GHOST OF ELVIS...IN JARROW.

Jarrow is a shipbuilding and coal mining town, in the county of Tyne and Wear, in the northeast of England. It is best known for the 'Jarrow March' of October 1936, in which a group of 200 men and a dog called Paddy, walked to London, some 291 miles to protest about unemployment after the closure of Palmer's Shipyard. It's also know for 'Jarrow Elvis'.

Joe Allen was just another unemployed shipbuilder, until one day he stood up in a pub and started singing Elvis Presley songs, in this a legend was born. Widely recognized as the very worst Elvis impersonator in the world, Joe, nonetheless thought he had a huge talent. His awful singing gained him a cult following in the 80s and 90s and he regularly packed out pubs with his performances. Soon his act attracted other performers and became known as the 'Road Show'. Busloads of fans travelled for miles to see Jarrow Elvis and the 'Road Show' play. Their main venue was the Victoria Park Inn where Jarrow Elvis and his friends generated 70% of the pub's weekly turnover, around £15,000 to £20,000!

Joe claimed one night, that he awoke to see the ghost of Elvis Presley floating on his ceiling. His mouth was moving, but Joe could hear no words. From that day on, Jarrow Elvis felt that his namesake was with him on stage each time he performed.

So what are we to make of ghosts? Most ghost encounters are not with apparitions that can be identified as a particular person, Elvis in Jarrow aside. If ghosts were the souls of dead people, with unfinished business that stay on Earth until their business is done, then the world would be awash with ghosts. Looking back through the cases above the reader will see that many of them do not seem human at all. A lot of these entities have more in common with the monsters we examined in the previous chapter. As we progress we will see similarities with other Fortean phenomena, that may provide clues to their nature. The restless souls of dead human beings just don't cut it as the answer to most ghosts.

CHAPTER THREE: UFOS

I am sure that cavemen looked into the night sky and saw strange lights. What did they make of them, gods or monsters? In 1561, in the skies above Nuremberg, Germany, were the scene of a bizarre aerial battle. Hanns Glaser, a letter-painter of Nuremberg recorded the strange events.

"In the morning of April 14, 1561, at daybreak, between 4 and 5 a.m., a dreadful apparition occurred on the sun, and then this was seen in Nuremberg in the city, before the gates and in the country – by many men and women. At first there appeared in the middle of the sun two blood-red semi-circular arcs, just like the moon in its last quarter. And in the sun, above and below and on both sides, the colour was blood, there stood a round ball of partly dull, partly black ferrous colour.

Likewise there stood on both sides and as a torus about the sun such blood-red ones and other balls in large number, about three in a line and four in a square, also some alone. In between these globes there were visible a few blood-red crosses, between which there were blood-red strips, becoming thicker to the rear and in the front malleable like the rods of reed-grass, which were intermingled, among them two big rods, one on the right, the other to the left, and within the small and big rods there were three, also four and more globes. These all started to fight among themselves, so that the globes, which were first in the sun, flew out to the ones standing on both sides, thereafter, the globes standing outside the sun, in the small and large rods, flew into the sun.

Besides the globes flew back and forth among themselves and fought vehemently with each other for over an hour. And when the conflict in and again out of the sun was most intense, they became fatigued to such an extent that they all, as said above, fell from the sun down upon the earth 'as if they all burned' and they then wasted away on the earth with immense smoke. After all this there was something like a black spear, very long and thick, sighted; the shaft pointed to the east, the point pointed west. Whatever such signs mean, God alone knows. Although we have seen, shortly one after another, many kinds of signs on the heaven, which are sent to us by the almighty God, to bring us to repentance, we still are, unfortunately, so ungrateful that we despise such high signs and miracles of God. Or we speak of them with ridicule and discard them to the wind, in order that God may send us a frightening punishment on account of our ungratefulness. After all, the God-fearing will by no means discard these signs, but will take it to heart as a warning of their merciful Father in heaven, will mend their lives and faithfully beg God that He may avert His wrath, including the well-deserved punishment, on us, so that we may temporarily here and perpetually there, live as his children. For it, may God grant us his help, Amen."

Weird lights and shapes in the sky have been interpreted as different things in different ages, all seen through the cultural filter of the time. The 1561 events were seen through the milieu of a cultures ruled and governed by religion.

Five years later, a similar phenomena occurred over Basel, in Switzerland. A local broadsheet recorded the sighting.

"It happened in 1566 three times, on 27 and 28 of July, and on August 7, against the sunrise and sunset; we saw strange shapes in the sky above Basel. During the year 1566, on the 27th of July, after the sun had shone warm on the clear, bright skies, and then around 9 pm, it suddenly took a different shape and colour. First, the sun lost all its radiance and lustre, and it was no bigger than the full moon, and finally it seemed to weep tears of blood and the air behind him went dark. And he was seen by all the people of the city and countryside. In much the same way also the moon, which has already been almost full and has shone through the night, assuming an almost blood-red colour in the sky. The next day, Sunday, the sun rose at about six o'clock and slept with the same appearance it had when it was lying before. He lit the houses, streets and around as if everything was blood-red and fiery. At the dawn of August 7, we saw large black spheres coming and going with great speed and precipitation before the sun and chattered as if they led a fight. Many of them were fiery red and, soon crumbled and then extinguished."

During WWI, strange flying objects were thought of as German airships. I believe it was Charles Fort, the great American pioneer of research into anomalies, who first postulated that flying objects and lights were craft piloted by intelligent beings from other planets. We think of these weird lights and objects as alien craft and the beings from within extra-terrestrial creatures. The truth is probably something much weirder.

STRANGE INVADERS

Ask the man in the street to draw an alien. Most people would produce a short, spindly limbed, grey skinned creature with a bulbous head and large dark eyes. These grey aliens were first reported in the abduction case of Betty and Barney Hill, in New Hampshire on the 19th - 20th of September, 1961.

Since that time the 'greys' have become ubiquitous. The problem with most so called 'aliens' reported by witnesses, is that they are far too humanoid. I call this the *Star Trek* problem. Most of the aliens in the TV show *Star Trek* are very humanoid. Two eyes, a nose and a mouth. A head, two arms and two legs. The idea of a life-form that evolved on another biosphere, looking anything like a human is astronomically unlikely. Humans only became the dominant life-form on Earth due to a series of happenstances. If we were to wind back evolution and start it again we would not necessarily get the same results. With something that evolved on another planet, all bets are off. The science fiction show *Doctor Who* the classic series of the 60s, 70s and early 80's (rather than the lack lustre, gender-flipped, woke modern reboot) features aliens that looked truly alien, with little or no human features.

However there are cases of supposed alien encounters, with things that are not human looking at all. Things that look like they crawled out of a scary episode of Doctor Who, or the slime festooned pages of a story by H.P. Lovecraft. In this first section, we will look at some of the strangest looking 'aliens' ever reported, a veritable cosmic menagerie.

THE HORROR OF BALD MOUNTAIN

On November 14th, 1974, a fiery object in the sky was witnessed crashing to Earth near Bald Mountain, in Lewis County, Washington, USA. Two days later, Ernest Smith, a grocer from Seattle was hunting deer in the area when he stumbled across a truly weird beast. He described it thus...

> *"...it was horse-sized, covered with scales and standing on four rubbery legs with suckers like octopus tentacles. Its head was football-shaped with an antenna sticking up...The thing gave off this green, iridescent light."*

The glowing horror was seen a couple of days later by Mr and Mrs Roger Ramsbaug, from Tacoma. According to the author Jim Brandon's report, the pair were driving along State Route 7, a nearly 60-mile stretch of road, between Morton and Tacoma, when they suddenly noticed *"...a dull glow near the side of the road."* The couple slowed down for a better look at what they assumed to be: *"...a neon sign in the fog."* It was then that the Ramsbaughs got the shock of their lives, when they spied the same horse sized, scaly, glowing, tentacled thing that Smith had reported just days before.

After reports of the creature found their way the local papers, the Sheriff of Lewis County, William H. Wiester, began an investigation. However a group of people

193

claiming to represent both NASA and the United States Air Force, stopped the Sheriff in his tracks. Wiesner said that the men were heavily armed and that their uniforms bore no insignia. And that's where the story dried up. There were no more reports of the creature or anything resembling it on Bald Mountain, or anywhere else for that matter.

VEGETABLE FROM THE VOID

Dangerous alien plants are a staple of science fiction, from the vast and monstrous Krynoids from the *Doctor Who* story *The Seeds of Doom,* to John Wyndham's creepy novel *Day of the Triffids*. But one man claimed a real life encounter with such an entity.

Bow hunting in woods just outside of Fairmont, West Virginia in July of 1968 the wonderfully named Jennings Fredrick had little luck in securing his quarry, woodchucks. The Air Force Veteran was about to head home when he heard a high pitched jabbering. The noise reminded Fredrick of a speeded up recording. Looking up he saw a tall figure of some seven feet. It was thin and resembled a stick of celery. The yellowish-green thing had long, pointed ears and yellowish, slit like eyes. It's arms that were no thicker than a quarter (quarter of a dollar, a coin just under an inch across) that ended in three fingers. The fingers terminated in what looked like suckers.

Fredrick thought the thing looked ill, and this seemed to be confirmed when he received a 'telepathic' message from the being as it's jabbering increased.

"You need not fear me. I wish to communicate. I come as a friend. We know of you all, I come in peace. I wish medical assistance. I need your help."

Then, rather in contradiction to its statement the vegetable man shot up one of its suckered hands and wrapped it around the startled hunter's arm. The suckers adhered and began to draw up blood. A jolt of pain shot up Fredrick's arm as the vegetable vampire began to feed on his blood. He could see his own blood being drawn up the inside of the thing's fingers, through its arm and into its body. As it fed on him, its eyes seemed to rotate and oscillate. This seemed to have a hypnotic effect that alleviated Frederick's pain, as it fed on his blood for the next two minutes.

Apparently satisfied, the creature disengaged its suckers and turned away. It sprinted up a hill with super-human speed as the pain in Fredrick's arm came back. Running away himself, Frederick heard what he interpreted as the creature's craft taking off.

BEWARE THE BLOB

Another staple of SF movies are blob like entities, the best known of which is Paramount's 1958 classic *The Blob*. Again, with life imitating art, two Swedish men had their own encounter with a group of hostile alien blobs just three months after the film's release.

The event occurred in December of 1958, near Domsten, Sweden. Stig Rydberg and Hans Gustafsson, who worked together at a dry cleaners, where driving home through fog after a night out with their girlfriends. At approximately 3:00am, the duo decided to pull over at an isolated spot on Route 45, which runs along the Strait of Öresund near the village of Domsten, in order to stretch their legs and relieve themselves. The men investigated a glow in the woods they initially thought to be a fire. The pair noticed a strange odour in the air. The glow turned out to be coming from a disc shaped craft twelve feet across, and three feet high that stood on three legs. Gustafsson said the craft looked like it was made of a peculiar shimmering light that seemed to change colour. Around the craft hopped four repulsive creatures, blue-grey in colour and about three feet tall. The blobs had a foul stench. Rydberg said...

> *"They were like protozoa, just a bit darker than most, a sort of bluish colour, hopping and jumping around the saucer like globs of animated jelly"*

Suddenly, three of the vile blobs flew at the men and fastened themselves onto them like leeches. Rydberg said that grappling with the blobs felt like being trapped in elastic dough. The blobs attempted to drag the men back towards the craft. Gustafsson described the struggle...

> *"The drag these thing exerted was terrific. And they gave off such a terrible smell-like ether and burnt sausage,"*

The men fought with their gelatinous assailants, but each time they struggled free the things would attach themselves again. Ryberg found his arm elbow deep in one of the horrors as he tried to punch it. Finally Ryberg broke free and rushed for the car. Looking back he saw his friend holding on to a 'no camping' sign. All three attacking blobs were now swarming over Gustafsson. Ryberg hit the horn of his car hoping to startle the blobs. The jelly bags dropped their prey and shot like lightning back to the ship, entering it from beneath. The craft took off with a high-pitched whistling. The whole encounter had lasted about five minutes, and left behind nothing except fear and a nauseating chemical smell. It took twenty minutes before Gustafsson's trembling would subside enough for him to drive home. Both men were so terrified that they wept.

Both men felt sick for three days after, but a doctor's examination found nothing wrong with them. The incident was reported to the Swedish Defence Staff. On January 9, 1959, the pair were called into the Helsingborg police station for questioning. The interview which lasted well over eleven hours was attended by police officers Captain Lennart Bunke and detectives Sven Rudolph and Ake Fernebrant, who conducted the examination. Michael Wächter, a German military psychologist and lab technician Sture Risberg. The investigators seemed hell bent on debunking the men's story, even going so far as to hide a microphone in the room when they were left alone in the hope that they would slip up and reveal it as a hoax. However Ryberg and Gustafsson stuck by their story.

Detective Rudolph confessed that they could not uncover anything that suggested a hoax. Despite that fact, the Swedish military publicly denounced the encounter as a hoax.

RAMPANT REPTIODS

Palaeontologist Dr Dale Russell speculated that if non-avian dinosaurs had not become extinct at the end of the Cretaceous, they would have evolved into a humanoid shape. Taking as his ancestral species *Stenonychus inequalis* (the most highly evolved known dinosaur, which was a close relative of the more familiar Velociraptor) he postulated an upright, bipedal creature with binocular-eyes and a large cranium, containing a large brain. The tailless beast was reconstructed with three long dexterous fingers, one of which was opposable. A reconstruction of the dinosaurid (as Russell christened his creation) is on show at the Canadian Museum of Nature in Ottawa, Canada. Russell believes that had such creatures evolved, their intelligence would have long since surpassed that of mankind. Mammals as a whole would not have developed beyond small, shrew-like primitives if this scenario had come to pass. It may be argued that Dr Russell was inspired in his theory by the BBC television programme *Doctor Who*. In the 1970s, The Doctor did battle with two races of super-intelligent dinosaur descendants, the Silurians and the Sea Devils. Both saw fruit well before Dr Russell's theories were published.

However, Dr Russell's reconstruction does bear a remarkable resemblance to reports of certain types of 'alien'. This has led some to conclude that reptoids come from a planet with a parallel development to Earth, but where dinosaurs continued their rule and developed into intelligent humanoids. The dinosaurid is totally hypothetical, and many say that it shows more about human narcissism than evolution. The tailless body, and human-like torso with the round face, all show a parallel evolution to humans. Would such a thing take place?

One class of 'alien' is called the reptilian or reptoid. This creature is a lizard-man resembling a human/reptile hybrid. The following encounter was said to have happened to Ron and Paula Watson, at Mount Vernon Missouri, in July 1983. During breakfast one day, the couple noticed flashes of light coming from a pasture across from their farmhouse. Looking at the field through binoculars, they were met with a strange sight. A black cow was lying motionless on the ground, whilst two silver-suited figures ran their hands over it. The animal seemed to levitate, and floated into a cone-shaped craft in a nearby clump of trees. The cone had a mirror-like surface that rendered it almost invisible. Standing next to the strange craft were two lizard-men, that were described as having green skin, eyes with vertical slits, and webbed hands and feet. On the other side of the cone was a hairy bigfoot-type creature (quite a menagerie all in all!) Ron wanted to have a closer look, but his wife would not let him go. They recounted what they saw to the owner of the pasture, when he mentioned that one of his cows was missing. The farmer would not listen to them, but his missing cow was never found. Later, under hypnosis, Paula claimed to have been abducted by the creatures.

In one strange case, a witness called 'Angie' claims to have been abducted by a race of 'lizard men' in Arizona, in March of 2000. Angie would collect rocks from the mountains and deserts. She was out collecting near the edge of a canyon, in the Superstition Mountains. On this particular day Angie noticed a cave. Looking up she was confronted by a humanoid creature with a lizard's head. Screaming in terror she passed out.

Later she woke up driving her car! She arrived home and showered for several hours due to the feeling of having to clean herself. Angie locked herself away for several days. She would find herself staring at the television screen, instead of watching whatever show was on, her mind completely scrambled and not at all at ease with itself.

Angie worked in a pet shop and when she returned to work somebody brought a lizard into the shop and she felt a wave of terror at the sight of the animal, something she had never experienced before. She realized that something had happened to her in the Superstition Mountains. A friend suggested contacting a professional hypnotist with experience of hypnotic regression.

Through several hypnosis sessions, she recalled what had befell her. She awoke to find herself in some kind of inner cavern or base. Strange sounds like barking or chirping filled the air around her. Any attempts she would make to sit up and look around would fail. She was held fast in some kind of jelly-like substance. Her captors were lizard men.

They were each adorned in a jumpsuit, that contained a logo or symbol of a curved dragon, with a seven-pointed star in the middle. Most of the creatures wore a black outfit, two others wore a white one, and one – who also had a paler complexion than the others – wore an orange one. The creatures were about seven feet tall. The one dressed in white seemed to be the leader. Something cold touched her head and she fell unconscious.

She awoke later in an oval room. Dangling from the ceiling were pipes attached to things like 'misshapen balloons'. The balloons moved as if something alive was inside them. She then noticed she was naked. Suddenly one of the reptoids entered, sans its jumpsuit. The thing approached her and shot a blue light at her rendering her unconscious once more.

Hypnotic regression is notoriously unreliable. It is not a door into the past. Memory is not like a video recorder. It is prone to mistakes. It's hard to determine if patients are truly accessing memories or creating false memories at the suggestion of a hypnotherapist. False memories are unique in that they represent a distinct recollection of something that did not actually happen. It is not like forgetting or mixing up details of things experienced; it is about remembering things they never experienced in the first place. However, it has been found that memory of traumatic or strange events can be much more accurate than memories of more mundane events. The trouble with hypnotic regression is the chance that the hypnotist influences the subconscious mind of the patient.

Angie has apparently stuck by her story for over twenty years.

The reptoids have become the super villains of modern day UFO folklore. A popular notion is that the reptoids can shape shift into human form and are in prominent places in world government. From there they control the human race. The British Royal Family and many politicians are thought, by conspiracy theorists, to be shape shifting space lizards! Chief pundit of this idea is David Icke, former footballer for Leicester, England, turned conspiracy theorist. He is convinced that such creatures are in control of Earth.

> *"When you get back into the ancient world, you find this recurring theme of a union between a non-human race and humans – creating a hybrid race. From 1998, I started coming across people who told me they had seen people change into a non-human form. It's an age-old phenomenon known as shape-shifting. The basic form is like a scaly humanoid, with reptilian rather than humanoid eyes"*

David Icke has been accused of being a leading producer of misinformation about COVID-19, as well as anti-Semitic content. Most serious researchers dismiss his claims and the shape shifting space lizards who rule the world theory.

KERMITS FROM THE STARS
As well as reptilian aliens there are reports of creatures that resemble amphibians, specifically frogs. The town of Loveland, on the Miami River, Ohio, has a tradition of such creatures dating back, once again, to 1955. In this year, a driver returning home from work at 3.30 in the morning spotted a trio of creatures some four feet tall, with leathery-skin, frog-like faces, and wrinkled heads. One of them held a rod like device that emitted sparks. He parked his car, and watched them for three minutes before calling Police Chief John Fritz. Fritz investigated, but found nothing. Almost seventeen years later, two of his colleagues had a better look. On March 3rd, 1973, at 1:00am, two Loveland police officers saw a creature, identical to the one described above, leap over a guard-rail, and scramble down into the river. Two weeks later one, of the officers saw the thing again. This time, it was lying in the middle of the road, and leapt up at his approach. Once again it scrambled over the rail, and into the river.

Miniature versions of the frog-men were reported from a UFO encounter in Orland Park, Illinois, in 1951. Twenty-four-year old steelworker, Harrison E. Bailey, would, in his spare time, attach advertising to a 4-foot-high, 150-pound tractor wheel and then roll the wheel on contract to make extra money. He had painted the wheel green and wore a matching outfit of green overalls and, for protection from stones thrown up by traffic, gloves, goggles and a cap. In late September, 1951, Bailey was working for the late baseball entrepreneur Bill Veeck to promote his St. Louis Browns baseball team. Bearing signs that said "*Let's come thru in '52--St. Louis Browns,*" Bailey's wheel was to be rolled from Gary to Chicago and then down to St. Louis. He was walking through some woods at 11am, when he experienced a strange cramping and burning sensation on his neck. Looking behind him, he saw what he took to be a small, grey whirlwind. He then noticed a silvery-grey, oval-shaped object, at the edge of a meadow. As he drew closer, he saw it was a craft of some kind with windows. Peering from the windows were two "men" in what appeared to be green-tinted face shields. The men asked him where he was from, and where he was going. The pain seemed to wear-off, and he walked away, glancing briefly back at the craft. The next thing Harrison remembered was awaking in the late afternoon. He was confused when a group of men asked him if he had come out of a flying-saucer on the previous day. He thought it was best to keep the story to himself.

In subsequent years, he began to suffer chronic health problems, and came to associate this with his encounter calling it his 'flying saucer disease'. After gall bladder surgery at Hines Veterans Hospital near Maywood, doctors told him his internal organs were the condition of a man twice or three times his age, possibly the result of an intemperate life. For clean-living Bailey, this was not an explanation. In 1975, under hypnosis, he recalled forgotten events that befell him that night. He remembered being swarmed by a horde of bipedal frog-like creatures with large eyes, brown-striped skin, and three fingers. The creatures stood some 1.5 feet (eighteen inches), tall, and were accompanied by swarms of black beetles that scurried around erratically.

The frog-creatures bit Harrison, and made him fall to sleep. He was taken on board the craft and given a telepathic message, by the men in face-shields, that they meant him no harm and wanted to communicate with mankind. They wanted Harrison as their spokesman .He said he was visited by the entities, that assigned him the mission of informing the American public, that otherworldly beings were among them and meant them no harm. So Bailey began spreading that word with his advertising wheel. On November 1st, 1978 he had his last visit from the creatures. He now lives in a retirement home in Pasadena and avoids discussing his UFO experience.

Down in Peru, a fifteen-year-old student had a run in with an alien batrachian, in the summer of 1965, on a rooftop in Lima. On Sunday, August 1, 1965, at 7pm Alberto San Roman Nunez, was taking in the washing from the line on his family's roof. Glancing skywards he saw a UFO descending onto the roof behind him.

The creature that emerged from the craft looked like a hairy, glowing toad, three and a half feet tall. Running for the stairs the boy was enveloped in a reddish light that seemed to freeze him. The glowing, hairy space toad bounded back into its craft and left. The UFO vanished over the Pacific Ocean.

THE AVELEY ABDUCTIONS

The most detailed case of alleged alien abduction in the UK took place in autumn of 1974 near the village of Aveley, Essex, just thirteen miles from London. The case is not only notable for its detail, but for the fact that other Fortean phenomena seem to bleed through into it. The saga was investigated in 1977, by UFOlogist Andrew Collins. Pseudonyms were used for the family of five involved. Collins called them the Avis family. John, 32, was a carpenter and joiner at the time. His wife Elaine, 28, had been an accountant but was now a full time mum for their children, Karen 11, Kevin 10, and Stuart 8.

The family had been visiting Elaine's mother in Harold Hill, and were returning at 9.50 in the evening. The drive to their home in Aveley would take around 20 minutes. Karen and Stuart had fallen asleep on the back seat, but Kevin pointed out an odd light in the sky to the east. The object looked like a pale blue, bright star, some 500 yards away and about 30 degrees above the horizon. It seemed to keep pace with the car as it travelled down the unlit country lane. They had an odd feeling that something was wrong as they approached a bend in the road. The car seemed oddly silent, they could not hear the engine or the wheels just the radio.

As they turned the bend, they saw a thick green fog blocking the road ahead. It was some 9 feet tall and was boarded on each side by bushes. The fog was around 90 feet ahead of them, as they drew close the car radio started crackling and smoking. John pulled out the wires. The car lights went dead as they entered the fog. The thick fog felt cold and they noticed a prickling sensation. Then there was a jolt as if the car had just crossed a hump-backed bridge. The car seemed to be about a mile further on than it had been a moment before, and as John came round from a kind of hazy feeling he could have sworn he was alone in the front of the car. When he looked again his wife was there.

Upon reaching their home, they found it was 1am, when they should have been home by 10.20pm. Nearly three hours seem to have been lost. After seeing an article on UFOs in a local paper several years later, John contacted a UFO group, who put him in touch with Andrew Collins. The story that unfolded grew increasingly strange.

The family underwent radical changes after the event. The family, with the exception of Stuart, became vegetarian and could not stand the taste or smell of meat. John and Elaine stopped drinking alcohol. They all became more interested in environmental matters. John left his job and got another looking after mentally handicapped children and Elaine returned to college. John, who used to smoke 60-70 cigarettes a day, stopped smoking.

Other things were stranger. The family cat would hiss at something that they could not see. They became convinced that they were being followed by three cars, a small red sports car, a blue Jaguar and a large white Ford Executive. Poltergeist activity broke out in the house. Small objects would vanish, only to turn up again days later in different places. A large radio was seen to rise up off a table and levitate. Doors would burst open violently and strange noises like a rustling and clicking were heard in the house. John tried to locate the sound, and it seemed like the sound would move away from him. Another noise, like morse-code was heard as well. Kevin awoke one night to see a clown standing next to his bed. Readers will recall occurrences like these in the haunting cases we looked at in the last chapter. Andrew Collins himself, whilst sleeping on the sofa in the living room one night, heard a violent crashing from the kitchen as if pots and pans were being smashed together.

Collins found out that the family had had other UFO experiences before the 1974 case. John and Elaine, whilst on holiday in Walton-On-The-Naze in 1961, had seen an object like a bright star whilst walking by the sea one evening. The object looped-the-loop and moved about in an erratic manner. Ten to fifteen other people saw it.

In 1968, John was a passenger in a car on the M1 Motorway near London. He and the others in the car, saw a blue light coming towards them. The power in their car failed as did the power in the car in front of them. The car stopped and the car John was in crashed into the back of it. Another car behind them also lost power and crashed into the back of them. All the drivers exchanged insurance information and phone numbers. When the police arrived they told the drivers that they had investigated a number of identical cases on the same night.

Then just before the main events, John was driving along the Aveley Road when he saw a metallic, cylindrical object pacing an airliner. The UFO seemed to be twice the size of the plane and was a matte silver. Then it overtook the plane and shot away in total silence,

After the 1974 case, the whole family saw another UFO. They were coming home from Walton-On-The-Naze to Aveley. and were driving between Gallows Corner and Brentwood. They saw a star like object. that followed their car for 40-50 miles, sometimes dropping as low as the treeline. It was oval in shape and effected the car's lights, radio and engine. The object finally shot off at great speed. It seemed like the Avis family was truly haunted!

Andrew Collins convinced John to undergo hypnotic regression. Elaine was too scared to take part. Both insisted that their son should be spared from the stress such a process may have caused. The couple had been having strange dreams and flashbacks. As they talked about them they recalled more. John remembered being operated on or touched by small ugly creatures. Elaine said she recalled being on an operating table and a small figure dressed in white.

John was regressed by hypnotist Dr Leonard Wilder, on September 25th 1977, with two follow up sessions on October 2nd and October 16th. John recalled a beam of light cutting through the green fog and hitting the car. He then passed out. When he awoke he was on a balcony looking at a blue car (his own car was white). Inside were an unconscious family that seemed to be himself and his wife and children. The car is in some kind of hanger. Beside him on the balcony were Elaine and Kevin who were also looking at the car. Behind them was a tall entity some six feet 8 inches tall. There was also some kind of 'air shaft' that John thought was used to ascend and descend.

John and the tall entity move to another room, using another air shaft to ascend. The entity touched John's shoulder and he blacked out again. When he came to, he was on a table with a bar shaped, moveable device seeming to scan him. Its underside seemed to be honeycombed. As it moved above him John felt a warm sensation. He noticed there were now three tall beings in the room with him, as well as two small, ugly creatures in white smocks. The smaller beings had brown fur and long, pointed ears. Their noses were arrow shaped and their eyes triangular. Their faces looked somewhat like those of a bat. They had three fingered hands with claws, and stood about 4 feet tall.

The creatures held 8 inch long, pen-like devices, that were attached by thin cables to the table. The examiners, as John dubbed them, seemed to be using the devices to scan him as well. When the larger beings approached, the examiners moved back making John think they had seniority over the smaller creatures.

John could not move until the examination was over, and the smaller creatures left. He asked if he could sit up and one of the tall beings told him that he could but to sit still for a while. The entity seemed to convey the message via telepathy. John felt weak for a short while.

The taller creatures dressed in once piece, white suits, that covered their hands, feet and heads. The latter was covered by a balaclava type hood. They seemed to be wearing masks with only their eyes visible. Their eyes were larger than a humans and had pink irises. Their skin was very pale, almost translucent. Each being had three fingers on their hands. Their limbs apparently had no joints and bend like those of a bendy toy or blow up doll.

They showed John a visor that they used to protect their eyes from light when they are outside of the ship. They went on to show John other rooms in their spaceship, including a leisure room and a laboratory with a device that displays holograms of small objects like a hi-tech microscope. There is also a room with couches that have bubble-like surfaces, that each have their own mini-atmosphere to counteract sudden stops and turns.

Whilst walking through a corridor, John saw a spherical device through a window. The metallic ball had an outer casing and two 'arms' that came from behind and linked in front. The outer casing was movable, and opened on a vertical axis like an eyelid. The aliens said that it was an ion drive, and the casing opened and closed at different speeds to control the flow of ions. This caused a vortex in the particles outside of the ship. The vortex was funnelled by a directional gun. In the atmosphere of Earth a magnetic drive was used, as the ion drive was too dangerous. They did not explain why.

The aliens said that they used magnetic fields to disguise their spaceship on Earth, and they could also channel magnetic fields into a powerful weapon, like a laser beam.

Finally, they reached a control room and John was placed on a couch, with a dish-shaped device above his head. Projected onto a wall, John sees many images of planets, star systems, plans, drawings and maps. There was an audible commentary that went with it. The words and images went so fast John could not recall them all, but the being told him that the information would be retained in his mind.

When he asked about computers, the beings told him that they had an organic computer that they controlled with their minds, rather than manually.

After this, they led him to a darker part of the control room, where he was shown a hologram of the being's home planet. They had ruined their world with pollution. The scene showed a collection of metallic cones, that he took to be a city, with mountains in the background and a sky layered in shades of green, red, yellow and blue. A hooded figure with pink eyes, holding a ball that glowed red and yellow, appeared in front of the scene. John was asked to touch the ball and felt a strange sensation flowing up his arm, he was told that it signified the energy being taken from their planet. John felt this had some form of religious meaning to the beings.

The leader told John it was time to go, and suddenly he felt himself back in the car. For a moment he noticed that his wife was missing from the passenger seat, but then suddenly she was there.

Elaine pieced her account together from recalling parts of it as she spoke with John. Like John she recalled the beam of white light and blacking out, only to awake on the balcony looking at the blue car with her own family in it, much as John did. A hole opened in the wall and two of the ugly examiner creatures were behind it. Kevin was taken away from her against her will by a tall creature, an examiner. She was herself led to a room and strapped to a table, that had two poles with metallic cones attached to them. Unlike John she struggled at first. The examiner creatures used their pen-like devices on her, and she noted that they emitted a purple light. The tall aliens told her the examiners were known as the Seb-a-teen and not to be afraid of them. One of the tall entities telepathically told her that they could not continue whilst she was in a frightened state. It put a finger on her forehead and she blacked out again.

When Elaine came around, she was still on the table but the straps were gone and she found herself dressed, in a long gown with a hood and puffy sleeves brought in at the wrists. She said it looked like crinkly cellophane. Three of the entities took her through some corridors to the control room. She remembered seeing John walking in the other direction on her way there. She was seated and offered some pink, disc shaped objects on a tray. She realized they were food but declined them fearing they may be drugged.

The leader of the beings asked another one to play some music. It twiddled its hands producing a music like a high pitched harp. The leader assured her that her children were safe.

Now feeling relaxed, she was taken to another part of the control room. The being showed her a star field on a panel. It zoomed in until she recognized the solar system, then Earth, then the UK and finally zooming in on Thames Estuary. Then, like John, she was taken to a couch with a dish shaped device above her head, and shown star charts, maps and plans. She said the effect was like having an encyclopaedia pumped into her head all at once. All she can recall clearly, was her own solar system, though she recalled seeing 11 planets in it.

Elaine was then shown the same hologram as John, with the hooded figure and glowing ball. She was told...

> *"This is the seed of life, our past and your future. Accept this from us for yourself, your children and your fellow kind."*

She was told it was time to go, and the next thing she knew she was with John and Kevin getting dressed. She drank from a bowl and told the others to hurry up. Her next recollection was being on a catwalk above the car. The car itself was in front of a horseshoe shaped wall. The leader and the alien that had played music, then came into view. The musical alien referred to the leader as 'Lyra' and the leader called the music man 'Ceres'. Looking at the car, she saw her husband and son were already inside it. The aliens said she could stay if she so wished. The room got brighter and the car dematerialized through the spaceship's wall, seeming to fizzle out. Elaine started to worry, but the entities told her not to worry as she could catch them up.

Elaine recalls seeing the car driving along the road by the woods. She impossibly got into the car, opening the door as it was moving and sat down. There was a jolt then the experience was over. All of her family were there.

Kevin only recalled being lifted up by the light, but he once said to his father *"They gave me a lot of things to do when I grow up but I've forgotten them all."*

It's quite a story, and a detailed one too, but we must remember that hypnotic regression is not an accurate way of accessing a person's memories. The fact is that Elaine was not hypnotised, but recalled events whilst in conversation with John, both before and after he was hypnotised. How much of this was just suggestion? Also, though I'm no physicist or engineer, the physics of the propulsion drive sound utterly made up.

I have no doubt that the Avis family experienced something very odd, but just what was it? The 'alien abduction' seems to have been pieced together from popular science fiction tropes. Dying worlds, tractor beams, medical examinations, unlikely propulsion drives, yet the case also calls to mind fairy lore we will be examining later. It would be interesting to track down members of the family and talk to them today.

DON'T MESS WITH HAIRY DWARFS
In the first chapter we looked at the association between UFOs and huge hairy ape men, but just as there seems to be large and small mystery primates of the flesh and blood kind, so there seems to be a smaller kind of paranormal hairy horror.

It was two in the morning, on the 28th of November 1954, when Gustave Gonzales and Jose Ponce left Caracas in Venezuela, in their truck to buy by foodstuffs from the town of Petare, some twenty minutes away. The pair found the road blocked by a glowing sphere ten feet across and hovering six feet above the ground. Stopping the truck, they both got out to investigate. They saw a weird creature resembling a troll. It was no more than three feet tall, had green, glowing eyes and was covered with hair. Its three fingered hands ended in claws. Gonzales, in an inversion of the usual alien abducts human, grabbed the creature, intending to take it to the police. Why he did this was unclear, perhaps he thought it was an animal that had escaped from a zoo. He reported that the little creature was very light, no more than 35lbs in weight. However the hairy dwarf was immensely strong, and with a single handed push, it hurled Gonzales fifteen feet!

Ponce lived up to his name and fled the scene, running to a nearby police station. Two more of the space trolls emerged, holding rocks and dirt in their hands, as if they had been taking samples. They hopped into doors that opened in the sphere, as their colleague continued its attack on Gonzales using its claws. The scared man pulled out a knife and tried to stab his attacker, only to find that the blade glanced off the creature's seemingly rock hard skin. Then, one of its friends emerged from the sphere and shot at Gonzales with a tubular weapon that emitted a light that dazzled him. All the space trolls scrambled back into their ship.

Ponce returned with the police who thought the two men were drunk. Back at the station they were examined and found to be sober. Gonzales also bore a livid scratch down his side. Both men were given sedatives.

Several days later, a doctor came forward saying that he had seen the hairy dwarfs attacking the men, but didn't say anything for fear of damaging his reputation. The lesson to take away is, if you see a hairy dwarf, don't try to kidnap it.

THE MEDFORD SHMOOS
The April-May 1984 issue of the CUFOS Associate Newsletter (Centre for UFO Studies), contains a letter from a woman who related a sighting of three strange creatures near Medford, Oregon, one night in the summer of 1953. The woman was with her husband and their twelve year old daughter. The car headlights lit up a trio of

strange creatures gliding along by the side of the road. She likened them to big geese with no beaks, wings or feet. The bowling pin shaped creatures were covered in silky white fur. The three creatures vanished into the woods, but all three family members got a good look at them.

> "They were white with 'very smooth satiny fur' and were shaped somewhat like peanuts. They lacked any features like arms, legs, wings, or beaks. They glided, in a sideways motion at first, then straight away from them till they disappeared into a wooded area. The last of the three was slightly smaller than the other two."

The baffled family asked their neighbours if they had seen anything odd, but none had. The writer of the letter said that her daughter still talks of the sighting all these years later. The family thought the creatures looked like the shmoos, a race of white, bowling pin shaped creatures, featured in the Li'l Abner comic strip by Al Capp.

There is little more information on this case, and the Medford Shmoos seem never to have been seen again but they were so strange that I just had to include them.

THE TUSCUMBIA SPACE PENGUINS
You thought it couldn't get any weirder than real life shmoos? Think again.

Claude Edwards, a 64 year old farmer, awoke on a cold February morning in 1967. He went to attend his duties on his remote farm near Tuscumbia, Missouri. As he marched towards his barn he noticed his cattle in the east field all staring at something. He looked in the same direction and saw a space ship in a grove of trees. The greyish-green craft was mushroom shaped, with a saucer like top, perched on a cylinder. Closing his barn door, he looked back at the craft and saw that a swarm of strange little creatures had emerged. The three foot tall beings were dressed in what seemed to be green hazmat suits with black, wide set goggles and beak-like protuberances that may have been breathing gear. The arms of the suit covered their hands, giving them a flipper like appearance.

The things, that he said looked like green space penguins, were some seventy feet away and separated from the witness by two fences. Annoyed at the alien trespassers, Edwards began to walk towards them. The creatures were scurrying around close to their ship with their arms moving fast. The farmer was angry rather than scared or baffled, and picked up a rock to hurl at the saucer. But as he got to about fifteen feet of the craft, he was stopped by a force field.

> "I thought I was going right up to it. I got up there and there it was. I just walked up against a wall."

He saw that the space ship was 18 feet across and 8 feet at it's apex. It was made of seamless metal he likened to shiny silk. Around the rim were a series of foot wide, oval portals from which bright light shone.

Backing off he hurled a rock, which bounced off the force field. Throwing another he saw it skip like a stone over water, this time landing behind the saucer. The panicked penguins scuttled back to the cylinder, and entered the craft, which then tilted a couple of times before lifting off and flying in a north-easterly direction.

UFO investigator Ted Phillips was introduced to Edwards by the farmer's brother. He agreed to be interviewed, as long as he could remain anonymous until his death. Philips interviewed the farmer not long after the events and was able to photograph marks in the field, where the craft had landed.

> *"When I arrived at the site the traces were still quite visible. It was one meter in diameter in a slightly irregular circle where the shaft had rested. The soil was extremely dehydrated in contrast with the surrounding soil."*

Phillips found that Edwards was reluctant to talk and wanted no publicity whatsoever. Why would a rural farmer in Missouri make up a story about space penguins in green hazmat suits?

THE CANARY ISLANDS GIANTS

One of the most spectacular UFO encounters took place on June 22nd, 1976, on Gran Canaria, in the Canary Islands. Dr Don Francisco-Julio Padron Leon had been called out by a young man, Santiago del Pino, to attend his sick mother. The Doctor and Santiago were riding in a taxi driven by Franscisco Estivez. They had just turned a bend in the road, at a place called Las Rosas, between Galdar and Agaete in the north west of the island. They were confronted by a huge bluish-grey sphere hovering a few yards above the ground. The taxi radio cut out and a wave of cold swept over them.

The sphere was as big as a two story house and transparent. There was a kind of platform inside and two gigantic beings. They had bright red, tight fitting uniforms and black, featureless, outsized helmets. They stood facing each other across the platform and seemed to be operating controls. Their hands seemed to be encased in black cones.

Estivez switched on the taxi's spotlight. Instantly the sphere rose up. A transparent tube inside emitted a blue gas that filled the sphere, which in turn swelled to the size of a 30 story building, though the occupants stayed the same size.

The terrified driver backtracked to some nearby houses. The people in the house that they went to told the Doctor that their TV had just stopped working. Together they joined the Doctor, the driver and the young man and all watched the object with the gas swirling around inside it. Then it shot off towards Tenerife. As it went, it changed shape to a spindle surrounded by a halo.

Later, Dr Pardon, was told not to talk about his sighting by the Spanish Air Ministry.

THE PENDELI EGG

In 1977, three witnesses, a Mr. and Mrs. "L.X." and a Miss "M.V.", were out hiking on Mt. Pendeli in the Attica region of Greece, when they came across a pristine car parked in a seemingly impossible location. On subsequent hikes the car was seen in exactly the same position, having not apparently moved. Around the car, in the snow, were strange oval tracks, some were in places where no man or animal could have trodden.

At one point Miss M.V had wandered off from the others. She encountered a *"dreadful creature, a dreadful white creature."* The entity was two feet tall and shaped like a smooth white egg. It had no features, save for two glowing, staring eyes.

Several days later, Mr. L.X. was about to start his car, when he began to scream and tremble without warning. When he finally came to, he claimed to have seen an unearthly spinning sphere, seemingly made of hair or black smoke, passing through the car window, and felt *"something trying to enter his mind."*

Again we have factors in the case which seem more like hauntings than alien encounters.

THE ALIEN OCTOPI

In the early hours of August 16th, 1968, Spanish farmer John Mateu and his dog were going to feed his cattle. His farm was located about 4 miles away from Tivissa, in the province of Tarragona. He noticed a metallic glinting, but assumed it was light reflecting off a car. Suspecting somebody had broken down and needed help, he walked towards the glimmering.

John was confronted by a hemispherical craft, that was floating several feet above the ground. Slithering towards the craft were a pair of strange entities that John described as looking like pale coloured octopi. The three-foot-tall creatures had four or five tentacles apiece. The whole area was lit up by light, and the craft shot up into the air. Time seemed to slip away for John and he passed out.

John was missing for a whole day and his brother Sebastian, an accountant from the nearby Village of Darmós, went out looking for him. John returned the next day with no memory of what happened, after seeing the space octopi slithering towards their ship.

John and Sebastian examined the area and found a circular area of burned grass. They also stumbled on two other, apparently older burned circles. On three separate occasions, Mateu and his brother noticed that their watches had stopped while they were wandering around in close proximity to the semi-circular scorch marks.

These circles were also observed by an Austrian couple, who had set up camp in the area. The man, identified as Hans Volkert, allegedly snapped and published photos of the circles, but, unfortunately, they seem to have been lost.

BUGS FROM BEYOND

A mainstay of 1950's B-movies such as *Tarantula, Them!, Deadly Mantis* and *The Black Scorpion,* giant insects and other arthropods are beloved of science fiction fans. The bugs usually grow to fantastic sizes after a dose of radiation. Indeed the very term 'bug eyed monsters' is associated with the genre. But there have been encounters with supposed extra-terrestrials that resemble magnified insects.

Mike Shea was a law student at the University of Baltimore. One night in that strange, strange year of 1973, he was driving to the town of Olney to meet a friend in a bar. About 15 miles outside of the city, he looked up to see a huge object in the sky. It had rings of alternating red and yellow lights. The object was shooting a beam of light onto a barn. Despite having his car window open, Mike could hear no noise. The beam of light faded and the object moved over his car, filling him with fear. Mike felt something like an electrical shock in his spine. And then the next thing he knew, he was driving into Olney feeling quite relaxed.

When he got to the bar, his friend was not waiting for him. The bar tender told him that his friend had been there at 7:00pm and it was now 9:00pm. Somehow he had lost two hours. He avoided using that road again.

A decade later, he was a lawyer with a wife and daughter. One day, in 1985, he was in a book shop during his lunch break and saw a book about UFOs entitled *Missing Time: A Documented Study of UFO Abductions* by Budd Hopkins. Whilst reading it, he came upon the case of a young man in rural Maryland, who seemingly lost hours of his life whilst driving on a lonely road. Later, under hypnotic regression. With mounting dread, Mike began to recall fragments of his own experience and confided in a friend at work,

Eventually, he contacted Budd Hopkins and underwent hypnosis. He recalled seeing four figures by the side of the road. At first he thought they were men, but they were not. They were insect like creatures clad in what looked like black armour. They looked like huge grasshoppers, and had four arms and two bowed legs. Three of them were large, the fourth small. Mike felt that this smaller one was in charge and that it was "ancient." Their heads were shaped like helmets, and the smaller being had on a silky looking suit with a zipper up the front.

There now seemed to be two crafts, a smaller one on the ground and a larger one hovering in the sky. Mike recalled been taken onto the ship and having samples taken from his body.

Mike continued to visit Budd, and even joined a group for people who had similar experiences. Many of them felt violated. One young woman said...

"When people come up behind me, even people I'm very fond of, I could kill them. It's an immediate adrenaline rush. And I wonder, why do I act this way? Always ready to turn around and fight, or run away. Always aware of who's coming into a room, who's going out, patrolling the area. And I realize, it all comes back to this."

Some of the victims recalled their encounters without hypnosis, others only with it. Most thought that they were being tagged and tracked in a manner that a zoologist would to a wild animal. Mike began to recall other events from his past. He recalled being given a silver sphere by the aliens, and hiding it in his attic as a boy. He consciously remembered his French tutor screaming at him once for arriving at a lesson hours late, and himself staring in disbelief at the clock. He knew he'd left home in time. He saw himself meeting the aliens in the woods on the way to the tutor's house. Another time, under hypnosis, he remembered seeing some symbols on a piece of paper aboard the craft. The ones he drew were nearly identical to symbols other abductees had remembered and drawn. But why would technologically advanced creatures from another world be using paper? One time, he recalled seeing one of the creatures looking at his sleeping daughter and seeming 'pleased' with her.

Mike agreed to go public and did several TV and radio interviews along with Budd Hopkins. He feared that the people he worked with would ridicule him, but one of the other lawyers confessed to Mike that he too had seen a strange object in the sky.

Mike was far from the only person to have encountered arthropod type entities. Project Blue Book, the US Air Force investigation into UFOs, ran from March of 1952 to December of 1969. One of the strangest cases it looked into was that of 23 year-old Canadian miner named Ennio La Sarza, who originally hailed from Italy. The case was first made public by a minister and ABC radio host from Buffalo, New York, Elder Charles Beck. The clergyman and a local reporter from the Sudbury Daily Star, Michael Bolton, conducted extensive taped interviews with the witness to this extraordinary event, less than a week after it transpired.

On July 2, 1954, at about 5:00pm, La Sarza, who worked as a miner and painter for a nickel mining company in Garson, about five miles from Sudbury, Ontario, saw something weird in the sky. The object hurtled down like a jet, but then hovered. It was 25 feet across, spherical, and had a ring of lights or portholes around it. Landing gear protruded from the bottom and antenna from the top.

Three creatures, each 13-feet-tall, got out of the craft. The things were greenish in colour and each had a single eye in their heads. Their bodies bore six pairs of limbs, with crab-like pincers, that snapped spasmodically. They had antenna sprouting from their heads. La Sarza thought they looked like giant insects.

As the cosmic horrors began to scuttle towards him, the scared minder tried to flee, but seemed to be frozen by the hypnotic glare of the creature's cyclopean eyes. The creatures telepathically ordered him to perform some form of act, that he refused to do, saying that he would rather die. La Sarza also refused to tell the researchers what the beings had asked him to do.

The things fixed him with their awful stare once more, and he passed out. When he came round the craft and the creatures were gone. The miner reported the event not only to his bosses, but also to the police and representatives of the Royal Canadian Air Force, who investigated. He also became so scared of the consequences of disobeying the space monsters and feeling their wrath, or obeying them and doing the awful, unknown thing they had asked him to do, that he asked to be locked in a jail cell for his own safety. Though quite how this would protect him from crab clawed horrors from the stars is unclear.

The Royal Canadian Air Force forwarded a copy of their final report to the United States Air Force's UFO investigative body, Project Blue Book, as well as the Air Ministry in London.

THE JELLY MAN
The village of Fragosa, Cáceres province, Spain, lies close to the border with Portugal. It is a small place, with a population of no more than 300. Strangely, it is also the sight of one of the weirdest entity encounters on record. One summer afternoon in 1965, Juan Dominguez and his sister Isabel were passing a rural garden wall. They were each loaded down with bundles of peaches, which they were bringing home, when they suddenly noticed a shadowy apparition skulking near the wall.

The humanoid figure was pacing up and down, and emitting a strange rattling sound. As they drew closer, they saw that the figure was not human. It was over six feet tall with very long arms and seemed to be created of translucent, gelatinous flesh. It seemed to be featureless and its long arms swayed in time to its body, as it paced slowly through the vegetation adjacent to the river.

Suddenly, the slimy, faceless horror stopped and changed direction, shambling towards the terrified siblings, who dropped their peaches and ran away.

No UFO was seen at the same time, and it could be argued that the 'jelly man' could just as easily have been placed in the previous chapter on ghosts, or the following chapter on miscellaneous entities. I would heartily agree, and that goes to show just how nebulous these cases can be. One man's alien is another's ghost, and in what passes for reality, it is probably neither, and both.

ATTACK OF THE SPACE BIRDS
Lyonel Trigano, a French UFO investigator, reported a highly unusual case in *Flying Saucer Review,* volume 14, number 6, in 1968. The events had occurred in 1961. The witness referred to only as 'Mr. S.' was "a solidly-built man in his fifties," who ran a garage in Heraul, France. Trigano took a witness statement from him.

"One evening in November, 1962, I was driving along a minor departmental road in Var. It was a dark night and raining in torrents so I was driving with my lights full on. Rounding a bend, I saw, 80 meters away, a group of figures clustered in the middle of the road. I slowed down to avoid the group and at the same time it split into two parts, suddenly and jerkily. My window was down and I leaned my head out slightly to see what was the matter; it was then I saw beasts, some kind of bizarre animals, with the heads of birds and covered in sort of plumage, which were hurling themselves from two sides towards my car. Terrified I wound up the window and accelerated like a madman and then stopped a further 150 meters further on. I turned round and saw these things, these beasts, these nightmarish kind of beings, which were heading, with a sort of flapping of wings towards a luminous, dark blue object which hung over a field on the other side of the road. It resembled two plates upside down and placed on one another. On reaching it these 'birds' were literally sucked into the underpart of this machine as if by a whirlwind. Then I heard a dull sound (clac!) and the object flew off at a prodigious speed and finally dissipated."

Who in their right mind would make up a story about aliens that looked like birds? It's a shame that we don't have a more detailed description.

THE MUMMIES OF PASCAGOULA

On the evening of Thursday October 11, 1973, 42-year-old Charles Hickson and 19-year-old Calvin Parker, co-workers at the Walker shipyard and natives of Gautier, Mississippi, USA, were fishing near a grain elevator on the shores of the Pascagoula River. Not having much luck, Hickson suggested that they relocate to the rusty iron piers of the dilapidated, old Shaupeter Shipyard.

At about 9pm the pair spotted a blue, glowing oblong shape. As it drew closer, hovering a few feet above the water, and only 100 feet from the witnesses, they began to become alarmed. They both heard the object emit a buzzing noise.

A hatch opened in the object and three deeply weird entities floated out. Hickson said...

"I jumped to my feet, looked over at Calvin, and he looked plumb strange. Then a door opened and this brilliant light came out of it. I couldn't figure what in the world was happening... and all of a sudden, right in the end of it, this opening was laid up there and three of them just floated out of the thing. They wasn't on no ground.

They were about five feet tall, had bullet-shaped heads without necks, slits for mouths and where their noses or ears would be, they had thin, conical objects sticking out, like carrots from a snowman's head. They had no eyes, grey, wrinkled skin, round feet, and claw-like hands... they didn't have toes. But they had feet shape. It was more or less just a round like thing on a leg– if you'd call it a leg.

I was scared to death, and me with a spinnin' reel out there– it's all I had. I couldn't– well, I was so scared– well, you can't imagine. Calvin done went hysterical on me."

The intensely wrinkled skin of the creatures made them look like mummies. One of them made a series of buzzing noises that may have been an attempt to communicate. Two of the creatures grasped the men with their lobster like claws and lifted them off the pier. Parker passed out at this point, another case of the nameless dread.

The creatures levitated, carrying the men with them and returned to the craft. Inside, it was brightly lit and featureless except for an eye like device. Hickson continued.

"Some kind of instrument I don't know what it was. I didn't see anything that I could call an instrument that I've ever seen before... it wasn't like no X-ray machine. There ain't no way to describe it. It looked like an eye. Like a big eye. It had some kind of an attachment to it... It moved right in front of my face. I saw dials and gadgets moving around. It went behind me, then came back over me... it went all over my body, up and down... then it disappeared back into the wall... I was just about out of my mind. I thought they were gonna kill me. Folks would think we fell off in the river and drowned, and nobody would ever know about this."

Parker could not recall what had happened to him while he was inside the ship, though, years later, he would begin to recall some hazy details during post-hypnotic regression sessions.

The wrinkled horrors then returned the men to the river bank. Calvin was crouched on the ground weeping and praying. The object rose up into the sky and vanished.

"When I got to Calvin, I had to slap him a time or two. I finally got him to where he could say something. He said, Charlie, what in the world was that? I said Son, I don't know, but they didn't kill us."

The men spent the next three quarters of an hour in Hickson's car, trying to calm down.

"The only thing I remember is that kid, Calvin, just standing there. I've never seen that sort of fear on a man's face as I saw on Calvin's. It took me a while to get him back to his senses, and the first thing I told him was, Son, ain't nobody gonna believe this. Let's just keep this whole thing to ourselves."

However, after further discussion, the men concluded that the creatures may prove a danger to their country and decided to tell the authorities, a brave decision on their part.

> *"I knew people would call us crazy and everything else, but I thought about it some more and said, 'What if it's a threat to our country?' That's when I decided to call Keesler."*

Hickson first telephoned Keesler Air Force Base, but was told that the Air Force no longer investigated UFO reports and that they should contact the local Sheriff.

Sheriff Fred Diamond and Captain Glen Ryder conducted an interview with the men, after they had taken the long drive to the Sheriff's office in Jackson County. The Sheriff noted that both men seemed very frightened by what had happened.

Captain Glen Ryder secretly taped the two discussing their abduction while they were alone in an interrogation room, following the initial interview. It was thought that the men would slip up and be caught talking about a hoax, but he was shocked to discover that in private, they seemed even more disturbed than they were while talking to the authorities.

Ryder said. *"I wasn't there with them, but I know you don't fake fear, and they were fearful. They were fearful."*

Fearing an exposure to radiation, the witnesses visited the local hospital, only to be told that they did not have the facilities to run the tests required. The men wound up, ironically at the Keesler Air Force Base whom had rejected their call a couple of days before. There, doctors examined them. Following the examination, the men were interviewed by the military intelligence chief of the base. Hickson recalled that the "whole base command" observed the proceedings and that an Air Force artist made a sketch of one of the creatures.

Dr. J. Allen Hynek, a consultant astronomer to Projects Sign, Grudge and Blue Book and James A. Harder, a U.C. Berkeley engineering professor, both interviewed the witnesses and performed hypnotic regression on Hickson. The man became so terrified that the process had to be halted.

> *"Under deep hypnosis, I discovered something that still gives me chills. There were people on that spaceship — living beings in another compartment. They never came in there where we were. And I'm telling you, they looked almost like us... only thing I can figure is that they couldn't live in our atmosphere, so they let the robots come out there and carry us inside."*

Soon after both men passed polygraph tests. This, together with the secretly recorded tapes convinced Dr Hynek that the men were telling the truth.

> *"... there was definitely something here that was not terrestrial... this was the first time I had seen for myself the profoundly disturbing effect of a UFO encounter on two ordinary human beings. It was impossible to be with Charlie and Calvin — or listen to that tape — and not believe that something terrifying had happened to them."*

THE RUSSIAN TRICLOPS

Voronezh is an industrial city located about 300 miles from Moscow, Russia. The city has a population of nearly one million. On the evening of September 27, 1989, at about 6:30pm Levoberezhniy's park in the city became the scene of events that resembled the plot to a 1950's B-movie.

A group of children, including Vasya Surin, Julia Sholokhova, Lena Sarokina, Alyosha Nikonov and Vova Startsev, were playing football in the park. Suddenly a glowing, deep red, ovoid object appeared. They estimated it to be approximately 45-feet wide and 18-feet high. Not only the children, but a group of 35-40 adults, who were waiting at a nearby bus stop on Poutiline St, saw the craft that made a disturbance in the grass as it hovered.

A hatch opened on the underside of the craft and a towering, 9 foot tall figure was seen looking out. It wore a silver jump suit and bronze coloured boots. The thing had a dome shaped helmet, through which three eyes were visible. The middle eye that was set above the other two glowed red and swivelled around like a radar. It also had a disc like object attached to its chest. The thing had legs, arms and a thick set body.

The craft descended, permanently damaging a poplar tree as it did so, and extended landing gear. The entity scanned the area then left the ship, followed by two more of its kind. Then what witnesses described as a box-like robot, with buttons on its front. An alien adjusted one of the controls on the robot's chest, enabling it to walk about in a mechanical fashion.

The entities then began to examine the ground near where their craft had landed and took some soil samples. One of the creatures made some sounds, which some eyewitnesses perceived to be orders. Then a beam of light emerged from the "chief" alien's chest. The beam delineated a number of luminous triangles and rectangles, about 1-foot by half a foot wide, as if indicating something to its comrades.

One of the boys screamed in alarm and one of the three eyed giants fixed him with a gaze that rooted him to the spot. Then the craft and its occupants seemed to dematerialize. About five minutes later, the ship and creatures reappeared in the exact places they had stood before. But this time one of them carried a tube like device in its hand. The creature pointed it towards a 16-year-old boy, and it shot a beam of light at him, causing him to vanish instantly. The creatures and their robot returned to the craft that raised its landing gear, rose up and flew away. The boy who had seemingly been vaporized reappeared at this point.

Soviet Government agencies, scientists and reporters descended upon the city of Voronezh. Those interviewed claimed they had observed this UFO, not just during the incident in question, but also many times between the dates of September 21 and October 2, 1989. The scientists also allegedly found anomalies in the local magnetic field and background radiation, as well as depressions in the ground in the shape of a large rhombus, that must have been caused by something weighing several tons.

THE BRAINS OF PALOS VERDAS
Alien brain creatures have featured in a number of SF films over the years, including *The Fiend Without a Face* and *The Brain from Planet Arous*. But with life imitating art, a duo of giant alien brains once confronted two men from Palos Verdes, California.

Peter Rodriguez and John Hodges, had been visiting a friend, and at 2am on the morning of August 17th, 1971, they were getting back into Hodges' car. As he switched the engine on, the headlights illuminated a pair of brain-like creatures, blocking the driveway. The 'brains' were blueish in colour and surrounded by a vapour. One of the brains was the size of a softball (3.5 inches across) but the other was as big as a human torso. The larger creature had a single red eye.

The scared men drove off as fast as they could. The men lived fairly close to their friend, so the drive was not a long one. Hodges first dropped off Rodriguez and then drove home. Once he arrived at his house he noticed that the time was 4:30am. Somehow, a 10 minute trip had turned into a 2 hour one.

In 1976, Hodges underwent hypnotic regression and surprise, surprise, remembered the events that had taken up nearly two hours of his life five years before. The brain creatures had sent him a telepathic message.

"Take the time to understand yourselves, the time draws near when you shall need to. You shall not remember this incident until we meet again."

After dropping off Rodriguez, he had found the brains waiting for him in the driveway of his own house. Hodges apparently lost consciousness and his next memory was of waking up in what he referred to as a "control room."

As well as the brain beasts, the room was occupied by 7-foot-tall, bald, thin-lipped, grey skinned beings with hands that bore six webbed fingers. Apparently, the brains were organic translators for the tall creatures. Hodges saw holographic images depicting a series of nuclear explosions and a television-like screen upon which were pinpoints of light. The larger brain explained to Hodges that the points represented places on Earth where there was "too much power." He was also shown images of another planet, that had been destroyed due to the fact that it also had "too much power."

The brain went on to tell him that the human race needed to stop abusing their power. After this he felt a buzzing sensation and found himself back inside the car.

Other so called 'aliens' have given out dire warnings about the abuse of nuclear power. But one wonders why they seem to choose anonymous members of the public and not world leaders to abduct and give these stern lectures to.

Hodges also indicated that he believed that the creatures had implanted what he referred to as a "translator cell" in his brain, in order to sustain contact with him. In the weeks that followed, Hodges claimed that he began to receive telepathic messages from the alien beings. They predicted that in 1983, a war would break out in the Middle East. They further prophesized that this war would swiftly spread into Europe, resulting in the use of atomic weapons. Following the war the creatures would establish a public dialogue between their species and the human race. None of this came to pass.

Ghosts, fairies, and messages from Ouija boards have all made similar predictions or promises, none of which seem to be accurate. The paranormal seems to have penchant for lies.

THE CATFISH MAN
A fish-like alien was reported on the evening of October 17th, in that golden year of 1973, in Europa, Mississippi. A 50-foot-wide, dome-shaped UFO hovered 2-3 feet above Highway 82 at twilight, seemingly suspended on a beam of light. The primary witness' car lights went out, and the engine died when it was 100 yards from the craft. Another car stopped just behind him. A second UFO hovered about 60 feet above the first, illuminating it with a light. Both craft were similar, like inverted cups, and had greenish blue flashing lights. A catfish-like creature came out from the top of the lower UFO, holding onto a handrail. It had grey, fish-like skin, a wide mouth, one glowing eye, flipper-like feet and webbing between the legs, like a flying squirrel. It had feather-like object on its back which opened and closed when it moved.

THE UGLIES FROM TEETONIA
Pier Zanfretta was a 26 year old night watchmen from Torriglia, Italy. On the night of December 6th, 1978, he was driving along the icy roads in his patrol car, en route to the currently unoccupied country home of a client, Dr. Ettore Righi, when the engine, radio and lights of his vehicle all simultaneously, and inexplicably, died. Electrical interference raises its head once more. You can guess what happened next.

Zanfretta saw four lights moving around the Righi house and suspected burglars. Gun at the ready he slipped out of the car, ready to tackle the intruders. Feeling a touch on the shoulder the security guard span round expecting to face a human intruder, instead he saw...

"An enormous green, ugly and frightful creature, with undulating skin... as though he were very fat or dressed in a loose, grey tunic... no less than 10 -feet tall."

The huge creature had greenish skin, points on the sides of their faces, rounded fingertips, monstrous, yellow triangular eyes and red veins across the forehead. The thing had spines on the side of its head. It also seemed to have some sort of breathing apparatus on its mouth.

The man ran for his car and a bright light shone from behind him. Looking back he saw a triangular craft. A wave of heat swept over him as he scrambled back into the car. He managed to get the radio working and contacted his security company's centre of operations in nearby Genoa.

At precisely 12:15am, when Carlo Toccalino, the security company's radio operator, testified that Zanfretta was speaking in a confused and agitated fashion. On asking Zanfretta to describe the men he replied...

"No, they aren't men, they aren't men... my God, are they ugly!"

Then communication was cut off and Toccalino sent another patrol car the check on the situation.

The icy road slowed the patrol car down. It was at about 1:15 in the morning when the two night watchmen, Walter Lauria and Raimondo Mascia, discovered Zanfretta lying prone on the frozen ground in front of the Righi house.

Zanfretta awoke and seemed terrified, shining his torch at his colleagues he raised his gun, eyes bulging with fear and seeming not to recognize them. Luckily Mascia and Lauria managed to overpower the panicked man. They noted that he felt hot despite having been lying on the frozen ground.

The Carabinieri, the Italian military police, were dispatched to the area and found two 9 -foot wide, horseshoe shaped marks in the frozen ground. The commandant of the Torriglia station, Antonio Nucchi said of Zanfretta ...

"I can state with certainty that he is a clear thinking man with no strange fantasies in his head. When we went to investigate the scene the next day, he almost didn't want to come, he was so scared. Only something exceptional could have frightened him so."

During the investigation, Nucchi also revealed that 52 Torriglia citizens had reported seeing a bright, glaring illumination emanating from the direction of the Righi house at exactly the same time Zanfretta testified to watching the triangular UFO.

On December 23rd, 1978, Zanfretta, encouraged by a supporting reporter Rino Di Stefano underwent hypnotic regression. It was conducted by Dr. Mauro Moretti, a psychotherapist, and member of the Italian Association of Medical Hypnosis. Underneath, he recalled that the huge creatures kidnapped him and transported him into a hot, luminous location where they thoroughly examined and interrogated him.

Using strange luminous device to translate what they were saying. During the same session, Zanfretta also indicated that the creatures came from the planet *"Teetonia,"* which was located in the third galaxy, and that they want to talk with us and that they will soon return in larger numbers.

Three nights after the first hypnotic session, at 11.45pm, Zanfretta was abducted a second time. Whilst driving his patrol car through the Bargagli tunnel, near the Scoffera Pass, when he suddenly lost control of the vehicle. He radioed in saying his car was out of control and driving itself. The breaks and steering wheel did not respond and the car drove down a steep incline through a wall of fog.

Finally the car screeched to a halt, and Zanfretta smashed his head on the steering wheel, as the car was bathed in light and heat. Zanfretta and his vehicle were discovered at 1:10am by another pair of security guards. The first man to spot Zanfretta was Sergeant Emanuele Travenzoli. Travenzoli stated that he found Zanfretta in a field near the road and, despite the continuing rain his clothes were warm and dry. Zanfretta was in a state of shock; quivering and weeping. He said...

> *"They say I must leave with them. What about my children? I don't want to... I don't want to."*

The Carabinier was called in again. Despite the cold, they found the man's car was hot as an oven, both inside and out. The police also discovered that the Fiat was surrounded by inexplicably huge boot prints measuring 20 inches long, by 8 inches wide. The Carabinieri then came across Zanfretta's revolver, a Smith & Wesson 38 Special, which had been fired five times.

The Carabinieri informed the Italian Department of Interior and other military commands, of the incident. In December of 1978, there were so many UFO sighting across Italy that Falco Accame, a former member of the Italian Parliament, asked both Italy's Premier, Giulio Andreotti, and Minister of Defence, Attilio Ruffini, to inform the Italian Congress about their opinion concerning the nature of the recent UFO sightings and what threats they may pose to the citizens of Italy.

Zanfretta's employers asked prominent neurologist, Dr. Giorgio Gianniotti, to examine Zanfretta. His diagnosis was: *"The man is in a state of shock, but he is perfectly sane."* Dr. Mauro Moretti once more put the witness under. During this session, Zanfretta recalled being stripped and forced by the creatures to wear a strange helmet, which enabled him to understand their language, but caused him great pain. One of the aliens also took his gun and tested it by firing the bullets into a panel,

The abductee told his captors...

> *"I know that you need me, but I don't want to. I like to be alone. I have two children. I feel good this way... and after all you are not human beings. You are horrible. I know you are trying to come more frequently... no, you can't come to Earth, people get scared if they look at you. You can't make friendship."*

Zanfretta may have been a first rate security guard, but he would make a lousy intergalactic ambassador!

On the night of July 30th, 1979, Zanfretta was on a motorcycle patrol in the residential area of Quarto, in Genoa, far from Torriglia. Again, two of his fellow guards were sent to find him. They found him after a two hour search on the summit of nearby Mount Fasce. The local people claimed that they had not seen the young guard or his motorcycle travel up the single road that led to the top of the mountain.

Another session of hypnotic regression was conducted at International Centre of Medical and Psychological Hypnosis in Milan where, on his own request, Zanfretta was injected with sodium penathol by Prof. Marco Marchesan. While under the serum's effects, Zanfretta claimed that he was lifted from the ground into to the alien spaceship by a mysterious green light.

Professor Marchesan confirmed that: *"No human being can knowingly lie while he is under Pentothal treatment, so I think it's very probable Zanfretta had these encounters."*

At about 10:30pm on Sunday, December 2nd, 1979, Zanfretta disappeared for the fourth time while driving in the suburbs of Genoa. This time, however, there were other witnesses. While driving in the hills of Genoa, searching once again for the serial abductee, four patrol guards claimed to have clearly seen a strange, cloud-like object floating above them. Suddenly, two beams of light seemed to emanate from within the large cloud, illuminating the patrol cars below. The car's engine stopped and the men got out to get a closer look at the cloud. Chief, Lt. Cassiba became afraid and shot his pistol into the cloud. The cloud and the lights vanished instantly.

One of the other witnesses Germano Zanardi was so frightened by the encounter he later committed suicide by shooting himself.

The following day things got even weirder. Zanfretta got out of his car at a petrol station in Genoa when he heard somebody calling his name. Turing around he saw a tall, humanoid figure with a bald, egg-shaped head, who was dressed in a chequered suit, that included a chest plate made of steel where the shirt should have been. The strange figure ordered him to drive his vehicle into a small cloud that was hovering just above the ground. He felt compelled to obey. The witness and his patrol car were levitated within a cloud, and deposited onto a huge spacecraft. On-board the ship the guard was allowed to explore with the company of the aliens.

On-board the craft, Zanfretta claimed to have seen large, transparent cylinders filled with a weird blue liquid. One of the cylinders was said to have contained a frog-shaped creature, which the aliens explained was: *"An enemy of ours from another planet."*

This creature recalls the frog like entities reported in other cases. Could they be the same species?

Zanfretta observed a large bird-like creature and another humanoid figure that he described as looking like a caveman. Could these be related to the bird like beings reported from the Var region of France and a bigfoot?

One of the giant beings attempted to give Zanfretta a transparent sphere, with what appeared to be an electrically charged pyramid inside. The aliens claimed that utilizing the sphere would enable human beings to comprehend who they were and how they live. Zanfretta tried to refuse the gift, stating that : *"had enough of all these strange encounters and wished only go back to his normal life."*

The creatures insisted he should take it, and told him to present it to a man he had never heard of, one Dr. J. Allen Hynek! The famed American researcher never received the device. For reasons best known to himself Zanfretta hid the object in the mountains of Genoa.

Zanfretta was abducted again on February14th, 1980, after which he was found by his colleagues in a state of shock and suffering from mild hypothermia. A villager living nearby stated that mere minutes before the rescuers arrived, he spied a huge, radiant mass in the sky.

During the next hypnosis session, Doctor Moretti found Zanfretta to be uncooperative, unlike in his past sessions. While hypnotized, he claimed he was contacting the aliens and began to speak an odd, unknown language. His voice became guttural and he uttered strange phrases like:

> *"Question with negative answer, tixel... you can't work out anything in a case like this. To believe or not to believe doesn't mean anything: each thing in its own time."*

Zanfretta vanished yet again on August 13, 1980, but was found before he could be taken by the beings. This was apparently the last time Zanfretta had his life interfered with by the creatures. The case is one of the strangest on record. As well as over 60 other witnesses, we must remember that one man was so disturbed by what he saw, that he took his own life.

THE CYBORG, CYCLOPS SPACE PIGS

Harrah is a small town of around 600 people. It is located in the county of Yakima in Washington State. On the morning of January 19, 1977, the sleepy little town had some very strange visitors. 9-year-old José Cantu woke up around 6.30am and decided to make himself breakfast rather than disturbing his parents. Whilst fixing himself breakfast, he glanced out of the window to see a 'little man' outside.

The boy rushed to tell his mother, Martha who, dismissing it as a prank, refused to get up and look.

The brave lad decided to go outside alone to get a closer look at the visitor. Hiding behind an old washing machine, he observed one of the weirdest 'aliens' on record. The thing had green skin, a single eye in the middle of its forehead, a mushroom-shaped antenna that grew out from under a head of bristly hair and a pig-like snout. It had vestigial arms growing from a round torso and no legs. The creature sat in a metallic chair connected via a metal tube to a wide base, which seemed to have wheels on it, allowing the creature to move about. The creature was about three feet tall.

José noticed two 'steely' craft, one parked on the lawn and another on the flat roof of his house. Each housed another space pig and had a ramp leading into it. The interiors were lit up by a brilliant light. José claimed that the access portals on the gleaming ships opened in *"two parts, like a cross,"* revealing an interior just big enough to contain two seats perched on tall bases.

The first creature roved about the garden before trundling up the ramp of one of the ships. The door closed and the craft rose up before vanishing in what seemed to be a puff of smoke.

Running back inside, he dragged his mother downstairs, telling her about the one eyed, green creatures. On finding no evidence, she packed him off to school. José told classroom aide, Diane Gomez, about his bizarre encounter. Later, when local reporters contacted Gomez for a story about these odd events the aide spoke highly of him.

> *"José is a serious boy. He's not one that tells stories or lies. What he told me, I took very seriously."*

Gomez took the story so seriously, that just after 10:00am, José's scheduled playtime, she and another aide decided to accompany the boy home and take a look at the alleged landing sites for themselves. The aides greeted his mother and they followed José through the yard as he indicated where he saw the creatures. The place where José insisted the creatures were rotating on their bases still bore two circular marks embedded in the gravel. The ladies also saw two, seemingly inexplicable, sets of indentations, consisting of three holes each, where the other being had stood.

After the aides returned to school with José, Martha called her neighbour, Irene

Sanchez, and asked her to come over. When Sanchez arrived at Martha's house, she claimed that both she and her brother had seen a strange, spherical impression 10 feet across in the Cantu's backyard from their own adjacent home. Sanchez and Martha examined the area, and discovered that the uncut grass in the centre of the circular pattern was swirled upward, looking like what in later years would be called a crop circle. Sanchez also testified to seeing the same marks that the aides had also observed.

The evidence was still clearly visible when Jose's father came home from work.

The next day the Cantu home was visited by Willard J. Vogel, a police officer for the Indian Reserve of Yakima, but also a UFO researcher. He was accompanied by fellow UFO researcher and electrical engineer, David Akers. They found the marks still in evidence. They told the family that there had been reports of strange lights in the area.

Frances Story, a journalist with the local *Toppenish Review* covered the story and interviewed the boy. He was impressed with the answers José gave him and commented *"I believe he saw what he said he saw."*

TRAIL OF THE STAR SLUG

Stan Gordon, who we met in chapter one on the track of Bigfoot and UFOs, recorded a very strange case in 2012. He was contacted on May 8th of that year by a man who had recently seen a UFO. At 11.10pm, on the previous night his living room had been illuminated by a white light. Looking out of the window he saw that the light was emanating from a weird craft. Stan interviewed the man by phone.

> *"One of the first questions the man asked me was if the object could be a 'weather plane' since there was lightning off in the distance. During our initial interview, the man suddenly brought up that until I had called him back to talk about what he had observed, he had blanked out the UFO sighting encounter and had forgotten about it, and this seemed disturbing to him. Upon seeing the object, the man attempted to video tape it as it moved off in the distance. Unfortunately, very little detail could be seen on the short video. Researcher and electronics specialist Jim Brown attempted to try to work with the footage to obtain more details but was unable to do so. Jim stated, 'There is simply not enough resolution to determine any detail to it."*

Stan's friend Keith Bastianini, an archaeologist and graphic artist, went to the location to conduct an on scene investigation and a first-hand interview with the man.

The UFO resembled an American football, with the pointed ends aligned vertically. It had two perpendicular light arrays in the middle. The witness estimated each light array measured approximately 30 feet in length. Each array supported seven round lights which reminded the witness of baseball stadium lights. A square-lined grill covered each of the lights. Three darker bands encircled the craft. The witness noted the object appeared to be of seamless fabrication.

The man grabbed a camera and ran outside to film the object. He shot video of the departing object until '*...it made a right turn like a car would...*' and shot from sight. The witness observed that the rear of the object looked different from the American football shape he first observed as it moved away. It was more like a rectangle with a triangle shape beneath it. Also of note, only five lights that blinked alternately red and white were now visible.

For 2 days afterwards, the witness experienced periods of confusion and forgetfulness during conversations or other activities. He attributed it to post-excitement, adding he's never experienced anything like it before. The witness also reported that he didn't sleep well that night expecting 'them' to come back for him.

The witness also reported a strange creature sighting earlier in the day. He speculated: '*They (the UFO occupants) left something – I think that's why they were here.*' The witness then led Bastianini to a rubble pile of concrete fragments adjacent to the area where the UFO hovered, and he recounted finding a large slug-like creature slithering in the weeds. The 'thing' was shiny black, like a slug, with no apparent eyes or appendages. It measured between 2.5 - 3 feet in length and 7 - 8 inches in thickness. As the witness watched the 'thing' slithering around in the weeds, it slithered into a hole in the rubble pile. He reported checking the area periodically since the sighting, hoping to capture a picture of the 'thing.'

Well, we have met a whole zoo-full of strange creatures. Space penguins and space pigs, vegetable men, evil blobs, disembodied brains, lizard men, and three eyed giants. Can we truly believe that all of these many alien races are visiting Earth? Their behaviour seems to make little sense, some act in an almost animalistic manner, others select random people over world leaders to give important messages to. Others make dramatic prophecies that never come true. Not what you would expect from supposedly superior beings. Maybe something much stranger than alien visitors is at work, as fantastic as that would be.

THE COMING OF THE ROBOTS
The word 'robot' comes from the Slavic term 'robota', meaning forced labour. In 1920 the Czech writer, Karel Čapek, penned a play called *R.U.R,* standing for *Rossum's Universal Robots.* The plot revolved around artificial humans created from organic material rather than metal, somewhat like the replicants in *Bladerunner.* They are used for slave labour, but gain sentience and revolt.

Even earlier, Lyman Frank Baum had featured metal, mechanical men in his series of *Oz* books, including The Tin man and Tik-Tok but the term Robot was not then in use for such beings.

After Čapek's play, robots began to be more widely used in fiction and were generally metallic creations of steel and chrome. Fritz Lang's *Metropolis* hit the screens in 1927 and cemented the idea of metal robots. These mechanical people became a mainstay of science fiction and remain so to the present day. They can range from endearing characters like C-3PO and R2D2 of *Star Wars* and K9 the robot dog of *Doctor Who* to menacing killers like the Voc robots, also in *Doctor Who,* and the brutal robot police in *THX 1138.*

It should come as no surprise then, that many so called extra-terrestrial encounters are with seemingly mechanical beings.

THE STEEL FLOWERS

The earliest example of which I am aware occurred in Florida in 1924, a scant four years after *R.U.R.* The story broke in 1974, when a beauty salon operator named Evelyn Wendt broke her silence about what had happened to her some fifty years before.

Wendt contacted *"The Weekday"* a West Palm Beach newspaper and claimed that, in 1924, as a girl she had a remarkable encounter that she still recalled clearly. Whilst playing on the lawn of St. Joseph school in Pasco County, Florida, Wendt claimed that she suddenly noticed a glowing egg-shaped object resting on the ground nearby. The object was so bright that she had to shield her eyes. The egg-shaped light faded revealing a pock-marked and leaden saucer. Then a hatch opened in the saucer and astounding little figures emerged.

> *"Little people emerged. I think they were robots. I tried to count them, but they charged about so. They were smaller than I was and resembled flowers with faces where the bud would be. Remember, I was just a bitty thing then, and kids don't fear flowers."*

The tiny robot flowers had some kind of weapon that they were aiming at the school science building. The dinky droids telepathically communicated with the girl, telling her that there were experiments going on in the science building, and they needed to be stopped before the place was destroyed.

For unknown reasons, the flowers took their weapon back to the saucer without firing it at the science building. One of the entities invited Wendt to come with them, but the little girl declined. The robot flowers promised to return in 35 years, an oddly specific length of time. They boarded their ship which rose up, turning a bright silver then vanished.

Sometime after her meeting with these plant-like robots, the science building that allegedly housed the experiments that had so angered the artificial blooms, had been left in what she called "shambles." Coincidence? But why would robot, alien flowers want to destroy a primary school science building?

In the 1970s, UFO researcher Stephen Putnam attempted to hypnotize Wendt to try and uncover more details, but he was unable to put her under.

THE ROBOT ASPARAGUS
Alien robots seem to like imitating plants. As well as flowers, we now have robot asparagus. In July of 1951, pilot Fred Regan was found next to his crashed plane. The craft had fallen from a height of several thousand feet, and had struck the ground so hard that its engine was embedded six feet under the earth. Fred, however, didn't have a mark on him. He also had no parachute.

Fred had been flying his Piper Cub plane over Georgia when it had been struck by a lozenge-shaped, pulsating UFO. His wrecked plane had begun to fall, but was caught in a tractor beam shot from the ship. He described it as a sticky, clinging force that drew him into the UFO. He found himself in the company of several three-foot-tall robots that he said looked like giant stalks of metallic asparagus.

The robots spoke to him in English, apologizing for the accidental crash. They gave him a medical examination and then told him that they had removed a cancer he didn't even know he had. They then put him to sleep, and deposited him in a farmer's field next to the wreckage of his plane.

Less than a year later, Fred Regan died in the Georgia State Asylum for the Insane. The cause of death, according to the report, was 'degeneration of the brain tissue due to exposure to extreme atomic radiation'.

It looks like the robot space asparagus gave him cancer rather than curing it. This is one of the rare cases of a human death associated with Fortean phenomena.

THE FLATWOODS MONSTER
Of all the robotic 'aliens' that have manifested over the years, the Flatwoods monster is by far the best known and most terrifying. The Flatwoods Monster has become a Fortean icon, lending its strange likeness to tee-shirts and toys. There is even a museum dedicated to it!

At 7:15pm, on September 12, 1952, three boys, Edward and Fred May, and their friend Tommy Hyer, witnessed a bright object cross the sky. The object came to rest on land belonging to a local farmer, G. Bailey Fisher. Once they saw the thing land, the boys went to the home of Kathleen May, where they told their story. May, accompanied by the three boys, local children Neil Nunley and Ronnie Shaver, and West Virginia National Guardsman Eugene Lemon, went to the Fisher farm to investigate.

One of their dogs ran ahead out of sight and started barking, and moments later ran back to the group with its tail between its legs. After travelling to the top of the hill, they reportedly saw a large pulsating "ball of fire" about 50 feet away. They also encountered a vile smelling mist that made their eyes and noses burn. Lemon then noticed two small lights over to the left of the object and directed his torch towards them. The lights were, in fact, eyes.

A dreadful thing glided towards the group. It towered 15 feet tall and appeared to be a robot rather than an organic being. It had a dark cowl in the shape of an ace of spades behind a red round head. Set in the head were two eyes, described as 'portholes,' glowing green-orange and the size of half-dollars. The 'body' was a metallic armoured structure, lined with thick vertical pipes. A green metal skirt formed the lower part of the body. Some of the witnesses said it had small arms, others failed to notice these.

The robot monster emitted a hissing noise and glided towards the group, who panicked and ran. Upon returning home, Mrs. May contacted the local sheriff and a news reporter. The reporter conducted a number of interviews and returned to the site with the farmer later that night, where he reported that "*there was a sickening, burnt, metallic odour still prevailing.*" The sheriff and his deputy searched the area separately, but found no trace of the craft or creature.

After the event, investigators associated with the research group Civilian Saucer Investigation obtained a number of accounts from witnesses who claimed to have experienced related phenomena. These accounts included the story of a mother and her 21-year-old daughter who claimed to have encountered a creature with the same appearance and odour a week prior to the September 12 incident; the encounter reportedly affected the daughter so badly that she was confined to a hospital for three weeks. They also gathered a statement from the mother of the local farmer in which she said that, at the approximate time of the crash, her house had been violently shaken and her radio had cut out for 45 minutes, and a report from the director of the local Board of Education, in which he claimed to have seen a flying saucer taking off at 6:30 on the morning of September 13th.

There is a myth that the May's family dog died after the encounter. In fact, though it vomited, the dog was fine afterwards and lived for many years.

THE RAILWAY ROBOTS
1954 saw one of the most intense UFO flaps in history and the focus was France. One of the best-known encounters occurred to Marius Dewilde, on September 10th. At 10.30pm Dewilde, a metalworker from Quarouble, in the Department of the Nord, stepped out of his house to see what his dog was barking at. Shining his torch along the railway tracks near his house, he saw a cigar shaped object some 23 feet away, resting on the tracks. It was 20 feet long. Hearing footsteps, he looked around and saw two 4 foot tall robot-like beings walking along the tracks.

They seemed to have legs but no arms and were clad in what looked like diving suits. Each had a large helmet. The light from his torch was reflected back as if the helmets were mirrored. As the witness moved closer to the tracks, a green ray was shot out of the craft, paralysing Dewilde. By the time he recovered, the two little robots were gone and the craft was rising and changing colour as it went.

Dewilde suffered from respiratory problems after the sighting, and his dog sadly died three days later. Three cows in nearby farms were found dead, and their post mortems revealed that their blood had been totally and inexplicably removed. Also, several local people claimed sightings of objects and creatures similar to the ones witnessed.

Investigators from the French Air Force soon arrived and interrogated Dewilde. They found cuts on the wooden railroad ties indicating that something weighing 30 tons had sat upon them. The gravel beneath this area was brittle as if calcified by great heat.

The local mayor, Lucien Jeune, and the village council sprang into action, to protect their vines, of course. They quickly passed a municipal decree to keep aliens out of the local skies and the vineyards.

Article 1. — The overflight, the landing and the take-off of aircraft known as flying saucers or flying cigars, whatever their nationality is, are prohibited on the territory of the community.

Article 2. — Any aircraft, known as flying saucer or flying cigar, which should land on the territory of the community will be immediately held in custody.

Article 3. — The forest officer and the city policeman are in charge, each one in what relates to him, of the execution of this decree.

THE BLOCKHEADS
On December 16th, 1957, near Old Saybrook, Connecticut, retired teacher Mrs. Mary M. Starr was awoken around two to three in the morning by a bright light. Peering out of her bedroom window, she thought at first that she was looking at a downed airplane, until she realized that the object was hovering above her clothesline and no more than 10 feet from her house.

Through lighted portholes in the side of the craft, she could see two weird figures moving along in opposite directions, each with their right arm raised. They were 4 feet tall with square, see-through helmets that contained a red glowing sphere. The thing's arms were featureless with a rounded end instead of a hand. They seem to have rubbery bodies, flared out like skirts. She could see no legs. A third figure appeared at one of the portholes. A sparkling antenna arose from one end of the craft and the entire hull began to glow for several minutes. Finally, the glow faded, the antenna retracted, and the object took a complicated path through the air before shooting straight up without a sound.

It would seem that the whole event was staged just for the witness, but to what end?

THE CISCO GROVE SEIGE

Three friends went bow hunting in Cisco Grove, Placer County, California, on 4th September 1964. Donald Shrum and his two friends Vincent and Tim. During the course of the afternoon's hunting, Shrum became separated from his friends. During his journey back to camp, Vincent claimed to have seen a 'slow moving meteor'. When Tim and Vincent got back to camp, they were not worried about Shrum, as he was an experienced woodsman and passing the night in the forest would be no great trial for him. Besides, it had become too dark to search for their friend.

As for Shrum, realizing he was lost, he decided to climb into a large tree to avoid bears. As night fell, he saw a white light moving below the horizon and thought it was a torch held by his friends as they came in search of him. Then he realized that the light was above the trees and thought that it was a helicopter, called by the others to rescue him.

He descended the tree and lit several small signal fires. But then he realized that the 'helicopter' was utterly silent. He quickly returned to the tree. The light was attached to an object that hovered over a valley. It was 150 feet long, oval-shaped, and had several illuminated, rectangular panels. It ejected a dome-like object with a flashing light attached to it. Soon after, Shrum heard crashing in the bushes that was drawing closer to him.

Soon a humanoid figure emerged. It was five feet tall and wore a one-piece silver suit that puffed out at the joints. Its face was dark and flat with a nose low upon it. The eyes were large and dark and reminded the witness of welding goggles. He had nightmares about the thing's eyes for some time afterwards.

The entity began to fiddle with a manzantia bush, and was soon joined by another identical being. A cooing and hooting began and the beings seemed to look back to the ship. The noises seemed to be some kind of communication.

Then the odd pair were joined by a different being. It looked like a metallic robot with jointed fingers and limbs. It had large, glowing orange eyes and large, square, mechanical jaws. The droid brushed away the remains of Shrums' signal fires. Then it opened its jaws and spat out a cloud of odourless white vapour. The smoke made the witness black out and when he awoke he felt sick. Shrum decided to fight back and shot an arrow at the robot. The arrow bounced off the robot's metal body causing a flash. The droid staggered back a couple of feet and its masters scattered.

Soon all three were back. Shrum shot two more arrows at the robot, but they had little effect. Having run out of arrows, the treed hunter lit a match and tossed it down. The entities fled back and Shrum saw that the ship shot upwards as well. This is an odd little detail. Why would a robot and two aliens from an advanced society fear a lit match? Why would the ship also withdraw as if all three were part of the same whole entity?

Shrum, seeing their reaction, began to set fire to every flammable thing he could, and threw it down towards the trio of weird visitors. This included his camouflage clothing, his hunting license, and his money. The fires scared the creatures away, but each time it waned they began to creep back again. The robot spewed more white smoke causing Shrum to black out again.

When he awoke, the two 'biological' entities were trying to climb into the tree. One was giving the other a boost by cupping its hands as its peer used them as a step. Shrum shook the tree causing them to fall. The robot spat more smoke and the hunter passed out again. When he awoke, the determined creatures were trying to clamber up the tree again, so he threw more things down. Now out of flammable items, he threw his flask at them, as well as coins, some of which the entities picked up.

Shrum says he saw another four of the biological creatures wandering the night time woods. Another robot of the same build as the first joined the trio. The robots stood side by side and arcs of light flashed between them. A cloud of vapour engulfed them and Shrum passed out again.

When the hunter awoke, entities, droids, and spaceship were gone. The besieged man staggered back to camp.

Shrum suffered PTSD from the affair and never sought out any publicity. He refused large sums of money offered for his story. Apparently, he was afraid of losing his job as a welder at the Airojet General Corporation. At the time this company held contracts with the US Government helping to construct rocket engines.

Sometime later he revisited the area and found his spent arrows, one of which had a blob of metal on its head. The Air Force took the sample away. He was haunted by nightmares afterwards and developed a phobia of owls.

THE COSMIC BEER CANS

James Townsend was out driving on the evening of October 23rd, 1965, near Long Prairie, Minnesota. He came upon an upright rocket ship, akin to the ones seen in old science fiction serials, like *Flash Gordon*. It was balanced on three fins. It was 30-40 feet high and 10 feet across. Realizing that nobody would believe him if he said he had seen such a thing, he decided to knock the rocket over so it could be kept as proof. At this point his car stalled.

Townsend got out of the car and began to approach the rocket on foot. As he drew closer, he saw three tiny robots, no more than 6 inches high and shaped like beer cans. They waddled along on two fins and when they stopped, a third fin was extended to act as a stabilizer. They had jointed arms that reminded Townsend of matchsticks. Despite not seeing any eyes on the robots, he had the distinct feeling that they were watching him. After a standoff, the cans scuttled back to the rocket which blasted off with a colourless light.

Townsend drove to the sheriff's office to report what he had seen. Knowing the witness had a good reputation, he investigated and found three strips of oily material on the ground at the site. Two hunters also came forward to say they had seen a bright object rise up in the area at the time of the incident.

THE ASHBURN BUBBLEHEAD

On October 19th, 1973, a woman was driving on Interstate 75, near Ashburn, Georgia, when her engine, power brakes, and steering quit. No UFO was seen, but a small, metallic man appeared after she had pulled the car to the roadside. It had a bubble-dome head with rectangular eye openings. The head moved like a robot. From the elbows down, the arms were narrow and wrinkled, like a chicken's legs. It moved around the car, then was gone. Afterwards she found the engine billowing smoke and the hood intensely hot.

This case brings to mind the effect the Bigfoot had on W.C 'Doc' Priestley's car in the case we looked at in chapter one.

THE PROSPECT SLABS

On January 26th, 1977, Lee Parish, 19, was driving his jeep at 1.05am near Prospect, Kentucky, when he saw a fire-coloured, rectangular UFO hovering over the treetops. It was 40 feet long and 10 feet wide. In a by now familiar scenario, his car failed and seemed to go out of control. It stopped beneath the UFO.

Next thing he knew Parish was back at his home. What should have been a 7 minute drive had taken 35. His eyes were bloodshot and sore.

Later, under hypnosis, he recalled that the soreness of his eyes had been caused by looking at the UFO. His jeep had been drawn up into it and though he had no recollection of opening the doors, he was transferred to a large circular room.

Three entities stood in the room. The largest was 20 feet tall, seemingly too large to fit in the ship. It looked like a giant, black tombstone with a small featureless head. Its texture was rough in some places, but smooth in others. It had one, single jointed arm with no hand.

The second being was shaped like a Coke machine and was red. It was less than 6 feet tall and had a single, unjointed arm. Parish somehow felt that it was afraid of him. It glided up to him and touched his head causing a cold, stinging sensation.

The third being was a glowing white in colour, and bulkier, it's front protruding like the boot of a car. It stood 6 feet tall and had a triangular head, shaped like a slab of cheese. It appeared to have two arms but did not use them. Parish thought that this one was the leader of the group.

The three robots merged with each other. The red one going behind the white and merging, then the white going behind the black and merging. Then the huge black slab moved backwards and vanished. Parish felt a warmth and found himself back in the car with a feeling that the slabs would one day return.

THE PACIENCIA MONOPEDS

It was 2.20 am on 15th of September 1977 and bus driver Antonia L Rubia was on his way to work in Paciencia, Rio Janeiro, Brazil. He spotted a 235-foot wide, hat-shaped UFO over a football field. As he turned to run, a blue light hit him and he was immobilized.

La Rubia found himself surrounded by three robots. The things stood 4 feet tall and had heads shaped like rugby balls stood on their ends. Their height was extended a further foot and a half by antenna that sprouted from the tops of their heads and ended in tiny swivelling dishes. Horizontally across the head of each was a band of mirrors or lenses, light blue in colour with one a darker hue. The robots had a bulky torso and were covered by aluminium-coloured scales, as were the heads. Two tentacles sprouted from either side of the broad torso. They reminded La Rubia of elephant's trunks and each ended in a flattened, finger like projection. They had belts with hooks that held syringe like objects. The robots lacked true legs and their torsos ended sat upon a single stiff pedestal with a wide, circular platform at the bottom on which they trundled about. This design resembles the cyclops pig creatures' chairs as reported by José Cantu in the same year.

One of the robots pointed a syringe at the witness who then found himself in a corridor inside the spaceship. Looking through a transparent wall, he could see they had taken off. In a hall area, he saw a huge, piano shaped device that held another two dozen of the robots. La Rubia says that he shouted at the beings, demanding to know what they wanted from him, but that they all promptly fell over as though injured by his voice. Perhaps they were vulnerable to loud noise. One of the robots shot him with a beam of blue light and he was rendered unconscious.

When he awoke, the robots projected a series of images on the wall. This clip show included scenes of La Rubia (some of them nude), a dilapidated train entering a tunnel, a flying saucer assembly line, and footage of a dog attacking one of the robots only to be shot with a blue beam. The dog turned blue and melted! During this sequence, one of his captors drew blood from his middle finger and squirted the contents at the wall, forming a strange pattern of three circles and an "L."

Suddenly, La Rubia found himself on the street opposite the bus station. One of the robots stood beside him, but when he turned around it had vanished. He saw an object like a lead balloon rising up into the sky. He experienced vomiting, pains, and dizziness for a few days.

THE JAPANESE BRAIN BLASTER

On October 3rd, 1978 Mr Hideichi Amano, who ran a snack bar called 'Juri' after his young daughter, drove to the top of a mountain near Sayama City, Japan. Mr Amano was a radio ham and wanted to get unobstructed transmission and reception of radio waves, while communicating with his younger brother who lives far away in Yamanashi Prefecture. The conversation began at about 8.50 p.m., and after they had finished, Mr. Amano chatted around with a few of his local "ham" friends before he decided, as he had two-year-old Juri with him, that it was time to go back down the mountain. It was at that point that the interior of the car became very bright. The interior was illuminated by a brightness tens of times more intense than he normally had from the fluorescent light that he had fitted into the car.

Amano stuck his head out of the car to look for the source of the illumination, but was amazed to see that the light was completely localized in his car. None of it was shining out through the doors or windows. Turning his head back into the car he was horrified to see that Juri, who moments before had been standing on the driver's seat beside him, was now lying on the seat foaming at the mouth.

Then he noticed that his stomach was being illuminated by an orange light. The beam was shining through the car window and seemed to be coming from a point in the night sky. At that moment he felt something metallic press against his temple. Looking sideways, he saw a robotic figure next to the car window. It was about 5 feet tall with a round, noseless head and eyes like two blueish-white lamps. There was a triangular depression on its forehead. It had pointed ears or fins on the side of its head and the head fitted directly onto the body. The metal tube attached to the witnesses' temple emerged from the robot's face, where the mouth should have been.

The thing made a noise like playback from a speeded-up tape recorder and seemed to be trying to transmit a message directly into his brain. When he tried to start the car, it would not work. All the time, the crazy "space message" continued to flow into his brain and he was immobilized, sensing all the time that his brain was gradually becoming more and more vague.

After an estimated four or five minutes, the figure of the robot began to dim out gradually until it had vanished; the round orange light on his stomach disappeared while the fluorescent interior light also returned to its normal brightness. Then, suddenly, all the things he had switched on began to work: the headlights came on, the car radio and stereo began to play. Later Mr. Amano's watch was found to have stopped at 9.57 p.m.

Amano drove straight back down the mountain. It was here that Juri stood up and said: *"I want a drink of water, papa!"* He drove on home and, having handed his daughter, who was now sleeping, to his wife who was still serving in the snack bar, crept into his bed complaining of a severe headache.

Some days later, professional hypnotist, Mr. Akio Morihe put Amano under hypnotic regression. He remembered having been asked by the creature to return to the same spot on the mountain top so that they could meet again. Apparently the second meeting never happened.

THE SCOTTISH SEA MINES

Bob Taylor was a forestry foreman who worked in Lothian, Scotland. It was on November 19th, 1977 that he had his run in with the strange. He was checking up on trees in a clearing at Dechmont Woods and took his dog with him. It was about 10.30am, broad daylight. He walked about a quarter of a mile into the forest.

The first he noticed was a smell *"like burning brakes."* Then he saw a domed craft 20 feet across. It was *"very dark grey colour"* and had an outer flange with arms on which were mounted propellers. The texture of the dome seemed to change as he looked at it, changing from smooth to rough and back again.

Then two spherical objects came tumbling out from beneath the craft. They were the size of beach-balls and each bore six projections. They reminded Taylor of old fashioned sea mines. The objects rolled towards the man making 'sucking' sounds as they came. The objects rolled right up to the forester and hooked his trousers with their projections pulling him off his feet. He felt the balls dragging him back towards the ship. Then the smell became so overpowering he passed out.

The barking of his dog woke him up some 20 minutes later. Taylor felt weak, sick, and had a sore throat. He struggled to his feet and staggered back to his truck. However, he was too uncoordinated to drive. He staggered home to Lothian on foot with his dog. His wife found him. Taylor's trousers had been ripped by something.

A doctor examined him and the police treated it as an assault and a criminal investigation followed, apparently making it the only UFO sighting in the UK that has been the subject to a criminal investigation. At the site, the police found lines indented into the ground and holes that may have been made by the spiked balls that attacked Taylor.

Taylor passed away in 2007, but there is a plaque commemorating the events erected in 1991.

BODIES, BOBBIES, BULB-HEADS, BEARDS AND BLACK DOGS!

Strap yourself in — this one is quite a ride. Between June and November in 1980, a bizarre series of events took place in the small town of Todmorden, Yorkshire.

For several weeks in June, reports of strange lights in the night sky were being received by the Todmorden police who, despite investigating them, never got to the bottom of it. But other things were afoot at the same time. On the 11th June at around 3.45pm, as Trevor Parker was entering his father's coal yard that was situated to Todmorden railway station, he came across the body of a man lying on his back and on top of a ten-foot high coal heap.

He telephoned for an ambulance which came around thirty minutes after his discovery, two police officers arrived to attend to the scene. Those officers were PC Malcolm Agley and P.C Alan Godfrey. The officers presumed that the man had died from a heart-attack, but the scene was puzzling. The coal heap was around ten foot high and none of the coal had been disturbed by someone walking up it. The body of the man also didn't have any signs of coal dust on his shoes or hands, which may have indicated that he had climbed up the coal heap. There was something odd about the man's clothing. It seemed he had rushed to get himself dressed as some of his clothing was not properly fastened. The zip on his trousers was down, his shoelaces untied. He was wearing a suit, but his shirt was missing.

Godfrey noticed strange burn marks at the back of the man's head and neck which also appeared to have a gel-like substance smeared over them. A post-mortem on the body took place that same evening, at around 9:00pm, over in nearby Hebden Bridge. It was found that the body was that of a Polish born 56-year-old miner by the name of Zigmund Adamski. He had lived in England for 40 years and worked at Lofthouse Colliery, but had wanted to retire due to ill health as well as wanting to be a full-time carer for his wife who was wheelchair bound.

Adamski lived in Tingley, near Wakefield, which is around thirty eight miles from where his body was found and it appears that he had left his home in Thornfield Crescent at around 3.30pm on Friday, 9th June, to visit a local shop. That was the last time he was seen alive. Nobody knew where he had been for the last five days.

The post-mortem revealed odd details. Despite being missing for five days, he had only one day's growth of beard. This suggests that he had been alive during the five days he was missing and that he had been somewhere that had allowed him the opportunity to shave. As for the burn marks found on his head, they may have been caused by a hot or corrosive fluid and seemed to have been caused at least two days before his death. And as for the gel-like substance found around the burn marks on the back of Adamski's head and neck, samples were taken and sent off for analysis, but it appears it could not be identified by forensic scientist.

One theory was that because he and his wife had allowed a family member to stay with them, as she had put out a restraining order against her husband, the husband had kidnapped and tortured Adamski, resulting in his heart attack. But no evidence was found to implicate the estranged husband and also, how would he get his victim's body to the top of a huge coal heap without disturbing it?

Adamski's wife, Leokadia or 'Lottie' as she was commonly known, stated in court that she thought her husband had been scared to death, such was the expression on his face when his body had been found. PC Godfrey echoed her sentiments, saying...

"Those eyes were staring up at me. I was looking down on him from a foot away. Those eyes sent a shudder down my spine. They were wide open.

*He had a look of someone who had seen something or someone that had
scared him to death."*

You may be thinking to yourself, *"what has a man's body got to do with robots or
UFOS?"* And indeed the Adamski case sounds more like something from the case book
of Sherlock Holmes than the books of Charles Fort. But the case was never solved and
speculation that Admaski had been kidnapped, not by an angry husband, but by aliens,
began to circulate.

Some months later, on the night of 28th November, 1980, P.C Godfrey had been sent
out to check on reports of cattle wandering around a local estate. Despite his searches,
and after driving up and down almost all of the avenues, there was no sign of any cows.
Godfrey decided to report back to the station, saying that the calls may have been a
hoax, but he then received another radio message, asking him to go back to the estate
because a third resident had now reported being awakened by a loud noise coming
from her garden. Peering through a bedroom window, she had seen five to six cows
moving about in the darkness. When she went downstairs to use the phone, she pulled
open the curtains to look. After ending the call to the police station, as she was putting
down the receiver, a very bright glow, like a car headlight that had suddenly gone onto
full beam, lit up the window, and within a second or so, everything went dark. When
she looked back out of the window, all the cows had vanished!

After speaking to the lady, Godfrey made his way back around the estate for one last
check to see if he could spot the roaming cattle, and despite his searching, he found
nothing. Annoyed, he drove back to the station. Later, after finishing his paperwork, he
decide to knock off an hour early, but he would have one last look for the vanishing
cows.

About half a mile or so along Burnley Road, and as he neared Ferney Lee Road,
Godfrey caught sight of something several hundred yards in front of him. At first, he
thought it was the staff bus that carried officers to and from the police station, but he
quickly remembered passing it in town and it was also too early for any of the actual
bus services to be running. He soon realised that whatever it was, it was blocking the
entire road.

The object, which was no more than 60 feet away, was hovering around five feet off
the road surface. It was diamond shaped, twenty feet wide, and fourteen feet in height.
It also had a row of dark panelling across the upper top of it. There seemed to be a
glow coming from a large dome on the top and the whole thing was spinning slowly in
an anti-clockwise direction.

P.C Godfrey switched on his blue beacon and hazards and tried to radio the station, but
in classic UFO form, the radio was not working. He then took out his notepad and with
a pencil, he quickly managed to sketch the object in front of him.

As he was drawing the object, a sudden flash of light appeared and he found himself about 300 feet further down the road than where he had first stopped his car and the object had disappeared.

Godfrey drove back to where he originally had pulled over, only to find the ground around him completely dry, despite it having rained. A bus travelling on the opposite side of the road soon appeared and, still in an obvious state of confusion, Godfrey spoke with the bus driver who also noticed the same thing: that one stretch of the road was completely dry.

Feeling very uneasy, Godfrey drove off to head back to the police station. Upon arriving, PC Malcolm Agley was just making his way into the station after ending his own patrol that morning. Godfrey told him to get into the passenger seat. Godfrey then set off back to where he had seen the bizarre sight in the middle of the road.

The road surface was still bone dry where the object had hovered whereas the surrounding areas were wet from the rain. There was also a strange swirl of leaves that seemed out of place. The two men touched the ground and felt it to be warm to the touch despite coldness of that late November air.

Looking around with their torches, the two policemen found the missing cows in the Centre Vale car park which was chained and locked.

Godfrey later found he had 25 minutes of missing time. He also found out that five other police officers, who were attending an incident on the moors above Todmorden, had seen a glowing light in the distance and near to or where Godfrey would, an hour later, encounter the strange object on Burnley Road. P. C. John Porter described it as being like a very cold, steel blue light. Another witness, school caretaker Leonard Smith had seen a bright light zig-zagging over the same area where Godfrey had his experience. After hearing of these reports, Godfrey submitted an official report.

Several months later, according to him, under pressure, Godfrey retired from the police force. He later underwent hypnosis.

Under hypnosis, Godfrey recalled getting out of his police car and making his way over to the craft, but upon getting close, a beam of light emitted from the craft, hitting him directly onto his chest. He then remembered feeling the sensation of weightlessness as he was brought into the craft.

At this point, everything went dark and when he recovered consciousness, he found himself face to face with a tall, bearded 'humanoid' figure, around 6 feet tall. The man was dressed in what Godfrey described as 'biblical clothes', very strange for a 'space man'. The figure then asked Godfrey to get onto a table. The man referred to himself as 'Yosef', then spoke to him telepathically, telling him to lie still and everything would be fine. Godfrey said he felt safe and as if he had met Yosef before.

Then a bunch of 8 seemingly mechanical beings entered. These robots had light-bulb shaped heads and proceeded to examine the prone policeman. The examination lasted for some time. Godfrey recalled seeing an animal like a huge black dog inside the craft. The reader will recall black dogs seen in association with UFOs before.

Godfrey began to get stressed and Yosef ended the examinations, telling him they feared he might have a heart attack.

Commendably, Godfrey does not think the memories uncovered by the three hypnosis sessions are gospel truth. He maintains he saw a weird object on that lonely, wet night, but thinks that Yosef and the robots may have been constructions of his own mind brought out by the hypnosis. We have already seen that this is not a 100% accurate ticket to past events.

The encounters with robotic aliens seem to make no more sense than those with beings of flesh and blood. Like their biological peers, they also seem to come in a bewildering number of shapes and sizes. One wonders if such encounters would have ever taken place if the idea of mechanical beings had never arisen in the minds of mankind?

FAY IN SPACESHIPS
The idea of hidden races, separate to mankind but sharing the Earth with them, is a universal one. Fairy lore is found in every culture under many different names. We will be examining modern day fairy sightings in the next chapter and examining the close links between so called 'alien' encounters and fairies in the final chapter. For the time being, suffice to say that the two phenomena seem to overlap and may even be one in the same. Presented here are some cases where the aliens seem much more like fairies of ancient lore than extra-terrestrials.

GOBLINS, HOPKINSVILLE AND OTHERWISE
One of the most infamous UFO cases took place in the hamlet of Kelly outside of Hopkinsville, Kentucky, and entered into Fortean legend. 11 witnesses on a remote farmhouse claimed to have been besieged by a clan of glowing, bullet proof goblins. On the night of August 21, 1955, a large extended family called the Suttons arrived breathlessly at the Hopkinsville police station in southwestern Kentucky and blurted out a weird tale.

At about 7 p.m. on the hot Sunday evening, Sutton family friend Billy Ray Taylor was fetching water from the backyard well when he saw a silvery object he described as *"real bright, with an exhaust all the colours of the rainbow."* As he later recounted, it came silently toward the house, passed over it, stopped in the air, and then dropped straight to the ground.

Taylor, 21, and his 18-year old wife, had come from Pennsylvania to visit Lucky Sutton, with whom he had worked on a travelling carnival. The Sutton family consisted of 50-year-old widow Glennie Lankford, her two older sons and their wives, a brother-in-law and the widow's three younger children (12, 10, and 7). They didn't take Billy Ray seriously, laughing off his UFO account. By the end of the night they would all regret their dismissive behaviour.

An hour later, the family dogs started to bark furiously. Lucky and Billy Ray went to the back door and saw a small humanoid creature. About 3 feet 6 inches tall, it had an oversized head, almost perfectly round, long arms that extended almost to the ground, large pointed ears and clawed hands. The goblin's huge eyes glowed with a yellowish light. The body gave off an eerie silvery shimmer in the light of the night's new moon,

Terrified, the two men grabbed a 20-gauge shotgun and a .22 rifle and fired at the goblin as it came toward the back door. They reported that it then did a "flip," scrambled upright and fled into the darkness. Shortly after, the men saw a similar creature appear in a side window, and fired through the window screen. Impervious to bullets, the creature again flipped, then disappeared.

> *"I went out in the hallway and crouched down next to Billy, when I saw one approaching the door. It looked like a five-gallon gasoline can with a head on top and small legs. It was a shimmering bright metal like on my refrigerator."*

Mrs. Lankford told UFO researcher Isabel Davis, who subsequently wrote a book on the affair.

Taylor stepped outside under the small overhanging roof and those behind him saw the claw-like hand of one of the goblins reach down and touch his hair. The group screamed and pulled Taylor back while Lucky shot above the overhang and then at another similar creature in a nearby tree. It floated to the ground and then scurried into the woods.

The family moved inside and spent several hours listening for movements, hearing mostly occasional scratches on the roof. At 11:00pm, the whole group ran for the cars and drove to the Hopkinsville police station at top speed. Sheriff Russell Greenwell noted that they seemed genuinely frightened.

After the local police chief called for backup, his team was joined at the Sutton farm by state police, military police from nearby Fort Campbell, and a photographer from the *Kentucky New Era*. There, investigators found shell casings from the gun shots, but no other evidence. Neither could they find proof that the Suttons had been drinking.

But the goblins were biding their time. Once the police and others left, the creatures returned between 2:30 a.m. and daybreak. Mrs. Lankford said she saw one glowing, repeatedly looking in at bedside window, its claw like hand on the screen. The goblins, when not floating, walked with a swaying motion as though wading through water. The family held out in the house till dawn. Finally, at daybreak, the creatures retreated. Isabel Davis's strong impression after meeting Mrs. Lankford was one of a sombre, no-nonsense matriarch who abhorred the limelight and had no reason to lie. None of the witnesses, Davis noted, had any history of making preposterous allegations.

Space goblins have been seen elsewhere and elsewhen.

In 1979 in Puerto Rico, a goblin-like creature was seen after a UFO sighting near the El Yunque rainforest. The case was investigated by respected researcher Jorge Martin. Martin visited the family in question and interviewed them at their home the Villa Carolina in Carolina municipality.

The family included Edwin Rodrigues Castro, his wife Carmen, and their two daughters. The three Castro women had seen a saucer-shaped, illuminated object hovering over their car as they drove to their farm in November of 1979. The girls became hysterical, scratching at their mother's throat, and screaming that the 'Martians' wanted to take them. This is clearly a case of 'the nameless dread. When they reached the farm, where Edwin was waiting, the whole family saw the saucer veer off and seem to enter the side of a mountain.

They were so shaken that the next morning they drove back from their farm to Villa Carolina. However, at about 9.00pm next night they heard beautiful music, like that of a flute coming from the back yard. Peering out of a window, they saw a goblin. It was standing on top of an electrical transformer box. It was three and a half feet tall and dressed in what looked like green, military fatigues that seemed to be iridescent. It had on a wide, dark belt and its boots ended in points. On its head it wore a tall pointed hat with 'eves' on the sides. The goblin's skin looked greenish and its eyes glowed with a hypnotic effect. Its ears were pointed. When Edwin ventured outside, it ran away. No traces could be found. Music is often associated with fairies in folklore and is often used to lure humans or put them under a spell.

The following year, again in Puerto Rico, there was a multiple-witness case involving goblin-like creatures. On March 3rd 1980 in Rio Piedras, two teenagers, Vivian and Jose Rodriguez, were woken by a barking dog at 3.30 in the morning. Looking through the window, they saw five strange creatures, with pointed ears and webbed feet and wearing tight fitting clothes. The creatures seemed to be interested in the family's chickens.

Next day it was found out that at the same time on the same night, two men had also seen the goblins. The witnesses had been sleeping in a parked car, resting after a long journey. Awaking, they saw a large domed object on the ground. They saw goblin-like creatures identical to those seen by the Rodriguez children. They were headed in the direction of the farm.

Ilkley Moor is a wild expanse of land in Yorkshire. It is well known for sightings of ghosts, mysterious big cats, and strange lights. There are stone circles and ancient carvings. The area has a spooky reputation.

On December 1st 1987, off duty policeman Philip Spencer (a pseudonym) was walking across the moor to visit his father in law. He had a camera with him, as he was an amateur photographer, and a compass in case he got lost. The moor can be a treacherous place with fog rolling in quickly and sudden weather changes.

As he walked up a hill, he saw a strange little figure. It was four feet tall with a round head and long arms and legs. It seemed to have green skin. The creature gestured to Spencer as if to tell him to stay away. He managed to take a photograph of the figure before it scuttled away. He chased the thing but lost it in the growing fog. Then he saw a glowing white craft rise up out of the fog. Spencer said it looked like two saucers one on top of the other. It made a humming nose.

Forgetting his original plan, he headed for the closest town to find two hours of his life had seemingly gone missing and his compass was pointing in the opposite direction, seemingly having its polarity reversed.

The photo, when developed, showed a green skinned, goblin like figure. Spencer made contact with UFO researchers Jenny Randles and Peter Hough. Spencer handed over the copyright of the photo to Hough and made it clear he did not want to make money from it. Experts from the Kodak laboratory in Hemel Hempstead who examined the photo said that they could not detect any evidence of tampering.

Spencer claimed that he experienced strange dreams. Following Hough's advice he attended a session of regressive hypnotherapy. This was carried out by Dr. Jim Singleton on 16 March 1988.

He recalled becoming paralysed upon seeing the creature. He was then lifted up into the craft where a voice told him not to be alarmed. A group of the green creatures performed an examination on him, inserting objects into his mouth and nose. They then gave him a tour of the ship. He was shown two films. The first consisted of floods, famines, and nuclear explosions. Spencer never revealed what he was shown in the second film, saying that the aliens did not want humanity to know. If that were the case, you have to ask yourself: why did they show him in the first place? The apocalyptical warnings are common in abduction cases. Then he was returned to the moor where he photographed one of the creatures. Apparently, it was waving him goodbye rather than telling him to stay away.

Folklorist Steve Jones tells me that in the early 19th century there was a bathhouse on the moor. One day the owner came to unlock it only to find the doors were already open. On entering he saw dozens of small, green coloured goblins frolicking in the water. On seeing the man, they scampered off across the moor and vanished.

THE EVERITSTOWN LEPRECHAUN

On the evening of November 6th 1957, paper mill worker John Trasco of Everitstown, New Jersey, went outside to feed his dog King. At the same time his wife, who was looking out of the window, noticed a shiny, oval-shaped object in front of a barn some 60 feet away. The glowing object was floating a few feet above the ground and seemed to be about 12 feet long.

Meanwhile, John had noticed that King was barking. Walking around the side of the house, he saw a three-foot-tall man in a green suit with shiny buttons, gloves with shiny objects at the tip of each finger, and a tam-o'-shanter hat. He had a face the 'colour of putty' and large, frog-like eyes. John described him as looking like a leprechaun.

The leprechaun spoke to the witness in a voice that was "sharp, scary, and broken."

"We are peaceful people. We don't want no trouble. We just want your dog."

Not wanting to give his beloved pet to a space leprechaun John shouted at the little man.

"Get the hell out of here!"

The little man marched back to the object and entered it, despite there being no visible door. The object shot off into the sky "like a flame." The next night, Mrs Trasco saw two lights that hovered over the house before vanishing.

TOLKIEN'S SPACEMEN

José Antonio da Silva, a military police officer, was fishing in a lagoon in Bebedouro, Minas Geriais, Brazil. It was May 4th 1969. He heard voices and turned to see two four -foot figures approaching him. They had helmets and aluminium-coloured suits. The beings grabbed da Silva and dragged him towards a craft that looked like an upright cylinder with black platforms at the side.

Once inside, one of the beings forced a helmet onto his head. It was rounded at the back and square at the front. It had sharp edges and was flat from the forehead down, but had a projection for the nose and inch wide holes for the eyes. The little men began to talk animatedly. He felt the craft begin to rise and was then blindfolded and taken into another room.

When the blindfold was removed, one of the men stood before him sans his helmet. His kidnapper resembled nothing so much as a dwarf from Scandinavian legend. He had a waist long, reddish beard, deep set green eyes larger than a human's, and ears larger than a human's. His nose was also large and pointed. He had a wide mouth and wide-set, thick eyebrows. The others removed their helmets and each looked very like the first. There were over a dozen of them.

The space dwarfs examined his fishing gear, which they had apparently brought with them. Later, the witness noticed on a low shelf the bodies of four human men and became frightened. Later still, the beings gave him a dark green liquid to drink out of a cubical stone glass.

The dwarf leader then began a conservation with the soldier, mostly about weapons, which was conducted entirely with gestures and drawings.

Da Silva understood that they wanted him to help in their relations with humans. When he refused, the dwarf snatched the crucifix from the rosary Da Silva always carried with him. As the soldier began praying, a Christ-like figure appeared to him, making revelations.

Shortly afterwards, he was blindfolded again and taken back to earth. As the craft landed, he felt he was being dragged and lost consciousness. When he awoke, de Silva found himself in the town of Vitoria some 250 miles from where he was abducted. He had, apparently, been missing for 4 days.

Not only did the 'aliens' look like legendary dwarves, but the encounter had religious overtones.

THE FINNISH PIXIE
Pixies are a kind of fairy associated with the southwest of England, infamous for trickery. To be under their spell is known as being 'pixie led'. You would not usually associate them with Finland, but a creature very like one was seen there in 1970.

Esko Viljo, a 38 year old farmer, and Arno Heinonen, a 36 year old forester, were skiing in Imjiarvi, Mikkeli, Finland. Both men saw a reddish light in the sky that emitted a buzzing sound. The light became enveloped by a cloud of mist that emitted smoke in puffs from the top. The cloud descended and the skiers saw it contained a metal sphere 9 feet across. On the bottom part was three hemispheres and a tube in the middle. The object hovered 10 feet from the ground.

The tube emitted a beam of light that lit up a three-foot circle in the snow. A small being materialized in the beam. It had a 'waxy' face with small eyes, pointed ears, and a hooked nose. It wore green overalls and darker green boots. It also had long white gloves that came up past its elbows. On its head was a shiny, conical hat. Including the hat, it was 3 feet tall.

It held a box with a round opening that shone a bright, pulsing light. The pixie pointed its box at Heinonen and as it did so a reddish grey mist billowed down from the craft. Then four-inch green, red, and purple sparks sprang up from the illuminated circle. The mist became so thick that it obscured the little creature who was standing just nine feet away.

The beam of light drew back into the craft and it rose and vanished. The pixie was gone. Heinonen felt numb down his right side and stumbled when he tried to walk. Abandoning his skis, he needed Viljo to help him walk. They returned to a cottage owned by Heinonen's parents. Heinonen felt ill, suffering a headache, shortness of breath, vomiting, body pains, memory loss and his urine turning black! Viljo's face became swollen and red, he had difficulty balancing, had a headache, and felt pain in his eyes.

A doctor who later examined them said they had radiation poisoning. Also, a reporter located two independent witnesses who had seen a UFO in the area at the same time.

THE MINCE PIE MARTIANS

On the morning of January 4th, 1979, 43-year-old Mrs Jean Hingley, from Rowley Regis, West Midlands, England, had some odd callers. The creatures that entered her house were subsequently known as 'the mince pie Martians'. She was later interviewed by UFO researcher Eileen Morris for the magazine *Flying Saucer Review*. Her account, in her own words follows.

"On the morning of January 4, 1979, a cold dark morning with snow on the ground, I had the strangest experience in my whole life.

I live in a small council house in Bluestone Walk, Rowley Regis, near Birmingham. The house is one of a number on a small estate surrounded by waste land and quarries. We are near Hailstone quarry and our road is named after the Bluestone Quarry. We, my husband Cyril and myself, have lived here for nine years. We have an Alsatian dog, Hobo, who is two and a half years old. I work at a factory making sound proofing for cars and my husband is employed at a Cement Works.

The house has a small front garden and a small lawn at the back about seventeen feet by eleven feet. There is a carport at the end of the lawn and a shed. A door opens to a road at the back of the house.

At seven o'clock on January 4th my husband was going to work by car and I stood at the back door to wave him off. Hobo, our Alsatian dog, was by my side. When my husband had gone I saw a light in the garden and thought, 'Cyril has left the light on in the car port.' I went down the garden to the car port but saw that the light was switched off. As we turned to go back to the house I saw an orange light over the garden which gradually turned white. It lit the whole garden.

We went into the back door of the house. Suddenly with a sound like Zee . . . zee . . . zee . . . three 'beings' floated past me through the open door. They glowed with a brilliant light and seemed to float about a foot above the floor. As they floated past me into the lounge I saw that they had wonderful wings. I was so terrified that I grabbed the steel sink in the kitchen. I couldn't speak. I was frozen.

I looked at Hobo. He seemed to 'hobble' to his drinking bowl, swaying from side to side. His hair was sticking out all over like a hedgehog's. Yet Hobo is afraid of nothing. He seemed as though he was drugged. He just flopped down and lay on the floor, stiff, with his eyes open. I felt as though all the blood in my body had drained out through my toes. I was paralysed. My mouth was wide open. I couldn't move or speak.

After a while the fear seemed to leave me. I felt as if I were lifted up. I wondered what was happening to me. I felt as if I were a different person; as though I was in Heaven although I was still at home. I seemed to float into the lounge. I held the door but my feet didn't touch the ground. The doors were wide open and it was a bitterly cold morning but I felt warm. All the downstairs lights were on as it was dark outside.

I could hear the little artificial Christmas tree shaking but the light was so brilliant that I had to cover my eyes. The three 'creatures' seemed to read my mind. It was like a light or an X-ray penetrating my mind.

When I took my hands from my eyes I could see. They seemed to have turned down the light that surrounded them. There was a glow round their heads. I could see them clearly. They were shaking and tugging at the little Christmas Tree. There they were - three little slim 'men' in silvery-green tunics and silver waistcoats with silver buttons or press studs. They were about three feet six inches to four feet high, all alike. Their pointed hands and feet were covered in the same silvery-green, and they had pointed caps on their heads of the same colour and with something like a lamp on top. They had transparent 'fish bowl' helmets over their heads which rested on their shoulders. There were no eyebrows or ears to be seen. Their faces were waxy white, corpse like, and they had 'black diamond' eyes. I don't know much about precious stones but that is how I would describe them. I didn't notice their noses. Their mouths were very thin.

Their wings were wonderful, large, oval-shaped and glowing with rainbow colours - red, violet, gold, blue, green - but more beautiful than our earthly colours. Their wings were covered in dots like 'Braille' dots. I thought of 'Joseph's coat of many colours.' Our colours seems like 'chemical' colours compared to them.

They were floating round the lounge touching everything - the Christmas cards, the clock, the radio and all the furniture.

At last I could speak. I said: 'Three of you and one of me. What are you going to do? What do you want with me?'

Each of them put their hands to their chests with their pointed hands and seemed to manipulate the buttons. A 'beep' sound came from each 'being's' chest and then the voice came from the chest. The mouths never moved. They all spoke together: 'We shall not harm you.'

'Where have you come from?' I asked, and they said: 'We come from the sky.'

They started to shake the little Christmas tree again and the little fairy fell from the top. I still seemed paralysed. I couldn't move to pick it up.

I said: 'We put up a tree at Christmas because we believe Jesus was born then.' They said, 'We know all about Jesus.'

They were looking at the Sunday papers on the table. There was an Honours list on the front page. I said, 'These people have been made lords.' They said: 'There is only one Lord.' They looked at a picture of the Queen, and I said, 'You should go to the Queen or go see a real lady.' I wondered why they came to me as I am just a working woman. They said, 'You are a lady.' We have a large lounge with a corner unit couch. They sat on the couch and bounced like children. I said: 'Be careful of my furniture,' and they stopped.

When I spoke sharply they put the light up, so I thought I had better be friendly with them. They were only small but they seemed to have power in their bodies that might have harmed me. My eyes still felt sore from the bright light but I felt happy with them. They looked at me with friendly eyes, I thought. I said: 'I can't call you 'creature' so I shall call you 'gentlemen.' I started to say, 'Nice to see you! Nice.' They replied: 'Nice.'

When they floated about the room their wings fluttered gently - there was no sound. When they moved through the open door to the hall they folded their wings behind their backs like pleated fans. They circled in the hall then floated upstairs. The doors to the upstairs rooms were closed and they floated down again.

They picked up the tapes for the tape recorder and looked at the packets of cigarettes on the sideboard. There were bottles of whisky and sherry on the sideboard as it was not long after Christmas. I asked: 'Do you want a drink?' They said 'Water' three times. I went to the kitchen and fetched four glasses of water. I put them on a metal tray. I thought I would bring one for myself as well - to show that it wasn't poisoned - and to drink with them to keep them company. As I came near them with the metal tray I could hardly hold it. The tray seemed to be magnetised towards them. I put the tray down on the table and each of them picked up a glass when I lifted mine. They seemed about to lift their masks but when they saw me watching they put the "power light" on. I didn't actually see them drink but when they put the glasses back on the tray the water was gone. They said: 'We have been to Australia, New Zealand and America. We come down here to try to talk to people but they don't seem to be interested.'

'Shall I tell people on earth about it?' I asked, and they replied 'Yes.'

They said: 'We have been here before,' and 'We shall come again.' Another thing they said was, 'Everybody will go to Heaven. There are beautiful colours there.'

They seemed to put a light on me to draw out the words they wanted to hear. I was stuttering with nervousness. I was talking about politics and women going to work and said, 'It's a man's world.' They seemed interested and excited as though they were listening and understanding. I told them I had not been to chapel for a year or two as chapels had pop groups and guitarists these days and I didn't like that kind of service.

'There is no need to worship in synagogues,' they said. I didn't know until my husband told me later that it was the name of the Jewish place of worship. I said: 'The bible is hard to understand,' and they seemed to know what I meant.

I told them that I had looked after foster children for seven years and at one time I looked after thirteen stray dogs. The neighbours' children used to come with me to take them for walks. Then I went to fetch a plate of mince pies for them. I put six on a plate and told them to help themselves. They each lifted a mince pie from the plate as though their hands were magnetic. I saw that they were looking at the cigarettes again. 'I'll show you how people smoke these,' I said. I struck a match and lit a cigarette. They leapt back as though they were frightened and began to float towards the back door. I stubbed out the cigarette and called out 'Come back! Come back!' I still seemed to be floating as I followed them and as they went through the back door I saw an orange coloured glowing 'thing' in the back garden - a 'space ship.' It must have been eight to ten feet long and four feet high. It had round windows or portholes. It seemed to be covered with a kind of shining plastic. I couldn't see through the portholes. There was something like a 'scorpion tail' at the back and a kind of 'wheel' on top.

They still held a mince pie each as they went towards the 'space ship' and entered it. They flashed the lights twice as if to say 'Goodbye.' Then they took off over the fence and away across the open ground towards Oldbury. The sky was still dark, with no stars, and there was snow on the ground. Hobo came to life then and wandered around the garden as though he was looking for them.

There was a deep impression on the back garden where the 'space ship' had settled. I felt warm and happy although it was such a cold morning. I felt 'good,' as though I had been blessed.

When I went into the house and looked around I realised that the clock and the radio had stopped. I spoke to my next door neighbour and she said: 'You should ring the police.' I rang the Oldbury police and they said they would come.

I rang my husband but he couldn't leave his job. I said to him: 'I have had visitors with wings.' He said: 'What do you mean? Birds?' I was shaking and crying. I said, 'No. Men with wings.' He laughed and said: 'Why don't you go and have your hair done and tell the girls about it.' I did that later on, and they were very kind.

The police came from West Bromwich Police Station. They looked around and said: 'You look pale as though you have had a shock.' They couldn't do much - they couldn't take fingerprints. They went away and came back later. They rang UFO Investigation Service in Birmingham. The people came and measured the impression on the ground in the back garden. It was eight feet by four feet. They also took soil samples for analysis. I haven't heard the results yet.

I have had a lot of nice letters from people who were interested after reading reports in the papers. Some people sent tapes to be returned to them with my own story.

The West Bromwich College of Technology - Television Production Course students have made a film of me telling my story. It was filmed at the Technical College on March 19, 1979. My eyes were sore for about a week and I had to wear dark glasses. I have had to have some weeks away from work as I haven't felt well and the doctor advised me to have a rest. My jaws ached after staring with my mouth open with shock when I first saw the 'beings'.

I have never read books about UFOs. I only read the papers. I don't look at a lot of television, but like the Crossroads programmes and Coronation Street, and love stories.

Some people have made jokes about me, but people who know me believe me as they know I am truthful.

Some people have written to say that they think the visitors were elves or beings from the Fairy Kingdom, or even robots, but I don't know what to think. I know I shall never forget them if I live to be a hundred.

A few days later we tried the 'tapes' that had been handled by the 'beings.' They were so distorted that they were ruined. Before January the 4th they were quite normal.

Wow, now that's a story and a half. It occurred on my 9th birthday and I recall very well the case being covered by the national newspapers in the UK. Unlike a lot of cases there was no menace or nameless fear here. However, there are some very interesting points.

Firstly, in actual legend, fairies do not have wings. The gossamer pinions seen in today's depiction of the little people are an invention of the Victorians (like the horns on Viking's helmets) and exasperated by the saccharine garbage of Disney. This may have had something to do with the witness' pre-conceptions or expectations.

The religious connotations are interesting too. These may have something to do with Mrs Hingley's subconscious or her own beliefs. In some ways it seems like the sighting of an angel or the Blessed Virgin Mary. However, in common with other UFO cases, we see things like possible radiation effects and animals acting in a strange manner. Also we must ask ourselves once more, why would technologically advanced creatures such as 'aliens' be so afraid of fire? It brings to mind the Cisco Grove case.

SPACE GHOSTS
We have already noted the similarities between 'extra-terrestrials' and fairies and Bigfoot creatures seen in and around UFOs. But there are also cases where the so-called aliens look and behave more like ghosts.

THE SPANISH SPACE SPECTER
In the early hours of November 12, 1965, at the Base Aérea de Talavera la Real (Talavera la Real Air Base), located in the province of Badajoz, Spain, two soldiers named José María Trejo and Juan Carrizosa Luján, claimed to have seen a creature that seem to have been as much of a phantom as it was an alien.

The men were patrolling the fuel storage area of the base. At 1:45 am, they were about 60 meters away from each other when they heard a noise similar to radio interference.

Suddenly, the noise became an acute whistle that lasted for 5 minutes. José asked Juan to follow him on an inspection around the area, fearing sabotage. The whistle ceased 5 minutes later. Soon after it, they spotted an intense bright light in the sky that lasted for 15 seconds. Another guard arrived with a guard dog.

The men were hit by a blast of air like a whirlwind. Together, they moved all across the area. When they arrived at a certain point, they heard the sound of eucalyptus twigs being broken. The dog was released and rushed towards the point where the sound of broken twigs came from. Soon the dog returned and was staggering as if sick. The soldiers urged the dog to return to the point several times, each time with the same result. Eventually, the dog started to walk in circles around the three soldiers (a technique taught to the guard dogs, meant to protect their handlers).

Suddenly they found themselves face to face with a human-shaped green light to their left. The being was very tall, about 3 meters, and floated in mid-air. It seemed to be constructed of small light points, and the brightness was more intense along its edges. The head was small and covered by something resembling a helmet. The arms were very long and were crossed. There was no sign of either hands or feet as if the figure was not fully formed.

Trejo fell down on his knees and fainted. His friends took the chance to open fire on the creature. Between 40 and 50 bullets were fired, and as the being was hit, its brightness became more intense and it disappeared. While Trejo was being assisted, the same whistling sound as before was heard, lasting 10–15 seconds. Next morning, some 50 men searched the whole area. They discovered that there weren't any bullet casings of the shots fired by the soldiers, nor any bullet holes in the wall behind where the creature had floated.

THE GHOST AND THE GOBLIN
On October. 19, 1973 at Albany, Ohio, 7:30 p.m. a witness saw a "ghost-like" figure floating about 50 ft above the ground at 1000 ft distance when she came home one evening. It was about 4 ft tall and thin *"like a person draped in a close-fitting sheet."* It was seen only briefly when she noticed a bright white object moving about, and approaching to within 200 ft, before going away. The object was about 20 ft in diameter and about 25-30 ft off the ground. Later, as she made supper, she saw a *"little blue-green thing"* about 2.5 ft tall and with a face with *"spiky things at the tops and the sides of the head,"* look in an open door; it had stumpy arms and it quickly disappeared from sight. UFO sightings occurred about the same time in nearby Athens and elsewhere.

SHADOW MEN FROM SPACE.
27 year old Kelly Carhill and her husband Bill were driving home from a friend's house in the Dandenong foothills near Belgrave, Victoria, Australia shortly after midnight on August 8th 1993.

They saw a round craft hovering above the road. It had windows and lights along its bottom. Kelly thought she saw 'people' at the windows. The craft shot away to the left and vanished.

Less than a mile further on, they were hit by a blinding light. Then they found themselves back home, having lost an hour of time and smelling of vomit! In the following days, she found triangular marks around her navel and was admitted to hospital twice with a uterine infection and severe stomach pains.

Through hypnosis, she said, she was able to unlock her 'missing time'. She recalled the 150-foot-wide craft being in a field beside the road. They stopped the car and got out. As they did so they noticed another car had stopped too.

Then they both saw a 7-foot tall, jet black, inhuman figure in the field. It had an oversized head and large eyes like those of a fly that glowed red. There were more of the creatures in the field, 'heaps of them'. One of the shadowy creatures moved towards them, gliding rather than walking. Another shadow man approached the people in the other car.

Kelly thought that the things were evil and lacked souls. The being grabbed her and she screamed at it to let her go. She also felt sick. Kelly heard in her mind its thoughts: *"Let's kill them."* The next thing she remembered was that she was back in the car.

She later dreamed of the encounter with the shadow alien bending over her.

> *"The creature was black, but not the colour black. It was as if someone had cut a hole in matter where he stood or as if he himself was a hole in space."*

Compare that description to Jon Downes' one of the Beast of Bolam Lake, *"a man shaped hole in reality."*

Sydney-based researcher Bill Chalker, of the UFO Investigation Centre, was one of the first people Kelly contacted after that night. And he in turn contacted Phenomena Research Australia (PRA), a group of loosely affiliated Fortean researchers.

The independent witnesses in the other car confirmed what Kelly and Bill had seen. They were interviewed by John Auchettl of PRA who had placed an ad in local newspapers in an effort to find the occupants of the second car. They detailed experiences inside the mystery craft where they were strapped to a table and examined by the beings.

According to the PRA, the women had the same triangular wounds near their navels, as well as other strange marks. There was even talk of a third car driven by a local lawyer, PRA discovered, whose story also lined up.

Kelly later wrote a book about her encounters but later retreated from the public eye after the other witnesses stayed anonymous. A 300-page report on the case by PRA was also canned for the same reason.

Why would aliens look and behave like ghosts? The terror Kelly reported puts me in mind of the fear felt by witnesses to the shadow man at Alanbrooke Hall. I don't for one moment think that these 'space ghosts' are the wraiths of dead aliens, but I do think they may be part of the same thing, a bigger phenomenological whole.

LIVING UFOS
Most people imagine UFOs to be machines, spaceships piloted by alien beings visiting Earth, but one man had other ideas. Trevor James Constable was a New Zealand military engineer, historian, and writer with very different ideas. They thought that the 'crafts' themselves were living creatures.

Constable believed that UFO were amoeba-like creatures that inhabited Earth's atmosphere and ranged in size from a coin to half a mile across. He theorized that these creatures were an 'offshoot of evolution' and that they had been on the planet since Earth was in a more gaseous form. The creatures, whom he dubbed 'critters', were sometimes invisible to the naked eye due to their ability to reflect infra-red light, but could be detected by radar. They could be observed with the naked eye when they changed colour. He also thought that radar provoked them into predatory behaviour and that cattle mutilations and even human deaths were the result of these sky beasts. In his book *The Cosmic Pulse of Life* he writes...

> *"Critters appear to be an elemental branch of evolution probably older than most life on Earth, dating from the time when the planet was more gaseous and plasmatic than solid ... They will probably one day be better classified as belonging to the general field of macrobiology or even macrobacteria inhabiting the aerial ocean we call the sky."*

One zoologist named the creatures *Amoebae constablea*, after their discoverer. Constable also claimed to have photographed the creatures using infrared film.

Another man who took a similar view was John Philip Bessor, an American *"psychical researcher and student of the mysterious"* as he once presented himself to a newspaper. Bessor thought UFOs were a form of living creature, which he called an ideoplasm. They were composed of a highly attenuated substance, enabling them to materialise or dematerialise at will, utilising telekinetic energy for propulsion. Bessor theorized that these entities must be capable of becoming visible, invisible, and changing colour, all very rapidly. He even submitted his thoughts to the U.S. Air Force, and, remarkably, was informed by them that they considered his notion to be *"one of the most intelligent theories we have received"* regarding the possible nature and identity of UFOs.

Countess Zoe Wassilko-Serecki was an Austrian astrologer and parapsychologist. The

Countess thought that UFOs may be enormous, glowing, stratosphere-inhabiting creatures resembling gargantuan bladders of colloidal silicones (suspensions of fine amorphous, non-porous, spherical silica particles in a liquid phase), containing a central core of insubstantial matter but otherwise composed predominantly of pure energy. She claimed that they appeared spherical when stationary but became fusiform when moving, and so diffuse at higher levels as to appear virtually invisible.

There are, indeed, some cases where the UFO observed seems more like a biological entity than a nuts-and-bolts craft. On the top of a remote mesa in Nevada, four amateur pilots had an encounter with what seemed to be some kind of organic UFOs back in 1925, years before the term had even been invented.

Don Wood Jr. and three friends had vintage, WWI aircraft known as the Curtiss JN-4 or, more commonly, as the "Jenny." The aviators used their Jennies to explore vast stretches of the Nevada desert and to land on hard-to-reach plateaus.

In the October 1959, issue of the ufological periodical *Flying Saucers*, a letter by Wood was published where he revealed what happened to him and his fellow pilots back in 1925.

> *"I must write to you of what happened to me in 1925, which I think solves most UFO reports. I have never told this to anyone, but can get a signed affidavit if needed. Four of us were flying old Jennies, over the Nevada desert. One plane was a two-seater, the one I was in. We landed on Flat Mesa, near Battle Mountain, Nevada. This mesa is about 5000 square feet and the walls are too steep to climb unless a lot of work is done.*
>
> *We wanted to see what was on top of this flat place. We landed at 1 pm. While walking about the top of this place we noticed something coming in for a landing. It was about 8 feet across and was round and flat like a saucer. The undersides were a reddish color. It skidded to a stop about 30 feet away.*
>
> *This next you won't believe, and I don't care but it's the truth. We walked up to the thing and it was some animal like we never saw before. It was hurt, and as it breathed the top would rise and fall, making a half-foot hole all round it like a clam opening and closing.*
>
> *Quite a hunk had been chewed out of one side of this rim and a sort of metal-looking froth issued. When it saw us, it breathed frantically and rose up only a few inches, only to fall back to earth again. It was moist and glistened on the top side. We could see no eyes or legs. After a 20 minute rest, it started pulsating once more, (We stayed 10 feet away.) And so help me the thing grew as bright as all get out, except where it was hurt. It had a mica-like shell body. It tried to rise up again but sank back*

again.

Then we saw a large round shadow fall on us. We looked up and ran. Coming in was a much larger animal 30-feet across. It paid no attention to us, but settled itself over the small one. Four sucker-like tongues settled on the little one and the big one got so dazzling bright you couldn't look at it. Both rose straight up and were out of sight in a second. They must have been travelling a thousand miles an hour to get so high so fast.

When we walked over, there was an awful stench, and the frothy stuff the little one had bled looked like fine aluminium wire. There was more frothy, wiry stuff in a 30-foot circle where the big one had breathed. This stuff finally melted in the sun, and we took off.

So help me, this was an animal. I have never told this before as we knew no one would believe us. I only write now because this animal would be one big 30-foot light if seen at night."

THE DISSOLVING FLYING SAUCER
"Flying 'Saucer' Just Dissolves." ran the headline In the September 27, 1950 edition of the *Philadelphia Inquirer.* The story involved veteran Philadelphia police officers Joe Keenan and John Collins who had seen an "object" plunging through the night sky the night before.

They were making their evening rounds in their patrol car when they turned down a desolate side street between Vare Avenue and 26th Street. As the two policemen rounded the corner, they noticed a large, glittering "mass" drifting toward an open field approximately half a block distant.

The bewildered cops wasted no time in racing toward the unidentified falling object that was sparking under their headlights' glare. Once their patrol car squealed to a halt, the officers swiftly retrieved their torches and rushed toward the quivering, domed mound of what they described as *"purple jelly."* They saw a strangely pulsating heap, which they claimed glistened beneath their torches. The officers estimated that the circular mass was approximately 6-feet in diameter and nearly a foot thick at the centre.

The edges sloped down from the peaked top of this blob tapering to a lip that was just a couple of inches thick. When they turned off their torches, the men discovered that the thing emitted a faint purplish, bioluminescent glow that lit up the darkened field.

The substance seemed to vibrate of its own accord. The fact that this thing moved at all seemed to indicate to the officers in question that this blob-like entity was almost certainly a living organism.

The two men radioed for backup and soon Patrolman James Cooper and Sergeant Joe Cook arrived at the scene. The men surrounded the thing and Collins tried to touch it. To his amazement the substance of the thing fell apart in his hands leaving nothing but an odourless scum on his fingers. The whole thing, whatever it was, dissolved in front of the four men's eyes and soon vanished completely.

In 1913 Arthur Conan Doyle wrote a story called *"The Horror of the Heights"* for *Strand Magazine.* The story revolved around an early aviator who was trying to fly higher than anybody ever had before. At 43,000 he encounters the inhabitants of the 'sky jungles', some of whom are predators.

Maybe Doyle was closer to the truth than he ever knew.

NIGHT LIGHTS
Sometimes UFO cases involve not actual, solid machines but weird lights that behave in an odd manner as if under intelligent control or intelligent in and of themselves. These 'night lights' can have weird effects. Light phenomena seem to go hand in hand with every kind of Fortean case and could even be called its core. We will now look at a few of the weirder cases on record.

THE MELTING MAN OF ARAÇARIGUAMA
It is rare that Fortean phenomena are fatal to those who encounter them. Once in a blue moon a monster is credited with killing somebody though conclusive proof is illusive. But in this x-rated story, a weird light seems to be connected to a truly horrific death. This really makes the hairs on the back of the author's neck prickle.

On the evening of March 4, 1946, Joao Prestes Filho was returning home from a day of fishing in Tieté River, Brazil. His wife had been attending a carnival, but Joao had no love of these events and had elected to spend the day fishing with a friend, Salvador dos Santos. Joao had given his friend a lift in his horse drawn cart and waved him goodnight before carrying on his way.

The village of Araçariguama was unlit, there being no electricity in those days, and mostly empty, with the inhabitants off at the carnival that he had shunned. On reaching home he found that his family, who were out enjoying themselves, had accidentally locked him out of the house. Joao put his horse into a corral then managed to climb into his home through a window.

Once inside, Joao got an eerie feeling of being watched. Looking out of the window he saw a glowing object floating in the air. A beam of yellow light emanating from the object struck him.. Filho stated that he covered his face with his hands and collapsed to his knees. The burst lasted only a moment before vanishing completely. He felt his body burning and that his beard, while short, was burning. Panicked, and unable to move his hands, Joao raised the door latch using his teeth and ran into the street barefoot.

Joao ran screaming to his sister María's house near the Araçariguama church. He dropped on a bed and said he'd been burned. The police chief, Joao Malaquías, went over immediately. Joao Filho told the chief there was no one to blame for what had happened, because his attacker *'had not been of this world''*

The man's flesh looked as if it had been boiled. Luis Prestes, Joao Filho's nephew, recalled the night.

When interviewed in 1997 he said...

> *"I was in Araçariguama when I learned that my uncle was dying at a relatives' house. I wanted to go in, but it was forbidden, since I was too young and Joao's physical condition could have caused a traumatic impression... My father said that Joao was only burned from the waist up, with the exception of the hair on his head."*

Vergilio Francisco Alves, Filho's second cousin, was also interviewed and his cousin said...

> *"When I got to María's house, I found Joao Malaquías, the sheriff, speaking with Joao. He was in bed and having problems using his tongue. His skin, which was fair, was toasted, reddish, as if he'd been roasted. His hands and face had the worst burns. The hands were twisted. His hair didn't burn, nor did his feet, nor clothing. He was only burned from the waist up. His feet were torn up from running barefoot on sharp rocks.*
>
> *Malaquías, the sheriff, wanted to take him to a hospital in Sao Paulo, but the road was in bad shape and they went to Santana de Parnaíba."*

Prestes recounted the moment that they took his uncle away.

> *"I managed to see my uncle when they removed him from the house to take him to Santana de Parnaíba by truck, where the nearest hospital was located. I remember that the sheets covering him were blackened, perhaps by the burns on his body... His appearance, according to my father, who escorted him to the hospital, was truly ruinous... He had serious burns all over his body. His flesh was dark and he presented no other bodily injuries."*

Foao was admitted into the hospital at once, where the baffled doctors, including Physician Luiz Caligiuri, failed to diagnose what had happened to him. In 1974, investigator Fernando Grossman, interviewed direct eyewitnesses to the event, including former Army medic and orderly, Aracy Gomide, who had seen the victim in his last hours.

He said that the man was literally disintegrating before his eyes, lumps of flesh falling from his body. The victim's face, his ears, nose, lips and eyelids incrementally peeled off, clinging to slender strands of skin before falling onto the bed. Weirdly, he seemed to be in no physical pain toward the end of his ordeal. Witnesses claim that he even chatted with the strong stomached Gomide and others until he lost too much soft tissue on his jaw to continue speaking.

Gomide stayed with Joao till the end and stated that his final request was for a glass of water. Not long after, at about 10 pm. on March 4, Filho expired and Dr. Caligiuri officially listed the cause of death as "*cardiac collapse.*"

Prestes continued...

> *"My father was a deputy policeman at Santana de Parnaíba and requested the assistance of the forensic unit to research the case, but I don't know anything about the results. The fact is that nothing burned in the room where Joao was when the fire appeared. He had no enemies or anyone who'd be interested in doing such a thing to him. Even as he died, he repeated that the light had attacked him and that it was 'otherworldly'."*

Alves said that the villagers at the time blamed some kind of supernatural attack. The inhabitants of a small, remote Brazilian village knew nothing of 'flying saucers' or aliens. They blamed the Boitatá, a fiery flying serpent from local legend that supposedly perused and killed people.

Jesuit priest José de Anchieta (1555- 1597) lived in Brazil and wrote of the Boitatá...

> *"There are also others (ghosts), especially on the beaches, who live most of the time by the sea and rivers, and are called baetatá , which means something of fire , which is the same as saying what is all of fire . You can't see anything other than a flashing beam running there; it quickly attacks the Indians and kills them, like the curupiras; what this is, we still don't know for sure."*

Maybe the villagers were right.

Alves continued...

> *"I think that the Boitatá was to blame, since it had attacked him once before, when Joao was a tropero [cattle driver], he was still young and lived with his father in Araçariguama. One day at sundown, as he led*

the donkeys over a hill, he saw a fire that fell from the sky, a fireball. He was near a chapel that had a cross, and he could feel the fireball passing him, almost knocking him down. Joao would tell me that at that spot you could sometimes see ten or twelve balls emerging from the sky. Some of them were red, others Moon-coloured. Sometimes five or six of them would fall to the ground and explode. People would call them the Boitatá lights."

Prestes said that other members of the family had been attacked by the Boitatá.

"Something equally scary happened to Emiliano Prestes, my uncle and Joao Prestes' brother. A few months after his brother's tragic death, Emiliano was walking through an Araçariguama forest, in Agua Podre... A fiery torch appeared above him, causing the terrified Emiliano to run to a canyon's edge when the thing fell on him. All he could do was kneel and pray for his life. He told us that he felt an intense heat, but luckily the fiery torch moved away and vanished.

The lights were seen most frequently between three and four in the morning and were three or four times larger than the Moon. People would feel their heat even at a distance, and they were able to move amazingly fast. My father stopped going to parties at night because of these lights."

Other legendary entities in the area were the assombraçoes or ghosts that dwelled in a long closed Morro Velho gold mine, located on Mt. Saboao. A Canadian by the name of General George Raston founded the mine in 1926 and lived on there until it closed in the 1930s.

Alves claimed to have seen these glowing apparitions floating behind the mountain.

"We also called those fireballs 'maes do ouro' [mothers of gold]. There was also the 'golden lizard', an elongated tongue of flame that moved in a straight line, slowly, without making a sound."

Dr. Luiz Braga who, while investigating the incident years later, noted that Joao's condition sounded like radiation burns. But there was nothing in those days that could shoot a concentrated beam of radiation. Even if there was, why use it on a peasant in a rural village in South America?

Filho's family refused to return to the site of the house. The police had it condemned and it was pulled down.

All these years later we are no closer to knowing what happened to poor Joao Prestes Filho.

SERPENT IN THE CIRCLE.

UFO and Fortean researcher Nick Redfern once spoke with a man called Malcolm Lees. Lees was in the British Royal Air Force in the early 1950s and retired in the late 1960s. In 1962 he received a posting to a RAF station in the county of Wiltshire and worked in intelligence gathering.

Most of his work was mundane, but he told Nick of one very strange story. Early one September morning in 1962, a call came into the base from someone who had seen a UFO hovering in the vicinity of the ancient standing stones in the historic village of Avebury.

The witness was a middle-aged lady who had lived in Avebury all of her life and who was fascinated by archaeological history. The name of the witness remains redacted in the official document - entitled *'Flying Saucer Incident, Avebury, Wilts., September 4, 1962'* But she was a 52-year-old woman would often stroll among the stone circle at night.

On the night in question, she had been out walking at around 10.30 p.m. when she was both startled and amazed to see a small ball of light, perhaps two feet in diameter, gliding slowly through the stones. Transfixed, she watched as it closed in on her at a height of about twelve feet. The ball then stopped fifteen feet from her and small amounts of what looked like liquid metal slowly and silently dripped from it to the ground. Then, the ball exploded in a bright, white flash.

For a moment, she was blinded by the flash and fell to her knees. When her eyes cleared, however, she was faced with a horrific sight. The ball of light had gone, but on the ground in front of her was what she could only describe as a monstrous, writhing worm.

The creature was about five feet long, perhaps eight or nine inches thick, and its skin was milk white. As she slowly rose to her feet, the creature's head turned suddenly in her direction and two bulging eyes opened. When it began to crawl towards her, she emitted a hysterical scream and fled the scene. Rushing back home, she slammed the door shut and frantically called the police who, being unimpressed with her story told her to phone the air base.

Lees and a colleague were sent to interview her the next day. The woman, barricaded in her house, was almost incoherent with fear, and only agreed to return to the scene after lengthy coaxing. Lees and his colleague found no evidence of the UFO or the worm. On the ground near the standing stone, however, was a three-foot long trail of a slime-like substance, not unlike that left by a snail. Lees' colleague quickly improvised and, after racing back to the woman's house, scooped some of the material onto a spoon and into a drinking glass.

A report was prepared and dispatched up the chain of command, along with the unidentified slime. For more than a week, Lees said, plain-clothes military personnel would wander casually among the stones looking for evidence of anything unusual. Nothing else was ever found, however.

Lees said that he was fascinated by this incident because it was one of the few UFO-related cases he had heard about that was taken very seriously at an official level and that had some form of material evidence in support of it. He did not know the outcome of the investigation, but he never forgot about it.

We have already seen balls of light associated with Bigfoot, lake monsters, and dragons. This makes the Avebury case all the more fascinating. Could the ball of light / energy be the raw form of the monster before it takes definite shape?

A PERSONAL ACCOUNT: NIGHTLIGHTS DOWN UNDER.
In 2017, I took my third trip to Tasmania in search of a creature known as the thylacine or Tasmanian wolf. This is a flesh-eating marsupial that superficially resembles a wolf or dog, with tiger like stripes down the hind quarters. It is an example of convergent evolution, where two unrelated species, often on opposite sides of the world, evolve to resemble each other due to filling similar ecological niches. The Tasmanian wolf was supposedly hunted into extinction in the 1930s by white settlers scared that the creatures would kill their sheep. However, since then, over 4000 sightings have been claimed, including one by a park ranger and one by a zoologist. The thylacine has been called the healthiest extinct animal you will ever meet. I personally talked to many witnesses with no axe to grind.

My friend and fellow thylacine hunter, Mike Williams, and I had been searching for the creature for a fortnight and were going to visit Mike's friends on their farm. The couple, a man from England and an Australian woman, had built their own farmhouse and raised goats. They lived near Bronte Lagoon, an artificial lake near the centre of the island. Their daughter, when she was young, had seen a Tasmanian wolf whilst walking home from school many years ago. Mike told me that there were weird light phenomena around the farm and in the woods between the house and the lake. Mike himself had seen these lights and were convinced that they were paranormal in nature. Apparently, they appeared every night unless it was raining.

I was pretty sceptical about the whole thing but after talking to the couple about their daughter's sighting, we went out after dark to look for the lights. Suddenly a yellowish white, oval light the size of a large grape appeared over a chicken coop. It floated in mid-air for a while then winked out. Shortly after, the lights began to appear in the woods. Some seemed to perch on branches or beside roots, others floated in mid-air. They were not glow worms or fireflies both of which I have seen before. They resembled Christmas tree lights. All were yellowish white and grape sized. When approached, they would wink out and vanish only to reappear further into the woods.

As we reached the far side of the woods, the lights would then appear behind us. They never allowed us to get closer than about 15 feet before vanishing.

Mike suggested that we drive to the other side of the lagoon in order to make sure that the lights were not linked to the lighthouse there, even though these were small, individual lights and not a beam. And so, we drove over to the lighthouse. We found out that its beam did not even reach the farmhouse due to it being on a higher level than the lighthouse. However, whilst over there and looking back across the lagoon Mike and I saw a much larger light. Glowing red and the size of a beach ball, it hovered in the tops of the trees.

By the time we drove back, the big light had gone, but the couple had both seen it too. The man told us that once, one of the tiny lights approached the house and floated along near a window box. Next morning all of the flowers had wilted.

The lights all vanished about two hours after they appeared. I felt like a show had been put on just for us that had a pre-determined length. Many Fortean cases covered in this book have a 'show' feeling to them, as if they have been put on just for the witness. Do Fortean events happen if nobody is there to see them? My gut feeling is not. It seems like these events are intimately linked to us,

Just as the souls of departed humans don't cut it as an explanation for most ghosts, alien entities do not cut it as the answer to the UFO phenomena. If the Earth was the size of an orange and sitting in central London, then our nearest neighbouring star, Alpha Centauri, would be in Moscow. It is 4.37 light years from Earth or, in other words, if it were possible to travel at the speed of light, it would still take you 4.37 years to reach the closest star to our sun. But it's not just the improbability of interstellar travel but the nonsensical behaviour of the 'aliens'. Are we to believe that the myriad forms of 'aliens' reported are travelling from other star systems simply to kidnap anonymous people and give them messages that make little sense and prophecies that do not come true? The apocalyptic messages we see again and again seem to have more to do with the subconscious fears of the witnesses. Also, we cannot ignore the links with fairy lore. It seems that the two phenomena are one in the same or at least aspects of a greater whole.

The white dragon of Oconto Falls

The Cult of the Moon Beast.

The giant ghost crab of South Africa.

Jon Pertwee's phantom tree.

The Domsten protozoa

Jean Hingley and the Mince Pie Martians.

The Ranton Pixies

The giant talking cat and the pervert.

CHAPTER FOUR: OTHER ENTITIES

"Time to show some sharper teeth
Time to grow a thicker skin
Time to drop the seventh veil
And let some madness wander in."

Faith and the Muse, "Sredni Vashtar."

In this chapter we will look at entities that do not fit comfortably in other categories. The notion of categories in Fortean phenomena may be a false one. The reader should, by now, realise how flimsy the walls between phenomena are and how they can run into one another like two colours of paint mixing. The walls we erect between phenomena seem to be a human construct. I have separated them out into chapters for ease of writing and ease of reading. The beings here could be slotted into the boxes of aliens, ghosts, or monsters, but they seem to have their own flavour. As we proceed, perhaps you will see what I mean.

FAIRIES

Most cultures have a tradition of another kind of human-like beings that share the earth with us. Usually unseen to humans, these people have their own society and culture quite separate to ours though they sometimes interact with us. The popular notion of a fairy as a tiny humanoid with butterfly wings that flits around granting wishes is utterly wrong. In former centuries faeries were something to fear. Many would not even call them by name, referring to them as good folk, little people, or "themselves."

Robert Kirk (9 December 1644 – 14 May 1692) was a Scottish minister who wrote a book on fairies called *The Secret Commonwealth.* He died before it could be published and it didn't see light until 1815, when it was published by Sir Walter Scott. Kirk describes fairies thus...

> *"These Siths, or Fairies, they call Sleagh Maith, or the Good People...are said to be of a middle nature betwixt Man and Angel, as were Daemons thought to be of old; of intelligent studious Spirits, and light changeable bodies (lyke those called Astral), somewhat of the nature of a condensed cloud, and best seen in twilight. These bodies be so pliable through the subtlety of Spirits that agitate them, that they can make them appear or disappear at pleasure."*

Kirk said the little people were broken into two courts, the Seelie Court, who were well disposed to humans, and the Unseelie Court, who were hostile to man. When Kirk was found dead on a hillside at the age of 44, some thought it was the work of the Unseelie Court.

Most people would relegate fairies to folklore, but, much like dragons, there are, surprisingly, still reports of them in the age of 'reason'.

PIXIE LED

The Transactions of the Devonshire Association record the experience of Mrs. G. Herbert in 1928. She recalled seeing a pixie close to Shaugh Bridge on Southern Dartmoor in 1897.

> *"It was like a little wizened man eighteen inches or possibly two feet high, but I incline to the lesser height. It had a pointed little hat, slightly curved at the front, a doublet and little, short knicker things. My impression is of contrasting colours, but I cannot now remember what colours, though I think they were red and blue. Its face was brown, wrinkled and wizened."*

These also record that in 1897 a farmer on Dartmoor had been suffering with sickness among his cattle. He sacrificed a sheep to the pixies on the moor and the illness stopped.

Nick Redfern recorded a pixie sighting in Ranton, Staffordshire, where the Man Monkey lurked. He interviewed an old lady back in 2000. The witness, now deceased, had her encounter in 1929 when she was a girl. She had seen a group of pixies dressed in green and prancing around the base of an old oak tree in her parent's back garden. The little folk tipped their hats and smiled at the girl, enticing her to come closer. As she did their demeanour quickly changed into something sinister and malevolent. Frowns twisted on their wizened faces and they now seemed to be stalking her, creeping forwards. The girl fled back into the house in terror. She never saw the sinister pixies again.

Pixies are a form of fay (fairy) usually associated with the southwest of England. They generally dress in green with conical red caps. They are known for deceiving and tricking humans.

GOULD'S GOBLINS

Reverend Sabine Baring Gould (28 January 1834 - 2 January 1924) was a noted author and folklorist. He had his own fairy encounter as a boy.

> *"In the year 1838, when I was a small boy of four years old, we were driving to Montpelier on a hot summer day over the long straight road that traverses the pebble-and-rubble-strewn plain, on which grows*

nothing save a few aromatic herbs. I was sitting on the box with my father when, to my great surprise, I saw legions of dwarfs of about two feet high running along beside the horses; some sat laughing on the pole, some were scrambling up the harness to get on the backs of the horses. I remarked to my father what I saw, when he abruptly stopped the carriage and put me inside beside my mother, where, the conveyance being closed, I was out of the sun. The effect was that, little by little, the host of imps diminished in number till they disappeared altogether."

Maybe the young Gould was suffering from sun stroke, but having had this myself I can say it afforded me no such strange visions! Gould's wife had an encounter herself as the Reverend records.

"When my wife was a girl of fifteen, she was walking down a lane in Yorkshire, between green hedges, when she saw seated in one of the privet hedges a little green man, perfectly well made, who looked at her with his beady black eyes...She was so frightened that she ran home."

Fairy sightings seem to run in the family as one of Gould's sons saw one too!

"One day a son of mine was sent into the garden to pick pea-pods for the cook to shell for dinner. Presently he rushed into the house as white as chalk to say that while he was thus engaged, and standing between the rows of peas, he saw a little man wearing a red cap, a green jacket, and brown knee-breeches, whose face was old and wan, and who had a grey beard and eyes as black and hard as sloes. He stared so intently at the boy that the latter took to his heels."

ALL MANNER OF FAY
The Reverend Dr. Edward Williams spoke of an event that befell him in 1757 near Bodfari, Denbighshire, north Wales when he was 7. He was out with his sister and two older friends, playing by a hedge close to their house. All of a sudden, the children saw a ring of figures dancing swiftly round and round in the field. There were 7 or 8 couples that Williams described as neither men, women, nor children but *beings*. They were dressed in what looked like red military uniforms and had red handkerchiefs with yellow spots on their heads.

Suddenly one of the dancers stopped and seemed to be staring at the children who grew afraid. The figure ran quickly towards them and the scared children scrambled over the stile. Edward, who was the youngest, was last at the stile and the fairy was hot on his heels *"having a clear view of him with his ancient, swarthy and grim complexion."*

Edward's sister managed to pull the boy over the stile and the fairy leaned over trying to grab him but did not follow them over the hedge. Running back to the house, the children told the adults there what had happened and a group of men went out to search the area but found nothing. In later life, he said that he was *"forced to class it among my unknowables."*

The controversial ghost hunter Elliot O'Donnell describes an incident not unlike that which befell the young Gould. He records it in his book *Byways of Ghostland.* O'Donnell says it happened to one of his relatives as he was driving his buggy down a road near Ballynanty in Limerick, Ireland.

> *"The horse had come to a dead stop, and was standing still, shivering, whilst the roadside was crowded with a number of tiny shadowy figures that were surging round the car trying to drag the unfortunate driver, who was quite frantic with terror, from his seat. Mr. B., however, concluding that what he saw could only be the fairies...of whose existence he had hitherto been very sceptical, seized the reins and urged the horse forward. Meanwhile his servant seemed to be still paralysed with fright, and it was not until they were well out of sight that the man found himself once again in possession of his tongue and normal faculties...Then he described what had befallen him...He was driving along quite all right, till the horse suddenly stopped, and when he looked down to see what was the cause of it, he perceived a crowd of fairies, who rushed at him, and tried to drag him off the car. He said their touch was so cold it benumbed him. But by praying hard he held on. The cause of the attack was apparent.*
>
> *'It was all because we came on them, sir, when they were dancing. They won't be disturbed when they are at their revels and enjoying themselves. Had they got me down into the road, maybe I should have lost my sight or my hearing or the use of my limbs, and in any case my soul.' "*

These stories give some insight into how feared faeries were in times past.

In 1907 an old blind man living with his wife in Ireland told Lady Archibald Campbell (wife of the Duke of Argyll and the subject of several paintings by Whistler) a strange story. He claimed that he had actually captured a fairy and kept it for two weeks before it escaped. The being was two feet tall, the man said, with dark skin and red hair. It was dressed in green clothes and boots, and wore a red cap.

> *"I gripped him close in my arms and took him home," the blind man said. "I called to the woman to look at what I had got. "What doll is it you have there?" she cried. "A living one," I said, and put it on the dresser. We feared to lose it; we kept the door locked. It talked and muttered to itself queer words...It might have been near on a fortnight since we had the fairy, when I said to the woman, "Sure, if we show it in the great city we will be made up. So we put it in a cage. At night we would leave the cage door open, and we would hear it stirring through the house...We fed it on*

bread and rice and milk out of a cup at the end of a spoon."

After the fairy escaped the couple's fortune took a turn for the worse. The man lost his sight and, it was said, eventually died, along with his wife, in poverty.

W.Y Evens Wentz, when collecting stories of fairy encounters in Ireland in early part of the 20th century when he was told a remarkable story by Neil Colton, then 73 years old. It had occurred to him when he was a boy, near the shores of Lough Derg in Donegal in the mid-19th century. Neil was gathering bilberries with his brother and his cousin at the back of his family home when all three children heard music playing close by. They were astonished to discover that it was the music of 6-8 small people who were dancing in a circle no more than a few dozen yards away.

> *"When they saw us, a little woman dressed all in red came running out from them towards us, and she struck my cousin in the face with what seemed to be a green rush. We ran for home as hard as we could and when my cousin reached the house she fell dead. Father saddled a horse and when for Father Ryan. When Father Ryan arrived he put a stole about his neck and began praying over my cousin and reciting psalms and striking her with his stole; and that way he brought her back. He said if she had not caught hold of my brother they would have taken her for ever."*

A far cry from wish granting, benign fairies of modern thought.

In 1918, whilst boarding at Claremont school, Eastbourne, Sussex, the novelist Pamela Frankau saw what she described as an albino dwarf scurry across her bedroom floor and out onto the landing.

Late one night in January 1931, 24-year-old farmhand Helge Eriksson had a bizarre experience. He worked on a farmstead called Dannebo, not far from Eslöv, in the province of Skåne, Sweden. It is a clear and beautiful winter evening when he decides to visit a friend on a farm nearby. He stayed a couple of hours and then started his way home around 10 P.M. Not far from Dannebo he was confronted by a very unusual sight. He was interviewed by Anders Liljegren,

> *"I was close to the farmhouse where I worked when I noticed 10-12 very small men approaching. They were at a distance of around 50 meters when I discovered them and they came in my direction at a fast pace. They were so small, they reached only to my knees, and I remember this strange sight frightened me."*

The little men were so close and the moonlight so bright that he got a good view of them. The small men wore dark green clothes with caps and their dress appeared to be made in one piece. All the men had beards and their heads were unusually large compared to their bodies. They had hard, cruel looks on their faces. A strong smell came from the gnome-like beings which lingered a long time afterwards.

The gnomes noticed Helge but they took no notice of him passing by, walking in an almost military fashion on the frozen ground. He could hear the sound of their footsteps continue out on a nearby field. Looking out on the field Helge saw a "strange, intense, blue violet light, hidden in some sort of fog." Suddenly there is a whining sound in the air and the whole sight is gone.

The strange smell and odd light phenomena have been noted in other Fortean cases. Later in life Helge described the entities as 'men from space' but he never mentions that in his early interviews. Clearly his world view and culture changed from when he was a 24 year old in rural Sweden to when he was a mature man. Yet clearly this has the hallmarks of a fairy sighting.

Mr. W. E. Thorner of Luton, Bedfordshire, had a strange experience on the island of Hoy in the Orkney Islands, Scotland, one winter's day during WWII. He was walking near a cliff top at Torness when he saw a group of small, wild-looking men dancing to and fro. He describes the encounter thus...

> *"One stormy day in winter I was walking or struggling along the cliff top at Torness. The wind was high and howled about, low-lying, swirling clouds part-enveloped the land in misty rain. At times the pressure was so great that I was forced to bend and clutch at the heather to retain a footing.*
>
> *On one such occasion, on looking up I was amazed to see that I had the company of what appeared to be a dozen or more ' wild men ' dancing about, to and fro. These creatures were small in stature, but they did not have long noses nor did they appear kindly in demeanour.*
>
> *They possessed round faces, sallow in complexion, with long, dark, bedraggled hair. As they danced about, seeming to throw themselves over the cliff edge, I felt that I was a witness to some ritual dance of a tribe of primitive men.*
>
> *It is difficult to describe in a few words my feelings at this juncture or my bewilderment. The whole sequence could have lasted about three minutes until I was able to leave the cliff edge."*

These beings seemed larger and more human like than the previous cases. The fay, like so called aliens, come in many forms.

Nandor Fodor in his 1964 book *Between Two Worlds* (not to be confused with Dermot MacManus' book of the same name) tells of a strange encounter with a tiny person who strayed too close to a woman washing her hair in Gloucester in the early 20th century.

> *"I was staying in an old house in Gloucester, and the garden at the back ended in the forest of Birdlip Beeches which covers part of the Cotswold Hills. It was before the days of the 'shingle' and I had washed my hair and was drying it in the sunshine in the forest, out of sight of the house. Suddenly, I felt something tugging at my hair and I turned to look.*
>
> *A most extraordinary sight met my eyes. He was about nine inches high, and the most dreadfully ugly, dreadfully misshapen, most wrinkled and tiniest mannikin I have ever seen.*
>
> *He was the colour of dead aspen leaves, sort of yellow brown-with a high squeaky voice. He was caught in the strands of my hair. He was struggling to escape, and he grumbled and complained all the time, telling me I had no right to be there, troubling honest folk, and, that I might have strangled him with my hair. Finally, he freed himself and disappeared.*
>
> *I mentioned my experience afterwards to a professor of Bristol University. He was not surprised and told me that Birdlip Beeches was one of the few places left where there were fairies, and no one could go there because of it."*

In 1948 Mr. E.J.A. Reynolds of London was staying in Horsham, Sussex, over the summer holidays. He was ten at the time. One night he was out setting rabbit tracks. He told the story to Fortean researcher Janet Bord.

> *"I decided to sit still and watch, being young I thought the rabbits would come out and I could see them being trapped. As I sat still and waited I suddenly realised that a small hairy man had stepped out from a blackberry bush. He was no more than 18 inches high and covered with hair. His face was bare but had a leathery look. The nose seemed sharp. I noticed it when it turned away in profile. It definitely had hands. Its arms seemed longer than a human being's. I did not notice his feet. It was definitely substantial, real. It did not notice me, or if it did it did not show it. It turned and disappeared back into the blackberry bush. When I told the couple I was staying with they laughed at me."*

The young Reynolds saw the same creature or another like it when he was on the upper deck of a bus. The creature was walking across a large garden in full daylight.

Irish researcher Dermot MacManus was told of an encounter on Dartmoor, a place infamous for sightings of the Little People. The witness Mrs. C. Woods was on holiday at Newton Abbot, Devon, in June 1952 and had visited Haytor on Dartmoor. On her return journey she saw a little man standing on one of the many boulders that are strewn around the area. He moved about six yards from the boulder and seemed to be watching her. The little man was elderly and wore a brown smock that hung to his knees and was tied with a cord. He had brown stockings and a cap. He stood between three and four feet tall.

> *"This was no momentary sight, as I watched him for some time standing there, and I wondered what he was. I had no idea at first that he was a little man; I thought rather of some animal until I got much nearer, and then I just said to myself, 'This is no animal, it is a tiny man in brown.' I felt, and still feel so convinced."*

Mrs. Woods returned to the summit of Haytor and returned with her son, but the tiny man was gone.

Another story recorded by MacManus concerned a girl of nine living in Cranagh, County Carlow. It occurred in the early 20th century. The girl was fetching her father's cows home and as she was waiting for them to come through the gate, she saw a tiny man walk in front of them. He had a buttoned up red coat and buff-coloured trousers. On his head was a black cap turned up at the front. One of the cows swung its head at him and he tapped it on the nose with a stick he was carrying. The little man walked past the girl so closely that he could have touched her, then stepped over a ditch and walked into a grassy bank. He walked through the earth as if it was not solid. This recalls the many cases of ghosts walking through walls.

Richard Holland, one time editor of *Paranormal Magazine,* recorded a story of a man walking his dog in Denbigshire, North Wales in 1961. The witness had struck the metal signpost several times to shake mud off his walking stick. Suddenly, a three-foot-tall goblin-like creature, dressed in green and with an ugly brown face appeared from nowhere. It seemed irritated at the noise and had an air of malevolence about it. The man's dog snarled at the creature and raised its hackles. The goblin seemed to vanish as quickly as it manifested.

Janet Board recorded another recent encounter from that fairy hotspot, Ireland. In the early 1990s, fifteen-year-old Brian Collins was on holiday in the Aran Islands off west Donegal. While out walking early one morning, he saw two little men fishing from a bank overlooking the sea. They were about 3 and a half feet tall, dressed in green with brown boots. One had a grey beard and a flat hat. They were laughing and talking in Irish, and suddenly they jumped over the bank. When Brian went to look for them, they had gone, but they had left a pipe behind. He took it back to the house where he was staying, but while there it disappeared from a locked drawer. When Brian saw the little men again, he tried to photograph them and tape-record their conversation, but nothing came out.

The vanishing evidence is something that turns up again and again in Fortean cases from photographs of monsters or UFOs not turning out to evidence going missing in the post or even from locked safes.

THE SLITTING MILLS TROLLS

Trolls are a race of creatures found in the folklore of Scandinavian countries. They are generally humanoid but hairy and ugly. They can range in size from dwarfs to giants. They are said to live underground in caves and tunnels and fear daylight. In terms of intelligence, they can range from club wielding brutes to beings that can use magic against humans. They were mostly thought of as hostile to man. You would not expect them to turn up in the English Midlands.

In 1975 a young couple, Barry and Elaine, are returning from a Christmas party in the early morning with their two children asleep in the back seat and driving towards their home in Slitting Mill, Staffordshire, England. The car stalled, as happens in so many of these cases. Barry coasted the car to the side of the road near Stonehouse by the edge of the village. He went to check on the engine but found nothing wrong. The car had simply stopped for no reason. When he returned, both he and his wife saw the 'trolls'. Elaine screamed as she saw a stunted figure scuttle across the road,

> "I just about saw it at the last second, and then another one followed it, and then a third one. The best way I can describe them to you is like a hairy troll or something like that. We had some moonlight and they were like little men, but with hunchbacks and big, hooked noses and not a stitch on them at all. Not a stitch, at all; just hair all over them. I'd say they were all four-feet-tallish, and when the third one crossed by us, you could see them at the edge of the trees – wary, or something, anyway."

Barry added...

> "We both know from memory that they came forwards, towards us, very slowly to us, and I've thought since that they were interested in us or wanted to know who we were. They came very slowly, and it was a bit like we were being hunted, to me. Elaine was hysterical; and with the kids with us, I wasn't far-off, either."

It was at this point that Barry and Elaine suddenly experienced a total lapse in memory - and the next thing they knew was that it was two o'clock and the car would now once again start easily.

Neither of them recall seeing the trolls leave. Barry later had dreams about the trolls surrounding the car and has been left with the nagging feeling that *'something happened to us'*.

The parallels with alien abduction cases are quite overt here despite the lack of a craft. It seems that fairies and aliens are interchangeable.

THE DWARFS FROM NARNIA

The user 'FineHair' wrote to the *Fortean Times* message board in 2004 with an account of what befell her one night in 1984. After a week away on a work seminar, she got home and unpacked her suitcase, hanging up her clothes in a walk-in wardrobe. Suddenly, she was overcome by sleepiness and decided to have a nap on her bed, fully clothed and with her shoes still on.

> *"The next thing I was aware of was the sound of several voices, all urging each other to 'Hurry, hurry' (I don't know if they were really speaking English or even if they were speaking aloud. Can only report the way it seemed.) It took a lot of effort to raise my head and look down at the foot of the bed, where the voices were. There were several small people tugging at my legs. They were trying to pull me from the bed and into the open wardrobe. It didn't seem odd to me. I wasn't afraid. I looked at them briefly and told myself that I was too big and heavy for them to move far. I decided it was safe for me to close my eyes for a few more minutes.*
>
> *Then I looked up and the small people were standing around me, looking at me silently. They were two and a half feet tall. There were males and females. They looked like 'gnomes' or 'dwarfs'. They were stocky. Their skin was very coarse and weathered, as if they spent a lot of time outdoors."*

The witness felt that they were waiting for her to fall asleep again and indeed she fell unconscious once more. Upon waking she found that she was in a reversed position, her head was where her feet were before and the little people had got her legs off the bed and were in the act of dragging her into the wardrobe. Again, she did not seem to think that the situation was strange. Nor did she feel any fear, but she did become angry. She kicked out at the little people and jumped up, standing in the middle of the room. She noticed that the light seemed very bright.

Realising that their victim was now fully awake, the dwarfs marched back into the wardrobe and vanished. They seemed to go down an 'incline' inside the wardrobe.

> *"I ran from the bedroom and into the living room. It was at this juncture that I fell apart in every way. It was all very sudden. I was just overcome with terror, shock, panic, hysteria. It was very acute. I couldn't breathe or think. I was close to being out of my mind with fear, but it had no real focus --it was just a hideous terror, I was reduced to the level of a very small, terrified child, in the space of a few seconds."*

She phoned a friend whom she had travelled back from the seminar with and he promised to drive straight over. Too afraid to stay in the house, the witness stumbled out onto the dark street. Her friend picked her up and drove her to his house. She was so frightened she spent the night in his spare room with the lights on and him in a chair beside her bed.

In the morning she told him what had happened, but he didn't want to speculate on it. Despite continuing to work together for a further 14 years, they never discussed the events again.

It would be tempting to dismiss 'Fine Hair's encounter as hypnopompic and hypnagogic hallucinations. These are hallucinations that occur when waking up or falling asleep. We will look at them more closely in the final chapter. But there is a sequel to the story. Some 10 year's later, the witness' daughter, then 20, revealed that she had seen the dwarfs as a girl and had been frightened of them. The little people came out of the wardrobe and tried to pull her in. She never told her mother as she had thought that she would not be believed.

This case is interesting as the witness felt no fear at the time, but was hit by a tidal wave of terror afterwards. It was as if she were under some kind of spell. Fairy lore is full of accounts of such bewitchment. Also of note, is the 'stairs leading downwards' in the wardrobe. Fairies and other entities are associated with subterranean dwelling places. The parallels with alien abduction need hardly be pointed out.

THE GREEN GOBLINS
In his 1977 phenomenological study, Timothy Bearsworth, studied 1000 submitted reports of experiences of a 'sense of presence'. One reference number 000235; involved what seemed to be an encounter with goblins.

> *"On the Friday a man came to clean the carpet and curtains in the drawing room. Later on there was a complete fusing of everything electrical. Clocks, radios, refrigerator, freezer, T.V. all the lights etc. In the evening I lay down on a sofa, closed my eyes and tried to relax. I then saw several little green men with very unpleasant expressions. They were looking at me. They seemed to be at a distance. I suppose "gnomes or goblins" would be an adequate description. I didn't like what I saw, and I was reminded of the time I had a rheumatic illness when I was seven, and had been very alarmed by the "little green men" I had seen then. Hallucinations, presumably."*

Maybe this was a hallucination linked with sleep and stress, but in our next case, the goblin left behind very tangible evidence of its visits.

THE RABBIT CRUSHER
The late, and sorely missed, Lee Walker was a Fortean researcher from Liverpool. He published an excellent magazine called *Dead of Night*. In one section called *"New Ferry after Midnight"* he recounted personal experiences from Merseyside. One of the strangest, concerned an encounter with what appeared to be a latter-day goblin. The story was reprinted in Lee's excellent book of the same name. The story occurred in the

late 1970s. Lee's family had two pet rabbits that they thought were both female. As it turned out one was male and had to be re-homed, but not before it had done what rabbits do best, and made its little friend pregnant. The rabbit gave birth to a number of young and all was well for several weeks.

Then one morning one of the babies was found dead. Its head had been crushed flat. It was assumed to have been an accident where the mother had stood on the youngster's head. Next day, another rabbit turned up dead. Once again its head had been literally crushed flat. It was decided that the babies, now weaned, were to be separated from their mother for their own safety. The young were put in a separate enclosure. However, the next day a third young rabbit was found dead in the same condition as the other two. The Walker family considered that some kind of predator was at work. But the hutch remained bolted, and what kind of predator crushed its victim's skull but did not eat them?

The next day, and the day after, the same thing happened. Finally the last rabbit was taken into a shed and put in a hutch with its lid weighted down with a brick. Lee's sister checked up on it and was amazed to see that something had moved the brick. She had a feeling she had disturbed something, just before it was about to attack the rabbit. But what kind of an animal could lift a brick off the roof of a hutch? She replaced the brick and went back inside the house. A short while later she had a feeling that he rabbit was in danger and rushed back out to the shed. Lee and his family heard her scream. She came rushing into the house in hysterics. She claimed to have seen a goblin.

Finally, they managed to calm her down and she blurted out a bizarre story. She had entered the shed, and once again saw that the brick had been removed from the lid of the hutch. She saw a dark shape crouching beside the hutch, and at first thought it was a huge rat. Then the thing stood up. She said it was a squat, humanoid figure clad in what looked like a monk's hooded robe. It was about three feet tall. As it turned to look at her, she saw it had leathery brown skin. It had a wide mouth, filled with sharp teeth, pointed ears, bloodshot eyes and a large, wart festooned nose. In one hand it clutched what looked like a lump hammer. She had run screaming into the house at that point. A search of the shed revealed no creature, but the last rabbit lay dead in its hutch. Its skull had been completely flattened.

LITTLE MAN OF THE LIZARD

I was once in a car with a friend Mohammed Bhula, who said he saw a small, brown skinned humanoid creature about 18 inches tall standing in the middle of the road. The event happened in the early 1990s, near Lizard Point, Cornwall. He saw the creature illuminated in the car headlights. It turned and ran off down a path that led to the beach. He could make out no facial features, nor could he be sure if the creature wore clothes. I did not see this as I was sitting in the back seat of the car, and he in the front passenger seat. We both walked down the path that the creature had run along, but saw nothing. There was however a distinctly eerie feeling.

NO RACE LIKE GNOMES

Wollaton in Nottingham, already the setting of a British UFO/Bigfoot encounter, also played host to one of the oddest modern fairy sightings on record. Wollaton Park consists of woods and lake grasslands, and surrounds Wollaton Hall, a mansion that has stood in for stately Wayne Manor in some of the Batman films. The park is slap bang in the middle of Nottingham.

On 23rd of September, 1979, 8 children between the ages of 8 and 10, were playing in the park. They included Angela Elliott, Glen Elliott, Julie Elliott, Andrew Pearce, Rosie Pearce and Patrick Olive. It was about 8.30 at night. They were alerted by a noise that sounded like a bell. The noise was coming from a marshy area of the park, fenced off with barbed wire. The kids found a hole and climbed through to investigate.

In the wooded area, they saw around 30 little bubble cars speeding around. The children estimated the cars were moving at 40-50 miles per hour. The tiny cars had triangular lights, and instead of a steering wheel, a circular device with a handle. Each car had two gnomes in it, one driver and one passenger. The gnomes had long white beards tipped with red, and wrinkled faces. They each had hats like old fashioned night caps with a bobble on the end. They also had blue tops and green or yellow tights. They were about half as high as the children, around 2 feet. The bubble cars made no noise and seemed to be able to leap over obstacles like logs. The children noted that they left no tracks in the mud.

Other gnomes were on foot, clambering out of the trees. The children emphasised how happy the gnomes looked, and how they were laughing. The kids watched the antics of the little men for about a quarter of an hour as they zoomed about, seeming to have the time of their lives. The gnomes then began to playfully chase the children with their cars. Again the witnesses stressed this was not aggressive, but a game. The gnomes would not follow them to well-lit areas, they noted. Finally, as it was getting late the children went home leaving the gnomes to their games.

When they returned to school the next day, their headmaster questioned them separately and taped the interviews. He also had them draw what they saw. The accounts and pictures dovetailed, the Headmaster's opinion was that the children were all telling the truth, although, as might be expected, there were minor differences of detail and emphasis between their different accounts.

This is not the only time little people have been seen driving cars. Fortean researcher and author, Janet Bord has this to add.

> *"Over six years before the Wollaton fairies were reported in the media, I had corresponded with Marina Fry of Cornwall, who wrote to me giving details of her own fairy sighting when she was nearly four years old, around 1940. One night she and her older sisters, all sleeping in one bedroom, awoke to hear a buzzing noise (one sister said 'music and bells').*

Looking out of the window, they saw a little man in a tiny blue and yellow car driving around in circles'. He was about 18 inches tall and had a white beard and a 'blue pointed hat'...he just disappeared after a while."

If that isn't weird enough for you, how about a fairy plane? In 1929, an 8-year-old boy and a 5-year-old girl were playing in their garden in Hertford, when they heard a noise like an engine. They saw a tiny plane, with a 15 inch wingspan, fly over their garden fence and land in the garden, narrowly missing their dustbin. The tiny pilot had a leather flying helmet on, which he doffed as he waved at them before taking off again.

FAIRY TREES, HILLS AND ROCKS
In some countries fairies are said to inhabit trees, hills or rocks and moving or damaging their dwellings is thought to bring disaster. In 1968, it was reported that the course of a new road had been altered in Donegal, Ireland, as none of the workmen would cut down a fairy tree. Even though he had just chopped down a wood, contractor Roy Green refused to touch the fairy tree, a gnarled old thing standing in the middle of a field.

"I refused to cut it down and I would not order any of my men to do the job. I have heard so much about these fairy trees that I would not risk it. There is something uncanny about it. The roots are not more than a couple of feet below ground-yet it defied a hurricane seven years ago."

The *Belfast Telegraph* of November 9th, 1959, records an even more remarkable tale. At Dunroe, not far from Cranagh, John Byrn was moving a large bush with a bulldozer. Suddenly, a 3-foot-tall man ran out from beneath the bush. Three other workmen saw him as he dashed across a field and over a fence. Under the tree the men discovered a flagstone that was seemingly over a hole. They couldn't budge the stone and when they tried to use dynamite on it the explosive failed to detonate. In the end, the farmer left the stone alone and removed all his machinery.

In 1999, in County Clare, also in the Emerald Isle, a fairy bush was to be cut down as part of the £100 million plan to bypass Newmarket-on-Fergus and Ennis. Folklorist Eddie Lenihan warned that the bypass plans could result in misfortune, and in some cases death, for those travelling the proposed new road.

Construction of roads also causes problems in Iceland, where fairies are supposed to live underground or in boulders and large rocks. In the early 1980s, a road was being built at Akureyri, in the north of the country. Machinery broke down and workers became ill. The road that connected Reykjavik to Kopavogur was diverted around a hill where elves were supposed to live.

Back in the 1930s, the first attempts to build a road through it were abandoned after a series of accidents. Plans to level the hill re-emerged in the 1980s, but problems reoccurred and workers refused to go anywhere near it. Even TV crews said their cameras failed to record anything when pointed at the rock. A road was finally built skirting round the protrusion, the house numbers skip a plot, in deference to the invisible neighbours, who ultimately had their way. Reykjavik's planning department even published a map, showing the dwelling places of the 'hidden people'.

A PERSONAL ACCOUNT: PIXE LED ON DARTMOOR

It was a drizzly evening in October of 2000, when, together with three friends, I set out to visit Scorhill, a Bronze Age stone circle on Dartmoor, Devon. We decided to visit on the spur of the moment and had not planned on going to the circle before that day. The bad weather didn't bother us and we set out for the moors in the early evening. None of us had visited Scorhill before, and only had a vague notion of where it was.

The circle actually has a rather rough carpark close to it, but we didn't know that at the time, as we had only decided to visit at the drop of a hat, with almost zero research. We arrived on Dartmoor, parked up, and walked in the general direction of the circle. If I recall correctly one of us had a map. We did have torches and boots but no waterproofs. We went tramping across Dartmoor in the drizzle, it was still light, but the skies were overcast and dusk was creeping up.

We entered a field enclosed by a dry stone wall, via a metal gate. There were a few sheep in the field. We began to cross the field to get to the gate on the other side. The field was not a large one, but we could not seem to find the gate. It was not yet dark and we didn't need to use the torches. We stumbled around the field for what seemed like ages, going round and round in circles. At times it felt like wading through treacle, a sort of heavy feeling on the limbs. Suddenly, we noticed it had turned dark, seemingly instantly. I remember my leather jacket being soaked. We all wondered how it got so dark so fast.

Then, beside the dry stone wall there appeared a circle of lights, standing vertically, and putting me in mind of a tiny Ferris wheel. The circle was about five feet across. It consisted of points of bluish-white lights, each about the size of a grape. I pointed it out and asked the others if they could see it. They all said that they could. It was visible for about ten seconds before 'winking out'.

As we all stood gaping, the lights re-appeared in a different configuration. This time the lights seemed to form the crude outline of a person. Again it was about five feet high. It stood still, with its arms and legs together. The head seemed to lack a neck and sat directly on the shoulders. If you imagine the logo used on public toilets to denote the men's, but with the head sunken down, you will get an approximation of the shape. Once more we all saw it, and once more it was there for around ten seconds before winking out again.

At this point I knew something was very wrong, and that something very strange was happening. I recalled the legends of travellers being 'pixy led' on the moors and that to break the spell you needed to turn your coat inside out. I said to the others *"Everybody stop. Take your coats off, turn them inside out and put them back on again."*

This is an utterly crazy thing to say to anybody, and I was fully expecting them to ask me what the bloody hell I was going on about. But the strange thing is, that nobody questioned me. They all just did as I asked. We all turned our coats and there, right in front of us was the gate! We walked through it to find that the Scorhill circle was right next to the field. There were a number of people there. They had lit a fire and some were playing instruments but we had not seen the fire, or heard the instruments, whilst in the field.

Another weird thing was that none of us talked about it afterwards. We just walked over to the circle, stayed a while, then drove home without discussing what had occurred. I told my housemate Jon about it when I got home but that was it.

These seemed to have been some missing time involved. Somebody looking at this with a modern viewpoint, would doubtless class it as an 'alien' encounter or possible abduction, but to me, it just seemed like a meeting with something of the fay.

I have visited Dartmoor countless times since, and never encountered anything even slightly odd.

'Aliens' supposedly abduct people, leading to missing time. They are said to put implants in their victims to stop them recalling the events. They are said to be interested in hybridizing with humans.

Fairies kidnap people and take them away to fairyland, leading to lost time. They sometimes place silver pins behind their victim's ears, to stop them recalling the events. They are supposed to have a weakening bloodline, and take human children for breeding, leaving a 'changeling' in their place.

It is clearly the same phenomena. The 'alien' mask is nothing but an update for a more technological age.

Back in the 1950s, 'contactees' like George Adamski, claimed to have met beautiful, Nordic looking aliens from Venus. Now we know that life on other planets is highly unlikely to resemble humans, and that Venus could not support life. Then, less human looking 'aliens' emerged, like the greys, who kidnapped Betty and Barney Hill and claimed to be from distant star-systems. It is as if the phenomena updates itself with mankind's knowledge as it grows. In less technological times, fairy sightings were abundant, in the space age it is aliens. Yet still we have the throwbacks, fairies seen in the modern age. Why is that? Other legendary creatures, like dragons, are still reported today but others not. For example, I know of no credible claims of modern day sightings of unicorns or griffins. So why should only certain types still manifest and not others? Fairies are still very much part of the modern world today.

DEMONS AND BEDROOM INVADERS

All cultures have a concept of actively evil spirits or beings, that exist to cause harm or bring fear. In the west we call them demons. In other cultures they can be termed as djinn, gallu, rakshasas, oni and a legion of other names. We generally think of a demon as a hairy, humanoid creature with hooves and horns. This mainly comes from Christianity demonizing the horned nature gods of other cultures, such as Pan or Cernunnos. Demons of this type are still reported, but these entities can come in an array of shapes and sizes, both in mythology and modern accounts. For example, the 17th century grimoire *Ars Goetia,* or *The Lesser Key of Solomon,* lists a number of weird looking *demons* .

Marchiosas takes the form of a wolf with eagle's wings and the tail of a serpent, Valefar had the body of a lion and the head of a thief. Bael was a man with three heads, one of a toad, one of a cat and one of a man. Even weirder were the demons supposedly summed up by the magician, Alistair Crowley, in his occult battle with MacGregor Mathers, head of the magickal group Order of the Golden Dawn. Crowley supposedly called on the attendant demons of Beelzebub, and they were recorded by his wife Rose. Nimorup was a dwarf with an outsized head, large ears and green and bronze lips that slobbered. Norminon looked like a huge red jellyfish, with a luminous green spot on it. Holastri was a gigantic pink beetle. Demons can take almost any shape.

We think of demons as the minions of Satan, but this is just a Christian world view. Many cultures that have no analogue of the Devil, still have these evil spirits. Like all Fortean phenomena, their interpretation is largely down to cultural factors.

Bedroom invaders are entities that manifest in the bedrooms of the witnesses. They generally, though not exclusively, bring feelings of menace. Again, they come in all shapes and sizes, and I have lumped them together with demons, because in centuries passed this is what they would have been identified as.

THE THING FROM OVER THE HILL

One of the weirdest cases was recorded *The Powers of Evil* by Richard Cavendish. It involved a couple he referred to as 'The Smiths'. One evening, in 1940, as she was chatting to her husband, Mrs Smith said, quite out of the blue *"It will come over the hill when it comes."* Afterwards she had no recollection of saying this, even though her husband insisted that she had. Shortly after she became nervous about being in the house after dark. About three months later, she woke her husband one night and told him that the thing from over the hill was nearly upon them. They heard one of the outside doors opening and a heavy, wet tread upon the stairs. As the Smiths clung together, the door swung open, and a hideous thing waddled in. It was bloated and naked, with skin blotched green, purple and yellow. It had a head that almost came to a point at the top, long earlobes that nearly reached its shoulders, webbed feet and a thick bull neck. It crossed to the window and vanished. Mrs Smith later commented.

"It was horrible and the absolute essence of evil...I have never

experienced anything so dreadful before or since, and I hope I never shall, God willing. I still experience the same horror when I talk about it, or write about it, as I am doing now. I have never been able to discover why I saw it, and I have never been able to find out what it was."

THE LIFE SUCKING TOAD-APE

J.W Brodie Innes was a lawyer, novelist, and one of the leaders of the Edinburgh branch of The Golden Dawn. He became convinced that an 'elemental' was preying on his wife, sucking up her life force and causing illness. He worked out a magickal ritual to manifest the demon visually. He says it appeared at first as *'a vague blot like a scrap of London fog'*. But then it gained a more tangible form ' *a most foul shape between a bloated, big bellied toad and a malicious ape'*.

Brodie Innes concentrated his destructive will on the thing. *'There was a slight feeling of shock, a foul smell, a momentary dimness and then the thing was gone."*

Afterwards his wife rapidly recovered.

THE THINGS UNDER THE BED.

In December of 1883, Mrs Charlotte E. Field wrote an article for the magazine *Merry England,* in which she described a haunting at a house in Montpelier Terrace, Cheltenham, England. Since her arrival in 1874, there had been crashing noises, running footsteps, talking and whispering sounds, and the appearance of two apparitions, a girl and a woman. As well as this, her two daughters had seen horrid creatures emerging from under their bed at night. The younger daughter stated...

> *"My sister always teased me for being afraid of ghosts, but, one night, she herself got really frightened. She was standing by my bed and mother and the maid were talking outside in the passage, when I heard something under my bed, and asked my sister to look. She laughed and said it was nothing, but when she looked, about a dozen black creatures, like toads, only with horns and a tail, came out from under the bed, and ran over my sister's foot and across the room, disappearing down the wainscot to return no more."*

The older sister confirmed the event.

> *"My sister was in bed, or standing, I forget which, and I was sitting by her, and our mother was outside with a servant. Suddenly my sister said to me 'There is something under the bed, do look' I did so and these little creatures ran out and one of them ran over my foot."*

The older sister tried to convince herself that the creatures were some kind of mechanical contrivance, put there by someone to frighten the family out of the house.

She failed to convince anybody, least of all, I think, herself. She wrote...

> *"Black things with horns and tails could not have come from a good place if they were supernatural."*

This is yet another case where a haunting features small strange creatures. We looked at a number of them in chapter two. These horned and tailed toads seemed particularly demonic.

SATYRS IN THE BEDROOM.

Dr Arthur Guirham was a British doctor and psychiatrist. In his autobiography *A Foot in Both World's: A Doctor's Autobiography of Psychic Experience,* he writes of seeing a demon in his bedroom as a boy.

> *"I lay on my bed and felt his presence. The air was crackling and electric. A wave of vibration came through the door of my bedroom. When the wave ebbed away quickly I was drawn towards the door...I knew he was calling and that the minute vibrations in the atmosphere were a summons to me I went from my bed through the air palpitating with a new cold and opened the door, and he was waiting for me...*
>
> *His face was hairy. It was covered, like his body, with a felt of blue-grey hair. He was a man in his features and in his almost upright, slightly leaning posture...His legs were different. I was not aware of them as human. They ended in a shaped stump of something like a hoof...There was a shining aura about him...*
>
> *I do not know how I went to bed...After he was gone, the night was empty."*

In *The Leicester Mercury* of August 28th, 1980, there was a story of a 10-year-old boy named Billingham, who shocked his art teacher by drawing a goat-headed humanoid with curving horns and cloven hooves. The boy said he often saw the creature standing at the end of his bed. His mother said that she and her husband had often heard crying, but on investigation had found that their children were quietly reading. Mrs Bellingham's mother, who was staying in the house one night, awoke to find herself being strangled by strong hands. Switching on the bedside light, she found nothing there. Some of their neighbours claimed they had woken up in the night, to see the figure of a monk standing over them.

The late Fortean researcher, Colin Wilson, linked the case to the nearby Humber Stone, a half buried boulder. It was thought, that anybody who tried to move or damage the stone, had ill luck. In the 19th century, a farmer whose plough had broken a piece off the stone, never prospered thereafter and died in the work house. A curate who attempted to cover the stone in earth was violently thrown from his gig.

THE DEMON SPHERE

Steve Cliffe, editor of the *Stockport Heritage* magazine, and paranormal investigator quotes a story in his 1993 book *Shadows A Northern Investigation of the Unknown*, that shows what strange forms demons can take and how they can be attached to certain objects.

His informant was a friend, whom he deliberately does not name. He and another man where sharing a flat in south London in the 1960s. Both men were students on degree courses. One of the men had acquired a curio, a wooden charm made in the South Seas. It has a stylized face, not unlike the Easter Island statues, with mother of pearl eyes. The matchbox sized carving brought with it an attached demon. Cliffe's informant stated...

> *"The actual psychokinetic effects included a feeling of cold, a subtle alteration in the quality of the light, and awareness of the approaching presence of something extremely evil knew when it was coming, approximately how long it would take to get there and from which physical direction it was approaching. While it was approaching I experienced a sense of terror and dread, which induced a kind of paralysis of thought and action.*
>
> *The appearance of the demon...A pulsating, translucent and only semi-visible entity a few feet in diameter. Globular in outline. Capable of flying-travelling through the air at will."*

The demon brought a wind with it, that banged gates and made falling leaves swirl. The men burned the charm, which changed into a weird yellow colour in the flames. When the informant's friend went to flush the ashes down the toilet, he received an electric shock that threw him against the wall.

THE DESERT GARGOYLES

Carlos Abett de la Torre, a sub-officer in the Chilean army, was driving through the Pampa Acha desert with his wife, three children and nephew one night, in July 2004. The family were visiting relations in the city of Arica.

At about 9.00 in the evening, and around 20 miles from their destination, Carlos' eldest daughter, Carmen, saw a pair of extraordinary entities through the back window.

> *"I was travelling in the back seat with my brothers, talking, and suddenly everything went dark. Then I told my brother what I was seeing and he told me to keep quiet, because Mom gets nervous. Later I looked through the window and saw some things that looked like birds, with dogs' heads and back swept wings. My father said they were like gargoyles."*

The creatures were about 6-feet-long, and Carmen and her brother watched them keep pace with the car. Their mother Teresa, soon caught sight of them too, describing them as 'dog-faced kangaroos'. Carlos stepped on the gas, but the gargoyles matched the speed of the car.

Suddenly, Carlos saw two more of the gargoyles on the ground, bounding out in front of the car. He narrowly missed them, and increased the car's speed, as four of the creatures, two flyng and two leaping, pursued the family. Finally, the car outpaced the creatures and the family arrived in Arica.

Carlos and his family told their relations of the encounter, but asked them not to tell anybody else, for fear of damaging Carlos' military career.

The family were terrified at the thought of driving back after the visit. but in the event, nothing happened. Subsequently another sub-officer, Diego Riquelme, claimed to have seen one of the creatures as well. After this the family came clean about the encounter.

Sceptics, rather pathetically, tried to explain the sighting as 'escaped ostriches' despite the fact that ostriches cannot fly, and look nothing like the description of the gargoyles.

THE BARCELONA DEVILS

In 1970, Margaret Fry visited Barcelona, in Spain. She had joined an international correspondence club and via this had been introduced to Francisco Deulander, a Spanish man who was trying to improve his English. Deulander had booked her into a hotel in Barcelona, and she arrived at about midnight. Fry was not pleased when she learned that Deulander was to be spending the night in the same room as her, but she was too tired to make a scene. Deulander said his intentions were honourable, but if he were to return home, he would not get there until three in the morning and would wake his elderly mother. In any case the beds were separate.

During the night, Margaret woke up to see moonlight streaming into the room through the patio windows and illuminating a weird sight.

> *"And there in the corner by a wash basin sat three little devils with horns.*
> *They formed a close circle and each had their heads bent, one arm linked*
> *and the other hand over their three quarters closed large eyes"*

Her screams awoke Deulander who saw the devils as well. Then their beds began to rock violently, so much so that Fry thought she would be tossed through the plate glass windows. She managed to leap onto Deulander's bed. He managed to switch the lights on and the beds stopped their wild zig-zagging, and the devils vanished.

The pair left the room and spent the night on chairs in the reception. The next day they checked into another hotel, and Fry spent the rest of her holiday being shown around Barcelona by Deulander, and nothing further happened of an occult nature.

Some year later, Margaret contacted Francisco again to find out if that night in 1970 was a fact or dream. His response was *"Don't mention that again, please, please it was the most horrible, horrible experience of my life"*

A DEMON AT NASA!

One windy night, sometime in 1986, Frank Shaw, a NASA archivist at Houston's Johnson Space Centre, met with a creature that looked like it had crawled out of hell. In 2004 Nick Redfern interviewed Shaw's daughter Desiree. She recalled how her father had come home late one night looking frightened, and after sufficiently composing himself, told his wife and daughter of the thing he had encountered.

After the end of his shift, Shaw had walked to his car. Glancing up he was horrified to see a jet black humanoid, perched on the edge of one of the Space Centre's buildings. It seemed to have a black cloak on its shoulders, from beneath which two bat-wings emerged. The thing looked down at the witness and Shaw felt that it was savouring the fear it created in him. Fumbling for his keys, the frightened man got into his car and drove off, too scared to look back.

His family advised him not to tell his superiors at work, for fear of him losing his job. But as the weeks went by, the sighting haunted him. Finally, he decided to confide in his immediate supervisor. Amazingly, he was told that he was not the first Johnson Space Centre employee to have seen the creature in the isolated areas of station. Shaw's boss revealed that a file had been opened on the creature just a few months prior to Shaw's sighting. The file had been opened after two guard dogs, large Alsatians, had been found torn apart.

Later, Shaw was questioned by NASA security, who were apparently flown in from Arizona. They advised Shaw that it was in his family's best interests to keep quiet about the sighting.

If the story is true and a NASA employee saw a demon, then it really makes you think!

THE BAT WOMAN OF DA NANG

UFO researcher Don Worley once interviewed a former US Marine, who had encountered a winged humanoid in Vietnam, in the summer of 1969. PFC Earl Morrison was on top of a bunker on guard duty with two other soldiers, when they saw something flying towards them.

> *"We saw what looked like wings, like a bat's, only it was gigantic compared to what a regular bat would be. After it got close enough we could see what it was. It looked like a woman. A naked woman. She was black. Her skin was black, the wings were black, everything was black. But it glowed, it glowed in the night...kind of a greenish cast to it.*
>
> *We watched her go straight over the top of us, and she didn't make any*

noise flapping her wings. She blotted out the moon once...That's how close she was to us. And dark...looked like pitch black but we could still define her because she just glowed"

The flying woman had long black hair on her head, and her whole body looked as if it were covered in fur. Her hands seemed moulded into her wings, and the wings rippled as they flapped. The three men watched her for about three minutes, before she vanished into the distance.

A BEAST CALLED BATSQUATCH

It was 9.30pm on April 16th, 1994, when 18-year-old student Brian Canfield, was driving southeast of Buckley, Washington. He was on his way to Camp One a settlement near Lake Kapowsin. It will come as no shock to the reader, that his pickup truck stalled. Suddenly, a huge winged creature landed in the road in front of him. The muscular creature was around 9-feet-tall and covered in bluish hair. Its head was like an outsized wolf, with white fangs and triangular, yellow eyes with crescent shaped pupils. Its hands and feet bore claws and on its back was a pair of huge, bat-like wings. It resembled a hybrid of wolf, ape and bat.

Brian felt that the thing was just resting, and indeed after a couple of minutes it unfurled it's wings and flew away again. As soon as it had gone, Brian's truck started again. He drove home in a panic and told his family. His father drove to the area to investigate but found nothing. Later there was a police investigation that again turned up nothing.

One of Brian's friends coined the name 'Batsquatch' and the moniker stuck.

THE SCREECHING MAN-BAT

Linda Godfrey, that tireless monster researcher from Wisconsin, heard a story of a demonic encounter from a member of the Cherokee Nation. The man - Wohali, 53 - was driving with his 23-year-old son along Briggs Road, at nine in the evening, on September 26, 2006. As they passed a wooded area, a winged creature dive-bombed the car. It had grey, wrinkled skin and was about six feet tall. Its bat wings spread to 14 feet, roughly half of the width of the road. Its horrible face had yellow eyes and a mouth full of fangs. Wohali recalled it had a protruding rib cage, like that of a bird that anchors the wing muscles. The legs were man-like but bore talons.

The witness thought that the creature was going to smash through the truck window, but it veered aside at the last moment, flying up into the trees and making a ghastly shrieking noise. The high pitched cry made both men feel violently sick, and they had to pull over and vomit. Both were ill for days.

They were convinced the entity was supernatural, as their home was subsequently plagued by weird noises that caused their dog to hide under the bed.

FLYING MONKEYS AND DEVIL APES

Like the witch's minions in *The Wizard of Oz,* winged monkeys of a seemingly demonic nature have been reported in real life. A woman called Jacki Hartley wrote to Karl Shuker, one of the world's foremost cryptozoologists, with a story of a series of encounters with a winged, simian horror in England. The first occurred when she was four, in 1969.

> *"I was in the back of the car when I suddenly heard an awful screeching scream. Mum and dad were in the front chatting and heard nothing. It was twilight and I looked out of the back window into the trees. I saw what I can only describe as a monster. It had wings which it unfolded and stretched out before folding back up again, red eyes and a kind of monster monkey face with a parrot's beak, and was about three feet high."*

You could argue that this could be a false memory, due to Jacki's young age, but she would see the creature twice again. She said she saw it again at the age of 11, whilst her family were driving near the village of Robertsbridge. Her most recent encounter was in 2006, when she saw the creature flying past the window of her house in Tumbridge.

She was convinced that, despite looking like a flesh and blood creature, the thing was supernatural. She had nightmares about the imp as a child.

Other demon reports involve a creature of a simian cast. Dermot MacManus describes how his father had an encounter with such a creature at the age of fourteen in Ireland.

He had been playing hide and seek with his elder brother among stable buildings during the Christmas holidays. He was alone in one of the granaries above the stables. Lifting up the trap door to jump down, he noticed that the horses below were snorting, trembling and stamping with fear. Then he saw the reason for their terror. Crouching on a manger, just 12 feet away from him, was a hideous creature squatting on its haunches. It was bony, humanoid and covered in greyish-brown hair. Its hands were long and bony, with sinewy fingers ending in claws. The creature looked at the boy with eyes glowing like coals of fire. The lad fled in terror.

A remarkably similar story was told about an entity in on old Essex farm, called Devil's House, on Wallasea Island, among the wild marshes around the River Crouch. The house got its name from its original owner, who was called Daville. According to records, it was known as Demon's Tenement as far back as the time of Charles II. The house seemed to have had a vile reputation for some time. A certain room was said to be haunted and anybody who slept in it experienced a paralysis, a creeping cold, a sense of deep terror and a noise like the flapping of huge wings close to the ceiling of the room. During World War One a sergeant, who scoffed at the stories, offered to spend the night in the room in order to debunk the 'ghost'. He was found sitting in a chair downstairs in the morning. He was deathly pale and flatly refused to talk about what he had seen.

One harvest time, a labourer working in one of the barns, heard his name being called and wandered through the house looking for the caller. Suddenly he felt an overwhelming urge to kill himself. He wandered like a man in a trance, until he found a length of rope and tied it in a noose about his neck. He found some steps and put them against a beam, intending to climb up it and attach the other end to the beam to hang himself.

He heard the same voice whispering *"do it, do it"* over and over. Looking up he saw a creature stretched out along the beam. It had black hair, yellow eyes and looked like an emaciated ape. The shock of seeing the creature snapped him out of the trance, and he threw off the noose, scrambled down the ladder and fled the house.

These creatures could have been Bigfoots on a crash diet, if it were not for the location of the sightings. Again, we see the walls we erect, and the categories we construct crumble.

Another encounter with an ape demon occurred much more recently.

In 2006, Jerry Glover wrote to the *Fortean Times* message board about a creature he had encountered when he was 16. He had awoke one night finding himself unable to move his arms and with a tight feeling on his chest. Opening his eyes, he saw an ape like creature squatting on his chest. The creature had brown fur and red glowing eyes. He likened the thing to the creature in Henry Fuseli's painting *The Nightmare,* only with a more simian appearance. It exuded an aura of evil, and Jerry had the impression that it had been waiting for him to wake up, so that it could gloat at his helplessness.

The ape-thing began to talk in a harsh, inhuman voice, mocking its victim and telling him how pathetic and helpless he was. With a herculean effort of will Jerry managed to overcome the force it was exerting, and it gradually faded away. The whole ordeal had lasted about 5 minutes. The case has a similarity to the 'Old Hag' or sleep paralysis condition that we will be looking at more closely in the final chapter.

Jon Hare, to whom this book is dedicated, is an old friend of mine and has accompanied me on a number of my cryptozoological expeditions. In 2020, he was subjected to a series of night-time attacks by a demonic entity of the bedroom invader type. He was living in a shared house in the Elephant & Castle area of London at the time. He wrote up the experience for me to use in this very book.

> *"I had the dubious honour of receiving a personal visit by the 'Old Hag' last night; the REAL thing, not just another one of my ubiquitous sleep-paralysis episodes.*
>
> *Motherfucker came sneaking out of the wardrobe in the corner of the*

room (I've seen things brewing in there out of the corner of my eye in the past) & proceeded to climb on top of me while I struggled helplessly to move, then snaked its way slowly up my body until it was lying on my chest with its face pressed right against mine, while I whispered "you aren't real.... you aren't real!!" at it in a useless attempt to convince myself that what was absolutely positively really happening, was not in fact actually really happening (even though it quite obviously WAS actually really happening).

It was grinning directly into my face, staring into my eyes for what seemed like forever... then it PULLED THE SKIN OFF ITS ENTIRE HEAD LEAVING JUST A WET LUMP OF DARK RED LEAKING FLESH, which it smeared brutally against my own gibbering face until I was finally able to scream myself into movement.

Delightful!

It was sleep paralysis, which I get at least a few times every week, but with the addition of an absolutely vivid visual & tactile hallucination (I hope) of an 'entity' that climbed on top of me & made absolute uncomfortably intimate personal contact with me, by making eye contact & sliming its way up my body until it was literally forcing its face against my own, keeping eye contact the whole time. VERY uncomfortable & scary, & ultra, ultra real. Definitely a 'hallucination' or 'vision' & not a 'dream', if you see what I mean. An 'experience', where something visited me that felt utterly unlike a dream.

I've had occasional scary hallucinations during sleep paralysis before, but this particular one has occurred a couple of times that I remember, & both of them were in the last few weeks. Plus I've had another couple of similar 'weird non-dream scary experiences' during sleep paralysis in recent weeks involving something moving around very frighteningly & realistically in my wardrobe—where the thing came from this time—although on those occasions it never emerged & I managed to shake myself out of the paralysis before anything else happened. But the last time this happened it was IDENTICAL: something sliming up my body & looking me in the eyes while it pressed its face into mine.

 When I first saw its face peering at me around the edge of the wardrobe door, as I lay in my paralysed state staring out across my room, I thought it was either a cute boy, or maybe my housemate Roz, but I quickly realised as it stepped out & strode inexorably towards me that this was a 'mask', & the thing became increasingly *nightmarish as it slathered itself all over me & started to force its face into mine, until what was pressing into me was just red peeled flesh: the skin literally fell away from its skull taking even its eyes with it until it was just a lump of meat.*

I actually woke up screaming a couple of months ago because I was paralysed & saw something moving in there... got REALLY scared when I realised that I was 'actually seeing it', if you see what I mean, & literally screamed myself awake. It was like a snake coiling around in the darkness... or maybe wet intestines sort of dangling there & moving around.

This time it was a little white smiling open face that peeped out into the room... I thought it was my housemate popping in to see if I was OK until I realised that it was looking out of my wardrobe: the door is on the other side of the room.

And then of course it stepped out in the room as I struggled to move & the whole thing became sort of a slow-motion waking nightmare.

This felt VERY VERY personal. I almost NEVER get 'entities'—the sense (the TRUE sense) that another being is in the space I'm in— during either sleep paralysis or drug/k/acid experiences. So the fact that this 'thing' was literally looking DIRECTLY INTO MY EYES & making things as personal as possible was really frightening. And I didn't just SEE it either: the sense of its body on mine was AT LEAST as 'real' as the seeing.

It felt UNCOMFORTABLY personal, is what I suppose I'm trying to say! I've seen eyes like that before on a night safari in Singapore, but these belonged to a leopard that emerged suddenly from beyond the circle of our lights & trotted directly & inexorably to within a few yards of me before it veered off & vanished back into the darkness.

It never broke eye contact either & they were similar eyes.

When I realised that this had happened before & that I'd also 'seen' things in that wardrobe before... makes me feel like I should be looking for some protection, frankly. Might have to start doing the LBRP again...

It's the fact that I was talking to it & trying to convince myself it wasn't there (futilely), even as it stared directly into my face & refused to not be... that's what really got me. I went from "oh this is weird" to "FUCK, THIS IS ACTUALLY HAPPENING!! ARGH!!" in about a minute.

All that happened was it just got realer & realer & ever closer & closer & more horribly intimate. & that was BEFORE it ripped its skin off!

I was definitely speaking to it last night (telling it that it wasn't real etc. in increasingly less-convinced ways as it advanced lol) but I couldn't speak above a whisper due to being unable to move, etc. It's why it was able to force itself into me without me struggling. If I can't SHOUT at it... do you just 'think' it away? How do you project that rage at it, when all your other normal methods of resistance are made impotent by your own body's paralysis? (By which I mean, this paralysis kind of creates a feedback loop of helplessness, since every time you try & resist you realise you can't... how do you get past that?)

I think it's the physicality that sort of blindsided me... it's just so SO completely 'there' that my paralysis—which is also physical—kind of made it seem like I was helpless in the face of a real person forcing themself on me, if you see what I mean.

In other words, I couldn't move, but IT DEFINITELY COULD... it didn't seem like a spiritual threat, but a physical one, & I can't deny that I was absolutely helpless: I literally couldn't move or speak above a whisper!

The 'reality' of the experience can't be understated, I think. SUPER-vivid & unlike the usual 'dream' that basically can be dismissed on waking. This really felt like something was in my room & lying on top of me. Not like any dream I've ever had, apart from the few other similar experiences I've been unlucky enough to have during sleep paralysis over the years.

(And not just 'something', either: an absolutely personal & malevolent 'entity' that was just as 'involved' in the whole thing as I was. & remember, I wasn't 'asleep', as it were; I was awake, paralysed & staring out at my room, when 'it' entered. If 'alien abduction' feels like that, then I can completely understand why people can't be rationalised out of the notion that something really happened. No joke though, as a person who I hope you agree is usually quite able to abstract my own experiences & 'not believe my own bullshit'... this REALLY FELT like some 'thing' was in my room & personally attacking my helpless body. Not remotely like a 'dream', which was very instructive actually.)

An interesting detail was that I can perfectly recall the feeling of the 'thing' on top of me. It felt like an area of pressurised air pushing down on me, like a balloon full of air if you removed the skin of the balloon; a faintly yielding force, rather than flesh & blood, giving a sense of pressure without texture. (A sense of textureless pressure.)"

It turned out that a guy in the flat directly below Jon's had been practising black magic. Whether this had any connection to Jon's nasty, nocturnal interloper is unknown. In a former time, this entity would probably be identified as an incubus, a sexual demon. The incubus and its female counterpart, the succubus, visit humans at night and lay with them. According to legend, they could swap gender, the succubus taking sperm from the men they lie with, then reusing it in their male form as the incubus.

I suggested to Jon that he left a small bowl of vinegar beside his bed. Vinegar contains acetic acid that is known to interfere with the etheric charge of some paranormal entities. This will be expanded upon in the next chapter. Upon taking my advice Jon was not troubled by the demon again.

THE TEMPLE DEMONS
Neil Geddes Ward, fairy and ghost researcher, was once told a story by a Persian gentleman he met on flight to Spain. The man was an architect and the two started a conversation. Neil mentioned his interest in strange things. The man, whom Neil gives the pseudonym Hussain, told him that many years ago he had been conscripted into the army, and was patrolling the streets one night with an army friend when a sandstorm began. The two took refuge in a Mosque. They took off their shoes as is customary.

The men decide to take turns on watch. His friend slept first as the Hussain sat up holding his gun. Hussain found himself nodding off and forcing himself to keep awake. After the third time he awoke, he found that there was a woman crouched down looking intently at his feet. Next to the woman was a girl of about 12. She was looking with the same interest at his friend's feet.

Hussain tried to wake up his friend by pressing his elbow into his chest, but no matter how hard he pressed, his friend remained asleep. Suddenly the woman looked up and right into Hussain's eyes. As soon as their eyes met he knew that she was 'not of this world'. At that moment, both the woman and the girl stood up and walked away. Hussain saw that both of them had cloven hooves instead of feet. Both of them vanished by walking through a wall like ghosts. In Islamic countries these beings are called 'djinn' and are thought to be shape shifting entities. We will take a closer look at them in the final chapter.

THE MOCKING BEAST
The writer, Joan Forman, worked at the village school in Goodhurst, Kent in the mid-1950s. She was spending the first few days of the summer holidays at the school. The

pupils and most of the staff had left. She was sleeping in oldest part of the school. On the second night she awoke at about 3.30 am and saw a creature crouching upon the floor.

> *"It was about two feet in length, I supposed about the size of a large cat or a small corgi. It resembled neither of these. It had a pair of huge nocturnal eyes like those of a lemur, and these were the clearest features of the apparition. I noticed them particularly because they were unwaveringly fixed on me. I think it was the most revolting gaze I have ever had to endure, for what emanated from the thing was a atmosphere of extreme malevolence. And obscenity. With all its exudation of evil it was at the same time mocking. It stared at me for what seemed half an hour (although I suspect it was only a few minutes in chronological time) and I stared back, playing rabbit to its snake. I could not move to switch on the light. And in any case the creature itself seemed to emit some kind of glow in which I could see the shape of its face and head and the huge eyes, and a dim suggestion of the rest of the body."*

Eventually, the room, that had been freezing cold despite it being summer, began to get warm again and the creature faded away. Once again you could argue that this was a hypnopompic or hypnagogic hallucination, but Joan's successor at the school saw the same creature in another bedroom.

The mocking aspect sounds very like the gloating ape-like demon, reported by Jerry Glover.

THE DEMON FOX
Richard Crule in his 1937 book, *Caravansary and Conversation*, wrote about a bedroom invader he encountered as a boy in Scotland. It was 2.00am on a winter's morning, and he was lying awake in bed, when he heard footsteps outside approaching the house. Then he heard them in the house coming towards his bedroom door.

> *"By this time I was sitting up in bed with my eyes glued to the door and with horror in my heart. The handle turned and in the opening stood a creature with the face of a fox, which walked on its hind legs. It was dressed in some sort of way, and would you credit it, wore a top hat, which added an indescribably macabre touch. But the face, I repeat, was fox-like, and it had a bushy tail. It was, perhaps, bigger than a real fox, but it was vulpine through and through, although I admit it had no rank odour.*
>
> *It gazed at me with a fixed and rather malign expression, but it did not speak. I shouted 'Go away!'-How well I remember the exact words!-and it turned round and went away. I heard its steps follow, in retreat, the precise route they had followed before, unhurried and steady as ever, until at last they died out on the road leading to the woods."*

This case puts me in mind of the phantom fox encountered by the father of Dermot MacManus in the 19th century.

THE HORNED CAT OF WORLD WAR II

We have already met demonic black dogs and demon foxes, so why not a devil cat?

One night in October, 1943, Howard Leland, an Air Raid Precautions volunteer, was caught in a bombing raid in south London. Fleeing into an empty house, he used his torch to locate a flight of stairs and sat on the bottom step waiting for the raid to pass.

Suddenly the house became colder, and Leland felt he was no longer alone. Switching the torch back on he played the beam up the stairs, where it fell onto an incredible sight. At the top of the stairs was a gigantic cat, with tabby-like markings of black and brown. Its eyes glowed red and sprouting from its head were a pair of horns!

Leland said he felt a palpable evil emanating from the creature. The thing leapt down the stairs at him, but vanished in mid-air. A hideous caterwauling began to ring through the room.

Suddenly, the sound of footsteps approached the building, and the yowling stopped. It was Leland's APR colleagues. He recounted what had just happened, fully expecting not to be believed. However they told him that a number of people had reported exactly the same thing in the house over the years.

Two days later, Leland visited a renowned clairvoyant, John Pendragon, Pendragon placed his finger on a map of the area, on the exact house and went into a trance. Later he described to Leland what he had seen.

Pendragon had a vision of a haggard looking man with a noose, about to hang himself. He was, Pendragon said, a black magician who had been sacrificing cats on an altar in the house. Ultimately he went mad and hung himself. The demon cat was a gestalt of the spirits of the dead cats.

THE JELLY DEMON AND THE FISH DEMON

Joshua P. Warren is a collector of weird tales. For nearly 30 years he has been recording strange events that have befallen ordinary members of the public. He published them, along with Andrea Saarkoppel, in a book entitled *It Was A Dark And Creepy Night,* in 2014. One of the weirdest was a demonic bedroom invader, seen by Fay Sennet in Seattle, Washington, in 1964. Fay, who was 18 at the time, had got up early to make packed lunches for her mum and dad. On returning to bed she heard a heavy breathing sound. Rolling over to look around, she was confronted by a horrific entity.

The thing standing next to her bed was as tall as a man, but cone shaped or 'like a Christmas tree'. It had no limbs and seemed to be composed of a grey, translucent jelly.

It had wart like growths on its body the size of walnuts, and slit like yellow eyes. Its gaping mouth were full of teeth that Fay thought looked like a dog's. She felt that the cone of jelly was 'pure evil".

The thing rolled on top of her and seemed to dissolve. The frightened girl thought the entity was a demon trying to possess her and shouted *"Please God Help Me."* The jelly monster instantly vanished. Understandably her parents did not believe her but to this day she still shakes recalling the event.

Another strange bedroom invader story that was related to Warren, was that of Judy Sikorski of Rossfield, Ohio. One night in 2000, she was sleeping alongside her husband when she felt something tugging at the bedclothes. Looking down she saw what looked like a huge fish with bulging eyes, a humped back and very sharp teeth. The fish creature bit into her foot and began to shake it. The pain was agonizing but Judy found that she could not scream. Then, suddenly, her attacker vanished. Her foot was throbbing but when she looked at it there were no marks.

It is tempting to write this off as some form of sleep paralysis, but some months later the fish beast was back and attacked her again. Afterwards she hung a rosary on her bedpost. The creature never returned.

VAMPIRES
In November, 1972, an inquest was held in the industrial, Midlands town of Stoke-On-Trent in England. It involved the death of a middle aged Polish pottery worker Demetrious Mykicura. Police constable, John Pye, was called to No.3, The Villas, a gloomy boarding house on the southwest side of the town. The landlady and several of her tenants had not seen the 56-year-old man in several days.

Forcing his way in, PC Pye found him dead on his single bed. Mykicura, who had lived in England for the previous 25 years, since the end of World War II, had been so frightened of electricity, that he had removed the light bulb from the light socket in his room, PC Pye was forced to carry out his examination by torchlight.

Crosses had been hung on the walls and salt was scattered over the furniture, and the bedclothes. Two small bags of salt were found in the bed, one between Mykicura's legs, the other on the pillow next to his head. On a flat roof close to the window sill, was an upturned washing bowl, under which was a mixture of human excrement and garlic. A subsequent post-mortem revealed that death had been due to asphyxiation, due to cloves of garlic getting lodged in his throat. The verdict of the coroner was that his fear of vampires had killed him, causing him to choke on the garlic.

Daniel Farson, wrote in his 1975 book, *Vampires, Zombies and Monster Men*:

> *'As a final desperate measure to ward off the vampires, this wretched man had slept with a clove of garlic in his mouth, and the garlic had choked*

him to death, So in a roundabout way, the vampires did get him in the end'.

The modern idea of a vampire as a beautiful, ageless, sexual predator is a construction of modern film and television. In legend they are far less palatable creatures.

In Egyptian lore a person had two souls, the *ba* and the *ka*. The *ba* could travel to the astral plane and its symbol was a human-headed bird. The *ka* however was linked to the physical body and could only travel short distances from it. The *ka* also needed the preservation of the physical body after death, one of the reasons for mummification. Regular offerings were left to the *ka* in order for it to feed. The *ba* travelled on to the afterlife but the *ka* remained in the mortal world. The former had to shed the latter to move on, but the latter could live on in the mortal world, where it needed a body as a home and food.

In the *Pert em Hru* or the Egyptian Book of the Dead, a funerary text written over 3550 years ago, it is stated that if food offerings are not left for the *ka*, it will leave the tomb at night in search of nourishment. These glowing phantoms were called *khu* or luminous ones. These creatures could enter the houses of the living to feed on their blood.

The English philosopher, Henry More (1614-1687), wrote up an early account of vampirism in central Europe. It concerns a shoemaker who committed suicide in the Silesia region of Poland, in 1591. Soon after, people in the town began to report that the shoemaker was appearing as a ghost after dark. It visited them at night, and lay on top of them almost suffocating them. It would pinch its victims leaving bruises. People began leaving candles lit all night and sleeping together. Yet still the phantom came. When people began to leave town the authorities dug up the man's body, seven months after it was buried. The body was found to be fresh, un-decayed and free of the smell of rotting flesh. The body was above ground for six days, during which his ghost was still seen.

The body was reburied under a gallows but it had no effect.

The shoemakers corpse was dug up a month later, and found to be equally fresh. The magistrate gave the order to cut off the head and cut out his heart. The heart was found to be as fresh as that of a newly slaughtered calf. The head, heart and body were burned and the ashes thrown into a river. After this the ghost did not return.

In 1725, two officers of the tribunal of Belgrade, Serbia, and an officer of the Imperial Army, all investigated a strange case in the village of Kisolova. A 62 year old farmer, Peter Plogojowitz, died seemingly of no cause. He had seemed perfectly healthy the night before. Three nights after being buried, Plogojowitz returned to his home and demanded food. His son put food on the table and the ghost ate it before leaving. The next night Plogojowitz returned demanding more food. The son refused and his late father took on a 'threatening aspect' and then vanished.

Next day the son died of seemingly no cause. Within a few hours friends and neighbours began to fall ill. In each case the victims seemed to suffer blood loss and exhaustion. Some said they had dreamed of Plogojowitz lying on top of them, biting their necks and sucking the blood. Within a week nine people died.

The apothecary gave details to the chief magistrate of the area, who sent them to Gradiska where the Commander of the Imperial Army was staying. His commander, together with two officers and the magistrate, had Plogojowitz dug up. He was found to be well preserved and his nails and hair still growing. A stake was driven into his heart and blood gushed up from his chest, as well as all the orifices of his body, flooding the coffin.

His victims were dug up, but showed no signs of vampirism, but as a precaution they were reburied with garlic and whitethorn in their coffins.

In the spring of 1727, Arnold Paole, a Serbian soldier returned from service in the military to settle in his home town of Meduegna, near Belgrade. He had been posted in Greece and had claimed to have been visited by a vampire. He cured himself, or so he said by finding the creature's grave and eating some of the earth from it.

Paole bought some land, built a house and married a girl who's father's land bordered his. He broke his neck in a fall from a haywagon. Sometime after his death several villagers claimed to have seen Paole in their homes at night. Then these people began to fall ill and die. Two military officers, two army surgeons, and a priest from the local church dug up Paole, and found his body fresh and un-decomposed. They drove a stake into the corpse, which is said to have screamed and vomited blood. Garlic was then strewn in the grave. The same was done with all of the villagers who had died.

All was quiet for a while. Then in 1732, a new spate of deaths began in Meduegna. A group from the village investigated, and found 11 corpses in the same state as the Paole corpse. One theory holds that Paole had feasted on local cattle as well as people. Then, the theory states, as time passed and the cows were killed for their meat, the vampire qualities were passed on to anyone who ate the meat.

Of course, the fresh state of the corpse is easily explainable with modern knowledge of decomposition. The scream it uttered when staked was probably caused by escaping gasses that had built up in the corpse. Modern undertakers have many such stories of bodies crying out, or even sitting up on the slab, due to contortions caused by internal gasses. Such pressures would account for the liquid vomited up from the mouth of the body. But this does not explain how the dead man visited other people in the village and why the deaths stopped after Paole's body had been staked. The same holds for other cases where the dead person is seen walking abroad.

The belief in the dead rising as vampires is still strong in Eastern Europe. Whilst travelling in Romania, American author Raymond T. MacNally witnessed the following.

> *"In 1969 I was passing through the village of Rodna,which is located near the Borgo Pass Noticing a burial taking place in the graveyard, I stopped to watch. As I talked to some of the bystanders, they told me that the deceased was a girl from the local village who had recently died by suicide. The villagers were afraid that she would become a vampire after death. So they did what they had to-and what I had read about for so many years. They plunged a stake through the heart of the corpse."*

In February 2004, a woman from the village of Marotinu de Sus in Dolj County, southern Romania, said that she had been visited by her late uncle, a 76-year-old Romanian man named Petre Toma, who had died in December 2003. Fearing the deceased might have become a 'moroi' (a local term for vampire), the woman's brother-in-law, Gheorghe Marinescu, organized a vampire hunting group.

They dug up the coffin of Petre Toma, and hacked his heart out. After removal of the heart, the body was burned and the ashes were mixed in water and drunk by Toma's niece, believing that this would put an end to the visitations. The local police later arrested six of the family members, who participated in the ritual, charging them with "disturbing the peace of the dead." They were sentenced to six months' imprisonment and ordered to pay damages to the family of the deceased.

The nearby village of Amărăştii de Sus, people drive a fire-hardened stake through the heart or belly of the dead, to prevent them becoming moroi.

In 2003, several villagers from another nearby village of Celaru dug up a dead man they suspected had transformed upon returning, they ripped out his heart, burned it, then drank the ashes with water. The case caused a stir, six people were found guilty of desecration of graves and were sentenced to prison, and to pay moral damages to the family. But the superstition has a hold. Dumitra, a 71-year-old woman told the Romanian online newspaper *Evz.ro* the following.

> *"Me, I've never been haunted by the dead because I stabbed them all in the heart, that's fine. My daughter was dug up five or six years ago. She had become an undead and she would come at night and shout: she would go into the barn (where the bread is kneaded), enter the closet and chatter and sniff the plates."*

The woman was later dug up and staked. The hauntings ended. Before you laugh at such superstitions, it its well to remember the case we mentioned previously of a farmer on Dartmoor, England, who had been suffering with illness among his cattle, sacrificed a sheep to the pixies. After he performed the ritual the illness ceased.

London in the swinging sixties seems like a world away from peasant villages in rural Romania. But was in a London graveyard that the best known case of a modern vampire occurred.

The *London Evening News* of 2nd November 1968, reported that...

> *"On the night of Halloween 1968, a graveyard desecration by persons unknown occurred at Tottenham Park Cemetery in London. These persons arranged flowers taken from graves in circular patterns with arrows of blooms pointing to a new grave, which was uncovered. A coffin was opened and the body inside "disturbed." But their most macabre act was driving an iron stake in form a cross through the lid and into the breast of the corpse."*

The event was linked to what followed the next year. In 1969, people began reporting a tall, dark, red eyed figure, floating around the graves in the famous Highgate Cemetery, in North London. It was also seen in the adjoining Swains Lane. Rumour spread that the thing was a vampire. One man who encountered it was David Farrant, when passing the cemetery on 24 December, 1969. Farrant saw a tall dark figure with red eyes floating above the ground. Farrant reported he had felt like losing control of his body and mind and the air suddenly turned icy. To escape from the terrifying situation, he had repeated some Kabbalistic incantation. Shortly after the phantom disappeared. David was a Wiccan and president of the Psychic and Occult Society. He subsequently wrote a letter to the *Hampstead and Highgate Express*, on February 6, 1970, David himself did not believe in vampire in the traditional sense of re-animated, blood sucking corpses but likened it more to an incubus, a demonic spirit that drained energy from its victims. After the article was published many other witnesses wrote to the paper, claiming that they too had seen the entity.

Enter Sean Manchester, a narcissistic publicity hunter who later became a self-proclaimed 'bishop' of the Old Catholic Church'. He made wild claims that the vampire had once been the leader of a satanic cult in Romania, and that he was going to find it and drive a stake through its heart. Farrant was dismissive of such an endeavour.

A rivalry grew between the two, and stories about them having a 'magickal duel' began to circulate. As you may expect, the media went bananas at this and stories in the local rags flew thick and fast. The situation reached fever pitch on Friday the 13th of February, 1970, as Thames TV ran a programme on the unfolding saga the night before the scheduled hunt. It culminated on a mob of at least a hundred members of the public descending on Highgate Cemetery, searching for the blood sucker one night in March. The mob burst past the hastily assembled police cordon.

It is said that Manchester and his companions broke into a catacomb, and found a sinister-looking black coffin that didn't seem to fit to the others laying down there.

Manchester performed an exorcism using garlic and holy water, and sprinkled salt around the coffin.

In the meantime, Farrant attempted to communicate with the Highgate Vampire using two circles, candles, incense and a medium. His first attempt was interrupted by the police, but the second attempt was successful. An entity appeared and grabbed him by the throat. Farrant had to break the circle to escape.

Some months later, it was found that a human corpse had been dragged out of a coffin, decapitated and set light to. The woman's body had been buried in the graveyard for over 100 years.

A few weeks later, Farrant found himself arrested in a nearby churchyard, carrying a crucifix. He later sued the vile and now defunct, right wing rag *News of the World,* for their intimation that he had sacrificed a cat in a ritual in the cemetery. Farrant was a lifelong animal lover and Wicca does not involve animal sacrifice.

Manchester claimed to have driven a stake through the vampire's heart, in a nearby house in Crouch End. Despite this claim the tall dark figure was still reported in Highgate.

Farrant was jailed for grave desecration in 1974, (charges he always denied). Having met David Farrant several times, I'm inclined to believe him. He says his occult group were trying to exorcise the phantom. His arrest prevented the ritual from taking place, and according to him, whatever the Highgate Vampire is, it's still at large. In 1991 Declan Walsh claimed he had witnessed a tall, very thin man, dressed in Victorian style as walking through a locked gate. Another witness took a glimpse of a figure floating from the east site to the west site of the cemetery in 2005.

In the 1978 General Election, David Farrant ran in Hornsey as the sole candidate for his own Wicca Workers Party, on a platform of free sex and nudity, restoring the Wiccan creed, outlawing communism, establishing state brothels. Sadly he was unsuccessful.

Manchester wrote a book on the subject, *The Highgate Vampire* in 1985. The book, such as it is, is a parade of narcissism in which the author depicts himself as a cross between Dr Van Helsing and Lord Byron, and battling a giant demonic spider and a vampire king. The whole book feels like a rejected script for a Hammer movie. There are also pictures of a young blond lady. "Elizabeth Wojdyla," whom Manchester claimed was killed by the vampire, and subsequently became a vampire herself. Manchester writes that he was forced to stake her too. David Farrant noted that

> *"...what we have here, is a man claiming to have killed his common law wife and buried her in a cemetery. Were it true, then he would have been arrested for publicly admitting his crime."*

Years later the girl who modelled for the shots, Jacqueline Cooper, turned up alive and well denying any attack from a 'vampire'. Cooper was in fact cited in Manchester's divorce, as he had been sleeping with her. She had been the "victim" in a film Manchester had made (where he played both hunter and vampire) titled *"The Vampire Exhumed."*

Far better is Farrant's 1991 book, *Beyond the Highgate Vampire.* It is an account quite free of boasting and wild flights of fantasy.

Don Ecker, the long-time head of research at *UFO Magazine,* contacted Sean Manchester to try and do an interview with him. Manchester flatly refused. Don a former criminal investigator writes...

> *"I had the attitude that the guy wasn't firing on all cylinders. Here was a guy that was screaming high and low that vampires were real, that they were an imminent threat to civilisation, and all the rest of it. But he didn't want to be interviewed by someone fairly well-known in the US?"*

As for Farrant he says...

> *"I really found him enjoyable. He was just a funny guy. My god, he had some hilarious stories."*

Ecker was later bombarded by abusive e-mails from the 'Friends of Sean Manchester' society (thought to be Manchester himself sending messages from multiple e-mail addresses). Manchester also told him that he was a direct descendent of King Arthur!

David Farrant found that he was supposed to have slain Seán Manchester during a different duel in France — with swords. Farrant claimed that *"Manchester himself circulated a photograph that showed his 'dead body'"* and a made-up obituary, to solicit funds to have his corpse shipped to Glastonbury, for internment. The police supposedly showed up to Farrant's flat demanding to know what he had done with his murder victim. It was of course all made up by Manchester.

The whole Highgate affair seems to be more about the rivalry between the two prime investigators, as it is about the vampire itself. Today Highgate Cemetery is under lock and key, and there is a hefty admission fee. The management and guides seem embarrassed by the whole affair an seldom speak of it.

The case was said to have inspired the Hammer film *Dracula-1972 AD*!

Ghost hunter Peter Underwood, had investigated a case of vampirism and wrote of it in his 1975 book *Further into the Occult.* Unfortunately, he does not provide a place and date though it must have been prior to 1948. It concerned a widowed mother and son. The father of the family had committed suicide. Underwood thought, at first, the case involved a poltergeist. The boy reported tappings at his window at night, and dark figures. Objects were moved about, but they were always religious objects. Religious themed pictures would fall from the wall, and a crucifix hung on the wall rattled so violently, it would fall and always land face down, and somehow make its way under the bed or out of sight. The Bible would quiver and open by itself, and the pages would turn on their own. The boy told Underwood that the book would stick to his hand. The only way he could get it unstuck, was to place it somewhere that it could not be seen.

One night he awoke to see something dark, blotting out the stars at his window. He screamed for his mother, who stayed with him until dawn. A few nights later the crucifix was flung from the wall, and the boy found it under his bed. It felt warm and had deep scratches in it. Looking to the window, he saw what looked like a cloud of black smoke pouring in. The smoke seemed to be taking on a more solid shape as it flowed towards him. Before he could scream he passed out. His mother found him pale and listless in the morning.

The boy was put in another room, and for a few nights slept undisturbed. But then the same events occurred once more. Peter Underwood sought out the help of Montague Summers. Summers (1880-1948) was a highly controversial character. A clergyman, author and occultist, was ordained as an Anglican deacon in 1908, but left the Church of England under a cloud due to his alleged interest in underage boys. Summers joined the Catholic Church, where such behaviour is not as frowned on as it is elsewhere. He called himself "Reverend Alphonsus Joseph-Mary Augustus Montague Summers" though it was disputed if he was ever ordained. He wrote a number of books on satanism, witchcraft and demonology in which he made some wild claims. He also translated, from Latin The *Malleus Maleficarum* or witch hammer, a 15th century treatise on witch hunting. He was also a member of the Uranians, a group of upper class men who wrote erotic poems about adolescent boys.

Summers had a plan of using the boy as bait to catch the vampire. The boy's mother was horrified and refused. Instead, she had a priest bless the room and obtained some holy water. This quietened things down for some time until Underwood received a frantic call from the mother. He found the boy in another room. He was very weak and could hardly tell Underwood what had happened. Looking at the room where the boy had slept the previous night, the window had been smashed inwards. The crucifix was under the bed, the religious picture had the wire it was hanging from twisted around and around, as if with pliers, and the Bible had its pages stuck together with 'an abominable substance'.

Summers gave him a brass medallion, that he said was a mid-European 15th century charm against vampires. Summers told a tall tale, worthy of a Hammer horror film, in which he fights a vampire count and his bride, in their castle in Italy. Using the amulet to keep the vampires at bay, Summers burns down their castle, so they can no longer prey on the innocent villagers. The whole story is chock full of clichés and impossible to believe. As it happened Underwood never had chance to use the amulet as the mother took her son to South Africa with her to live with her sister.

Asia too, has stories of blood sucking monsters. Commander RT Gould (1890-1948) was lieutenant-commander the British Navy, and an expert on the science of time keeping devices. He was also a cryptozoologist and Fortean researcher. He wrote books on sea serpents and the Loch Ness Monster. In his 1928 book *Oddities: A Book of Unexplained Facts* he writes of uncovering papers from the respected Asiatic Society of Bengal, written in 1896, by the wonderfully named Ethelbert Forbes Skertchely. Skertchely lived in Hong Kong and was an officer in the British Navy.

He was visiting Cagayan Sulu Island, that lies north of Borneo, and is part of the Philippines. He writes..

"In the centre of the island is a small village the inhabitants of which owe allegiance to neither of the two chiefs. These people are called "Berbalangs," and the Cagayans live in great fear of them. These Berbalangs are ghouls and must eat human flesh occasionally or they would die. You can always tell them because the pupils of their eyes are not round but just narrow slits like those of a cat. They dig open the graves and eat the entrails of the corpses; but in Cagayan the supply is limited, so when they feel the craving for a feed of human flesh they go away into the grass, and having carefully hidden their bodies hold their breath and fall into a trance. Their astral bodies are then liberated in the form of heads with the feet attached to the ears as wings. They fly away, and entering a house make their way into the body of one of the occupants and feed on his entrails, when of course he dies in fearful agony. The Berbalangs may be heard coming, as they make a moaning noise which is loud at a distance and dies away to a feeble wail as they approach. When they are near you the sound of their wings may be heard and the flashing lights of their eyes can be seen like dancing fire-flies in the dark. Should you be the happy possessor of a cocoa-nut pearl you are safe, but otherwise the only way to beat them off is to cut at them with a kris, the blade of which has been rubbed with the juice of a lime. If you see the lights and hear the moaning in front of you, wheel suddenly round and make a cut in the opposite direction. Berbalangs always go by contraries and are never where they appear to be. The cocoa-nut pearl, a stone like an opal sometimes found in the cocoa-nut, is the only really efficacious charm against their attacks and it is only of value to the finder, as its magic powers cease when it is given away. When the finder dies the pearl loses its lustre and becomes dead. The juice of limes sprinkled on a grave will prevent the Berbalangs from entering it, so all the dead are buried either under or near the houses and the graves are sprinkled daily with fresh lime-juice."

Most of the natives were too scared to guide Skertchely to the village. The son of Chief Mohomet, a man called Matali, agreed to lead him to within half a mile of the village.

Matali gave Skertchely a kris knife sprinkled with lime juice.

> *"Taking the kris and limes and leaving Matali praying for my safety, I soon arrived at the village. It consisted of about a dozen houses of the ordinary native type; but with the exception of a few fowls and a solitary goat there was no living thing to be seen. I was surprised at this and entered several of the houses, but all were alike deserted. Everything was in perfect order, and in one house some rice was standing in basins, still quite hot, as though the occupants had been suddenly called away when about to begin their evening meal. Thinking perhaps that they had run away I halooed but received no reply, and though I made a thorough search of the vicinity could discover no one. I returned to Matali, and on telling him of the deserted state of the village, he turned pale and implored me to come back at once as the Berbalangs were out and it would be dangerous to return in the dark"*

As they hurried home, the pair heard the beating of huge wings, weird moaning and saw red lights. As they approached the remote house of a man called Hassan, the lights and sounds faded away, and Matali thought that the berbalangs had gone to Hassan's house. The pair made it back to Chief Mohomet's village, but Skertchely suggested that they check on Hussan the following day.

> *"Accordingly, shortly after day-break, I started off alone, as I could get no one to accompany me, and in due course came to Hassan's house. There was no sign of anyone about so I tried the door but found it fastened. I shouted several times but no one answered, so, putting my shoulder to the door I gave a good push and it fell in. I entered the house and looked round but could see no one, going further in I suddenly started back, for, huddled up on the bed, with hands clenched, face distorted, and eyes staring as in horror, lay my friend Hassan-dead ."*

Apparently Hassan was drained of blood.

In the 1970s, another vampire case with its roots in Asia, was emerging in the USA.

The Hmong are an ethnic group of hill tribes from the mountains of Laos. The Hmong were recruited and armed by the CIA as a guerilla force, to counter the Communist Pather Lao insurgency. After the Pather Lao victory of 1975, over 100,000 of the Hmong moved to the USA.

In 1977, dozens of perfectly healthy young Hmong men began to die silently in their sleep, with seemingly no detectable cause. The epidemic became known as Sudden, Unexplainable Nocturnal Death Syndrome or SUNDS for short. One factor came to the attention of Dr Neal Holtan, an epidemiologist in charge of research into SUNDS. Dr Holtan described this as a 'transient nocturnal event'. This consisted of a feeling of fear, paralysis, a pressure on the chest, the sense that something was in the room with the person and a sensory factor such as seeing, hearing or touching something. One witness, the wife of a victim, awoke one night to find her husband seemingly in a violent struggle with an invisible attacker, despite the fact that he was still asleep. He suddenly went limp and was pronounced dead when the ambulance arrived.

The effects were similar to an experience called 'The Old Hag', recorded by David Hufford, who was researching in Newfoundland. We will examine the Old Hag in the next chapter. This effect was called the '*tsog tsuam*' (pronounced 'cho chau') among the Hmong. They attributed it to a spirit called the *dab tsog* (pronounced 'da cho'). The dab tsog is held to be an ancestor spirit, that dwells in caves by day, but ventures forth at night. These potentially deadly spirits were held in check by offerings and rituals, performed by the elders of the family. If the rituals were not performed, the angry spirit would sit on top of a sleeping man and suffocate them.

A survivor from a *dab tsog* attack, Vang Xiong, described it as a pallid, hideously wizened old woman that crawled up on top of him, as he lay parallelized. The hag-thing sat upon him and crushed his lungs making it hard to breath, all the time leering down with an expression of hate. After about 15 minutes, the entity crawled off him and vanished.

When the Hmong relocated to the USA, families were broken up, and rituals disrupted. Elders with the knowledge or the rituals were often left behind or scattered across the country, and without them, the dab tsog was free to carry out its night time depredations. Procuring animals for sacrifice was hard in America, and the city apartments that the refugees lived in lacked the central pillar of the traditional architecture, that was important in the rituals.

50-60 percent of Hmong living in America experienced attacks, though not all were fatal. Of the Hmong that had converted to Christianity, abandoning all of their native rituals 72 percent experienced dab tsog attacks.

Within a few years the Hmong communities moved to more rural areas, and elders rejoined the groups to use and teach the rituals that foil the dab tsog. Whereas the SUNDS deaths rose sharply from 1977 to 1981 they have decreased since.

Apparently, the affair of the *dab tsog* was the inspiration behind Wes Craven's 1984 horror film *A Nightmare on Elm Street*.

THE WENDIGO
In the legends of the Algonquin people, of the east coast Canadian forests, and the plains and Great Lakes region of the USA, no other creature is as feared as the wendigo. Also known as the wetiko, the windigo, the wīhtikow and several other similar names, it seems to be an avatar of winter and hunger. It is seen as a giant, humanoid figure but skeletally thin. Its pale skin is often caked with hoarfrost. The

wendigo has teeth and claws like daggers of ice, and large eyes that gleam like indigo stars on its skull like face. Sometimes it is shown with a wild mane of white or silver hair.

The wendigo is eternally hungry, and the more it eats, the more its appetite grows. It favours flesh and prizes the flesh of humans more than any other. The icy giant can fly by 'walking on the winds'.

But the spirit has another, more insidious power. Should a human, caught in the wilderness in winter, be forced into eating human flesh in order to survive, a real possibility in the wilds of Canada or the northern US, the spirit of the wendigo could possess them. Once in the thrall of the wendigo the victim has an overwhelming craving for human flesh and becomes a cannibal. The victim then begins to change physically, until they change into a small version of the wendigo itself and go on a killing rampage. Such a state is called 'going wendigo' and those suspected of it were generally killed by their peers out of fear. It caught early enough, the process could be reversed, by binding the victim and suspending them over a fire. It was thought that those going wendigo had green ice growing around their heart. If the green ice could be melted the curse would be lifted.

It is a creepy legend but is that all it is? Modern psychology acknowledges a condition called 'wendigo psychosis' wherein the sufferer believes they are possessed by the wendigo, and crave human flesh.

The pioneer and surveyor David Thompson, recorded a case of wendigo possession in 1796, whilst living with a group of Cree between Rainy River and The Lake of the Woods. One morning a young man woke up filled with an urge to kill and eat his own sister. Despite his family's efforts the youth's morbid urges grew stronger. At a tribal meeting the medicine man decided that a wendigo had possessed the boy. A sentence of death was passed on him and he was quite willing to die. He was strangled with a cord and his father burned his body to ashes.

At Smokey River in 1899, a man called Moostoos was bludgeoned with a hatchet, and stabbed by members of the party he was with. He had warned them that he was about to go wendigo. He said

> *"I don't want to do anything to my children. It is better that they should kill me. How would it do if I should eat my little ones, and especially their noses?"*

Even babies could go wendigo. Anthropologist Ruth Landes records such a case in her 1938 study *The Ojibwa Woman.*

> *"The infant son of the Shaman Great Mallard Duck was viewed by his mother's co-wife and by his half-sisters as a windigo and therefore*

killed. This happened during a period of starvation, when seven out of Duck's family of 16 persons died of hunger. The baby that was nursing was just crazy, He was eating his fingers up and biting the nipples off his dead mother's breasts. They knew he was to become a little windigo. His eyes were blazing and his teeth rattling. So the old woman killed the little boy."

The best known case of wendigo psychosis occurred in Alberta, in 1879. It gained notoriety in newspapers across Canada and the US. A Plains Cree hunter called Swift Runner, walked into a Catholic mission in St Alberts. He claimed that his whole family had starved during the harsh winter. The priests became suspicious when Swift Runner began to scream in his sleep, and was plagued with nightmares. He also looked well-nourished for a man whose family had all starved. He told them that he was being tormented by a wendigo.

The priests reported their suspicions to the police. The police took Swift Runner back to his camp, in the forests of northeast Edmonton. Here they found the bones of his whole family, with every scrap of flesh gnawed from the bones, and the marrow sucked out. Swift Runner had not eaten his family out of desperation. There was plenty of food at the Hudson Bay Company post, only 25 miles away, no distance for a Cree hunter, even in winter. He confessed to killing and eating them all. He told Father Hippolyte Ledue *"I am the least of men and I do not merit even being called a man."*

A contemporary photograph of Swift Runner shows a haunted man staring blankly as the snows swirl around him. He was hung on 20th of December.

One could argue that wendigo psychosis is simply a matter of mental illness. A number of serial killers have also been cannibals, including Ed Gien, Peter Bryan and Jeffery Dahmer. Yet none of them claimed to have been possessed by a monster. Victims of the wendigo seem to know there is something evil at work within them.

And then we have the physical manifestations recorded in some cases. A man named Napanin arrived at Trout Lake, an outpost near Wabassca, Alberta, Canada, in 1896, claiming that the wendigo was within him. He had been travelling with his wife and son when he began to see the child as a moose calf, and wanted to eat him. The wife and son fled and Napanin went looking for help for his affliction. He suffered from a constant, bone chilling cold, as if the monster's heart of ice were growing inside of him. Even when wrapped in six blankets, he felt as if he were freezing. Witnesses said that he made a noise like a wild bull and grew terribly swollen in both body and face. The men of the village decided that he must be killed before he began to eat the people of the village. A medicine man slew him with an axe and his body was buried with a massive pile of logs on top of it, to prevent him rising from his grave. His severed head was buried separately.

Here we have a physical change as well as a change in body temperature.

In another case, Dr A. J Hallowell, in field notes made in the 1930s on the Berens River, Saulteaux, an old woman was killed by her sons on her own request, as she thought she was turning wendigo. They cut open her body and found that her heart had already begun to turn to ice. Dr Hallowell records another case where a Saulteaux woman believed to be possessed by the wendigo, she developed a mass of ice on her back. She recovered after her family melted the ice with boiling water.

Cryptozoologist Derek 'Tex' Grebner, once wrote to me with an account of what seems like a wendigo, seen by a lady he knew.

> *"I have recalled something that I heard a year or two ago from a friend and this sighting is what I believe to be the wendigo because it was hairless and beastly and upright and she said it was about 7-feet tall. She drove past it in the ditch on a country road and stopped to see what it was. When it stood up she screamed and drove away but the beast kept pace with her car, as she drove the forest lined road, and it screamed at her. She said she could see it had claws. It swung into the trees and followed the terrified girl, screaming as it went, covering the distance quickly.*
>
> *The case finally ended when she left the trees and came into open territory, where she could get up sufficient speed until she reached a well-lit residential area. Her last sighting of the beast was it standing atop a street light at the edge of the residential area. She did not think what she saw looked at all like a bigfoot—it was all but hairless she said, and had a bit of a snout like face, and grey skin."*

Another manifestation was seen by paranormal investigator, Jan Thompson, on Christmas Eve, 1983. The area of the sighting was far south of the creature's usual range but it sounds very like a wendigo. Whilst driving along the back roads between Sturgis and Morganfield, in Union County, Kentucky, during a snow storm, Jan saw a monstrous creature crossing the road in front of her car.

> *"It appeared to be a naked man, relatively tall: about six-and-a-half feet in height, with milky white flesh that blended in with the snow. There was long, thick, dishevelled patches of dark brown hair protruding from its body in various odd places, mostly on its upper torso. The head was covered in the same untamed mess that hung like a lion's mane over its shoulders and down its back. It just stood there staring at me through the windshield. Its eyes were like fibre optic lights. With no whites around the pupils and they seemed to have the same iridescent glow that an animal has when caught in the beams of headlights. There were large clouds of warm air coming fast from its mouth as it momentarily rested from running through the field. Its nostrils were large and flared and seemed out of contour for a human. Between thick exhales of breath I*

could see some colour around the mouth, chin, and neck that went down to his chest. It too was red but this glistened in the lights, gleam as something wet would.

His arms were as muscular as his thighs and were abnormally long and slender, and his hands ended in spindly fingers with long ragged nails. Its mouth showed boxy, outsized teeth in what appeared to be a hideous grin, mixed with saliva and blood and stained pink. Its lips were opened wide in warning and revealed two large canines that were thick and longer than the other teeth. Then it licked its bottom teeth with its tongue, running it across his bottom lip.

It still maintained its unblinking stare with eyes that seemed to pulse inside its sunken cheekbones as if pondering what action to take. It closed its mouth, snorted a hefty gust of air, then just as quickly as it had jumped onto the road a few moments before, jumped away to the other side with an uncanny grace, barely touching the ditch and running through the opposite field."

Jan christened the entity the 'Sturgis Vampire', though its description sounds more like a wendigo, as does it's manifestation in a snow storm.

The most comprehensive and well researched study of the wendigo, is Chad Lewis and Kevin Lee Nelson's book, *Wendigo Lore: Monsters Myths and Madness*. Chad has spoken of how the Native Americans still fear the wendigo. In one case, when he was speaking about native legends, some tribal elders asked him not to mention the wendigo, such as the dread of the entity that still persisted and with good reason, as sightings still persisted.

Donald Kakaygeesick was playing baseball with a group of friends, near Warroad Minnesota, in the 1970s. Donald was 12 or 13 at the time. The ball had been knocked into some woods and Donald went to retrieve it. Bending down to pick the ball up he saw a pair of white feet. Filled with fear he ran back to his friend. The kids bravely decided to tackle the monster, and ran into the woods as a group. They heard the creature crashing through the undergrowth, and then it emerged to cross a road and they got a decent look at it. Donald said it was ten feet tall and had white skin, covered in dirty white hair. The monster was slender in build and had a weird gait, like somebody walking on stilts. On its forehead it had a star shaped light. The entity vanished back into the woods.

The star shaped light and strange gait seem like odd little details, but exactly the same features were reported by early pioneer Jake Nelson, in an obscure book called *Forty Years in The Roseau Valle*. Nelson, who lived in the village of Ross, Minnesota, and several other witnesses, mainly native people, reported a tall, slender, white entity with a star-like light on its head.

Hammerson Peters is a Canadian Fortean, who has written a number of excellent books on the mysteries of the Great White North. He was contacted by a miner called Don, from the remote town of Hay River, in the North-western Territories, who claimed to have seen a wendigo. Don worked in a two week rotation at the mine, staying there for a fortnight then having two weeks at home. One night in August, 2018, during his time off, he was alone in his truck driving along North-Western Highway 2. A pale figure came into view, crouching by the side of the road. The thing was greyish-white and hairless, it had dark eyes and a cruel looking mouth. Its limbs were spindly. Bearing its teeth, the creature began to lope alongside the truck. Don, afraid of the creature hit the accelerator and left it behind in the dark.

In the months that followed he tried to identify the strange creature he had seen on the remote road that night. He also became afraid of the woods at night, even those close to his house. Taking out garbage at night made him feel uneasy. As somebody who had a lifelong love of the wilderness and outdoor recreation, Don resolved that the only way to conquer his fear, was to find the creature and kill it. Three weeks after the encounter, he returned to the area and found the skid marks his truck had made, and strange tracks in the earth. He later showed them to his friend and work colleague Tim. The tracks seemed to lead back to an old gravel pit. After returning to the pit for a number of days he found more tracks, seemingly of two creatures. He found more trails later and noticed that they followed wolf tracks, leading him to speculate the creature fed on the remains of wolf kills.

Don showed photographs of the prints to trackers and zoologists, none of whom could identify them.

In 2019 Don sent Hammerson Peters a picture, taken from a trail cam he had set up. He claimed to have caught the creature, lurking in the background amidst some trees. The picture is far from clear, and what to some may look like a pale, leering face may be nothing more than pareidolia, the tendency of the human mind to see faces and shapes in random things like vegetation, clouds or fire. Indeed, Hammerson himself thought that the 'face' was nothing more than branches covered in snow.

Another witness who contacted Hammerson was a French-Canadian called 'Pierre'. The 40-year-old miner owned a remote cabin in the St Lawrence Delta. He brought the cabin from a man in 2017, who sold it because of 'some beast' that haunted the surrounding woods. In January 2020, he was in the cabin with his dog. He was awoken by loud sounds of banging on his walls and garage. He went out with his shotgun. He saw a 7-9 foot tall, grey creature crawling on the wall. As it crawled past the motion detector, the lightbulb exploded. Following it around he saw another, similar, but smaller creature leap from the garage roof into a tree. Running back inside, he put on a bullet proof vest and retrieved a semi-automatic rifle. Again, he saw the smaller creature leap from the roof into a tree, then bound into another tree. Pierre's dog began barking at the thing in the tree.

At 1.12am, the noises from the garage stopped, and a strange light appeared in front of the cabin. The light vanished at 1.25am. He checked his game camera and saw that it had stopped at about 10pm, about an hour before the noises started. Again, note the presence of a strange light, along with the other phenomena.

In 2015, John Crowder was working in Brasher State Forest, St Lawrence, New York. John was a contract worker who cut and marked trails through the forest in preparation for a public opening, later in the year. As an out of state worker he often found himself staying alone in a cabin at the trail head, whilst the other workers went home.

On the evening of 12th of December it had begun to snow. John heard a thump on the roof and thought some snow had fallen off the tree branches, and hit the cabin. Shortly after, there came a scratching. He dismissed this as a racoon. An hour or so later, he had just turned in when he was startled by a series of loud thuds on the roof. John went outside, but found nobody on the roof. He did, however, have an overwhelming feeling of being watched by something from the trees. Turning back inside, very uneasy, he locked the door and managed to fall asleep.

John awoke at sunrise and looking outside, saw bare, humanoid footprints in the snow, as if something had peered through the widow, watching him as he slept. When his colleagues returned on Monday, he told them what had happened. Most laughed at him but other took the events seriously, warning him that he had been visited by a wendigo.

Andrew Latimer, wrote to the *Fortean Times* about something that happened to his father Charles in 1983. Charles was a hunter from Delta County, Michigan, and was on a trip to the Gwinn Forest in Michigan's Upper Peninsular. It was December, and three days into the hunting trip, when he shot a stag in a forest clearing. Following the trail of blood he got halfway across the clearing, when the air seemed to turn thick and dull and everything got quiet. This sounds very like the Oz effect, that has been noted in other curious encounters. He found the dying deer and walked towards it. Behind it he noticed a mass of white that he thought was a pile of snow. But as he drew closer the 'pile of snow stood up'. It was a white, humanoid creature with blood about its mouth. It was very tall, thin, and its skin was almost as white as the snow. It stood in a defensive posture over Charles' kill. It let out a high pitched shriek that caused the hunter to flee. Charles Latimer never hunted in those woods again.

THE TOKOLOSHE
From the freezing wastes of Canada and the northern USA, to the burning wastes of South Africa.

Another culture bound horror is wreaking havoc today. The tokoloshe is a supernatural being, who is evoked to explain a number of strange phenomena over the countries of Southern Africa. They are generally described as short, ugly and hairy. They are often said to be the servants or familiars of witch doctors or more properly sangomas. The sangomas are thought to summon up these spirits in rituals or alternatively create them.

In the latter tradition, the sangoma digs up a human corpse and performs magickal rituals, that shrink the body to a dwarfish size. It sprouts hair and is re-animated as a tokoloshe. The sangoma can then send the hairy horror out to do his bidding.

Poltergeist like activity is blamed on the tokoloshe that can turn itself invisible, as are incubus like visitations on women. The tokoloshe is said to rape woman, and sometimes men, in nocturnal visitations. It has a massive phallus (tokoloshes are always male) that can extend to huge degrees, like a penile python and slither around, creeping through windows, to rape women from a distance.

The same sangomas, in recent years, claim to be able to enchant TV remote controls, so that men can make their own members invisible and detachable. They then use the remote control to make their flying, invisible penises rape women from a long way off! Africa is a continent where such ridiculous superstitions are commonplace.

But back to the tokoloshe. The belief in the creature is almost universal in South African countries, and it is even brought in in legal cases in court. In 1916, Flora Nthshutshe, then a teenager, was walking home in the Northern Cape Province town of De Aar, when she says she saw a tokoloshe.

"He was short and fat with a fur cape of animal skin round his shoulders. His eyes were like lights, yellow and shining brightly and looking evil. He looked hard and very angrily at me but otherwise took no notice...he was evidently going to something much more important and was in a great hurry. I shook with terror as I watched him hurrying down the street. I wondered if any evil would befall me and my people, but I think his thoughts were elsewhere and he was too concerned with thinking about something else to worry about me."

But it is not just black Africans who have seen the tokoloshe. Mrs Minnie Martin, wife of a British colonial officer living in Basutoland, now Lesotho, wrote of her own sighting in her 1903 book *Basutoland: Its Legends and Customs.*

"Some years ago, before I knew of the existence of the tokoloshe, I was obliged to go out to our cowshed rather late one evening to investigate disturbances amongst the cows. The moon was nearly full at the time and shining brightly. The shed was at the bottom of the garden, some way from the house. I went accompanied by my native nurse and a big black retriever. Nothing occurred until we were returning when suddenly I heard what I took to be a dog running from the residency through dead leaves in the garden towards us. I had barely said "What's that? "When we heard the 'ping' of the wire fence and saw, crossing the path not a dozen yards from us, a little black creature about the size and shape of a boy of six. The night being clear and bright there was no mistaking the fact that it was a human figure of

some sort. It ran with a peculiar shuffle, moving its head from side to side, straight through our garden into the darkness beyond. When my girl saw it she caught hold of me in terror but uttered no word. The dog to the contrary gave vent to a sound half growl, half howl and tore off to the house, where we followed as quickly as possible, and found him under my little son's bed from whence he refused to stir. This to my mind was conclusive proof that I had not been 'imagining things', as was said to me when I had described what had occurred, for the dog is a real plucky one and I had never seen him afraid before. My girl told me we had seen the tokoloshe."

She continues with a description:

"He is not much bigger than a baboon, but is minus the tail and is perfectly black, with a quantity of black hair on his body. His hands and feet are like an ordinary mortal but it is never heard to speak."

In modern times the tokoloshe is still feared and blamed for all kinds of things. For example the *Zimbabwe Mail* for December 23rd, 2019, reported on a woman being sexually molested by a tokoloshe.

"A South African woman, Abigail Monyethabeng (59) from Soweto has given praise to popular Prophet Paseka "Mboro" Motsoeneng for bringing her relief after she was tormented by a spiritual husband (tokoloshe).

"Almost every night it came to the window and when I fell asleep it had sex with me with body fluids,"

"Abigail, who has been married to her husband for over 30 years, said the trouble first began in the early 90s when she wanted to have a second child.

"I couldn't conceive but doctors said they couldn't find anything wrong. We tried for six years but things got worse as years went by.

"I fought with my husband about anything. We hardly spoke to each other and when we bought groceries some of it would disappear and frozen meat would suddenly rot.

"Something would come in my dreams and have sex with me. Gradually it became worse. It waited till I fell asleep. I never saw its face but felt it."

She said when she tried to have sex with her husband she was never satisfied.

"It haunted me that I couldn't satisfy my husband and it made me feel less of a woman," she said.

Abigail said she decided to visit Mboro's Incredible Happenings church and since September her sex life had improved. Her husband Lucas (63) said: "She complained about a strange thing having sex with her while I slept beside her.

"Meanwhile, I was getting stabbed with needles in my dreams. I tried to assure her that I was okay with our sex life but she was not convinced."

Mboro said the tokoloshe has been removed.

"Prayer has destroyed the tokoloshe. We will keep praying for the family."

Lazarus Seitsho, a lecturer who collected this tale, while researching witchcraft beliefs in Mpumalanga province. It appeared in the *Weekly Mail & Guardian*, 27 January 1995.

"People from the district of Ohrigstad say they chased a man from their village after finding evidence that he kept a Tokoloshe. The witch then approached a white farmer and suggested that he fire all the black labourers on the estate as he could do their work alone. This the farmer promptly did, but his curiosity was aroused. One night he visited the fields, and found a large number of small creatures tilling the land, with his new employee acting as the supervisor."

Another story involved a South African woman who had not seen the creature herself but had a friend who did.

"I live in SA and although I haven't seen it myself, I know two people that claim they have...

One was a black girl I went to school with (Zinchle) she said that you are not supposed to say the word Tokoloshe repeatedly out loud otherwise he comes to see you in the night. Anyway her & her sister, obviously non-believers at the time were defiantly chanting his name and that night she awoke to strangled noises coming from her sister. She looked over and apparently the tokoloshe had his hands on her sister's throat trying to kill her. She screamed and her parents came running and switched on the light...apart from marks on her sisters neck there was no sign of the bastard. Naturally, I asked her for a description of the creature and she said that it was a half a meter tall, very old black man, with red eyes and a long beard.

The other witness was my Italian friend's mom - although she didn't call it the tokoloshe (she had no idea what it was), she described the exact same entity, but it did not hurt her - it was just a sighting in her home."

In a number of cases, tokoloshes have caused mass panics at schools. In November 2010, stories began circulating in the Botswana press, of the so-called 'creature-school'. Boipuso Primary school in the town of Palapye, where a tiny hairy man with blazing red eyes was terrorizing children. The creature was said to haunt the toilets, locking the toilet doors when children tried to use them. When a girl's face was scratched, the tokoloshe was blamed and panicking parents demanded the school was closed down. The local education office, not believing a word of it, came to a compromise asking local churches to pray in order to cast the evil spirits out of the school.

Another school effected in the previous year, was Kalmare Primary. Victims described a long clawed, hairy creature, dressed in colourful clothes. The local papers ran with the headline *'Children Attacked By Hairy Sex Dwarf'*. The attacks were sexual in nature and involved girls of around 12. The girls were found comatose around the school and were often unconscious for up to 4 hours.

Lenyatso Galebetwe told the Botswanan newspaper *The Voice...*

"My child will not go back to that school until that evil thing has left. She has had enough! She was the first to be molested and has suffered repeated attacks since then. On the first day of school a certain schoolboy ran into my yard crying saying the teachers had sent him to summon me...When I got there my child was lying in a seemingly lifeless state with her eyes wide open and whitish thick saliva dribbling from the corners of her mouth. It was a pitiful and painful sight."

Kereng Peloentle, another mother of a pupil who fell victim to the attacks wrote...

"My child who is 12 and in standard five was attacked last month while in class. I took her to the New Good Shepard Church where the prophets told me the thing that had attacked my child was a tokoloshe. That's when I decided to withdraw her from school...She will only go back when that thing has been dealt with. I think only churches, especially the Zion Christian Church, can chase this creature away and give our children a chance to go on with their education."

A sceptical person might suggest that it was in the interests of the prophets to push the tokoloshe angle. But examination of the girls showed that they had been raped.

Over in Zimbabwe, the headmaster of St Mark's Secondary School, in Mhondro, was forced to flee after been accused of having an army of invisible tokoloshes that he was using to rape pupils. About 30 girls claimed to have been sexually assaulted by invisible beings in their beds at night. Also a number of female staff said that they awoke in the morning, and found their bedding wet and suspected 'foul play.'

The 'pinky-pinky' is a weird sub-species of tokoloshe (as if the standard ones were not odd enough!). An albino tokoloshe is pink or white, and lurks exclusively in school toilets. The best known pinky-pinky scare occurred in 1994, at Morretele Primary School, in the black township of Mamelodi, north of Pretoria, South Africa. The pupils phoned the police and Lieutenant Elias Maswele arrived, having been told that there was a case of child molestation to investigate. He was swarmed by children, telling him that the pinky-pinky had raped and slapped girls and stolen socks off boys. He investigated the toilets but found nothing. The children explained that the pinky-pinky would not appear to adults and it had to be conjured up. The kids went into the toilet and chanted a rhyme to summon the pinky-pinky.

> "I am the Pinky-Pinky.
> I live in a toilet
> My father was a sangoma
> My mother was a witch."

The kids came running out of the toilets claiming that their summoning had worked. Maswele armed himself with a monkey wrench and investigated. He himself claimed to have been touched by an invisible entity whilst in the toilets. The school narrowly missed being burned down by a gang of irate parents soon after. But like the poltergeist, the pinky-pinky seems to have a limited lifespan and the disturbances simply faded away.

SPRING HEELED JACK

In 1837, the year Queen Victoria ascended to the throne, a weird figure emerged in the suburbs of London. One who would become an urban legend. One that resembled a super villain from the comic books, that would emerge in America a century later. The public christened him Spring Heeled Jack. Jack's first attacks were in places such as Barnes, St John's Wood, Hammersmith and Isleworth and descriptions of him differed. News reports of an attacker in the guise of a 'ghost, devil or bear' were reported in local papers.

A 'devil' appeared in the west of London, attacking people with iron claws and climbing over the walls of Holland Park and Kensington Palace at midnight, to dance in the gardens. The phantom scared the daughter of Mr Plutarch Dickinson, of Dulwich, it was in the appearance of a ghost 'enveloped in a white sheet and blue fire'.

She was so scared that she 'was nearly deprived of her senses' and was taken to bed 'in a very dangerous state.' The white outfit and blue fire were to become trademarks of Spring Heeled Jack, along with his fantastic agility from whence his name came. In another case, a servant girl named Polly Adams on Shooter's Hill was attacked. Jack breathed fire into her face and tore her clothes from her body.

A committee of citizens was formed to try and catch the menace. On the night of January 20th, 1838, at Old Ford in the east of London, the bell of a lonely cottage was rung loudly and 18 year old Jane Alsop answered it. An angular, cloaked figure with some sort of head gear stood in the shadows.

'I am a policeman, for God's sake, bring me a light, for we have caught Spring-heeled Jack here in the lane.' He said.

Jane ran to get a candle. The stranger took the light and drew it to his face. His face was hideously ugly; its eyes blazed red and its pinched, tight features were topped by a peculiar sort of helmet; the body, meanwhile, was encased in a tightly-fitting, white shining suit, and a strange object, resembling a lamp, was strapped to the chest. The 'policeman' was Spring Heeled Jack himself, and he tore at Janes face and dress with iron talons, whilst spitting blue flames into her face. Jane pulled herself free but was grabbed again, Jack ripping out clumps of her hair. Her sister Mary, hearing her screams ran to help, and saw Jack herself as did her other sister, Sarah Harrison. Between them they managed to drag Jane back inside and slam the door on the monster. Jack banged on the door before bounding away into the night.

Next day, Jane went with her father and two sisters, to report the attack to Lambeth-street police office. One of the investigating officers was James Lea, who was thought of as the best detective in London in the 1830s, having worked on the infamous Red Barn murder of Maria Marten in Polstead, Suffolk, in 1827.

The phantom was seen in Bow-Fair fields, and was closely pursued by a number of men in the employment of Mr Giles, a coach-master at Bow, but, by the most extraordinary agility, he escaped.

Five days after Jane was attacked, Jack appeared again in the East End of London. This time he knocked at the door of 2 Turner Street, when a servant-boy answered the knock, Jack threw down his cloak and 'presented a most hideous appearance'. The shocked boy screamed and Jack leapt away.

On 28 February, at 8.30pm, Jack attacked Lucy Scales and her sister in Green Dragon Alley, Limehouse. He stunned the unfortunate girl with a flash of blue flame. The assault left Lucy temporarily blinded and 'in violent fits'. Lucy was taken home and a surgeon was called to attend her. Lucy described the attacker as 'tall and thin, looked 'gentlemanly and wore a head-dress rather like a bonnet, and a large cloak'.

The police force had no luck in catching Jack. The Duke of Wellington, victor of the battle of Warterloo, even came out of retirement, despite being 69 at the time. He patrolled the haunts of the phantom on horseback, armed with a sword and pistols. He had no better luck than the cops.

Jack seemed to vanish from London for years, only to return to his old tricks. The editor of the *Camberwell and Peckham Times* related his own encounter on 6 November, 1872.

> *"While returning from a friend's house at Brixton-hill last evening, (via Herne hill) I was accosted by that malapropre fellow the ghost. I had just arrived at the point in Herne-hill-road, where the footpath runs from the side of St. Paul's into Half Moon-lane, when the figure came forth from beside the stile. I confess I was momentarily frightened, but speedily recovering my presence of mind, was on the point of making an onslaught with my umbrella, when the object turned sharp round, and clearing the low railings at a bound, made off across the country. Being now over forty, it was useless thinking of pursuit, but I, however, satisfied myself that he is clad in a black suit, which, by some means, he transposes into white when needful. He also has spring-heeled or india-rubber soled boots, for no man living could leap so lightly, and, I might say, fly across the ground in the manner he did last night.'*

Jack took a holiday to Yorkshire in 1873, showing himself in Sheffield. Journalist David Clark recalls his grandparents speaking of what their own parents had heard, of a tall figure in white with a cloak and steel springs in his boots. Jack attacked passers-by at night and leapt over walls and bounded from rooftop to rooftop, leaving women in a state of shock.

The police pursued him through the Pitsmoor area of the city, finally cornering him in the wonderfully named Bumsgrove Cemetery. The agile fiend escaped by leaping over the graveyard gates.

Jack began turning up in the nearby town of Rotherham, and gangs of vigilantes hunted him on the streets at night, to no avail. One of his favourite haunts was an area imaginatively called 'The Park', which was an old hunting ground. The *Sheffield Daily Telegraph* of May 31st, 1873, describes Jack as follows.

> *"He was described as tall, gaunt and of an unearthly aspect, as skimming over the ground with supernatural swiftness and as making bounds in the air. He has been seen stalking in stately fashion through St Paul's Churchyard. He has curdled the blood of the inhabitants of Upperthorp and has frightened the residents of Daniel Hill into fits"*

One witness claimed to have seen Jack leap over a wall that was later measured at 14 feet 3 inches!

In 1877, Spring Heeled Jack turned up at the army barracks in Aldershot, terrifying the soldiers on duty. He bounded over sentry boxes, and despite being shot at by soldiers on duty, he escaped. The *Illustrated Police News* reported it thus.

> *"His method of proceeding seems to be to approach unobserved some post, then climb the sentry box, and pass his hand (which is arranged to feel as cold and clammy as that of a corpse) over the face of the sentinel. The sentries had lately been ordered to fire on the ghost, and were loaded with ball, but this precaution had lately been given up. 'Jack' pursued his old tactics on 31 August 1877. He managed to reach unseen the powder magazine in the North Camp. Here, having nearly frightened the sentry out of his wits, by slapping his face with his deathlike hand, he disappeared, hopping and bounding in to the mist."*

In 1884, Jack turned up in Richmond. As John Bradley and his servant were driving along Reeth Road one night, a white clad phantom emerged from near a convent and chased the horse and cart at inhuman speed, before turning off into a field.

Henry de la Poer Beresford, the 3rd Marquis of Waterford, was suspected of being Spring Heeled Jack. Known as "the Mad Marquess" he was infamous for rowdy and dangerous pranks. For example, in the early hours of Thursday, 6 April 1837, the Marquis and his friends arrived in Melton Mowbray at the Thorpe End tollgate. They had been at Croxton races, and were very drunk. The tollkeeper asked to be paid before he opened the gate for them. Sadly for him some repairs were underway, and ladders, brushes and pots of red paint were lying nearby; the Marquess and his friends seized these and attacked the tollkeeper, painting him, and a constable who intervened, red.

They then nailed up the door of the tollhouse and painted that red before moving into the town carrying the stolen equipment. They ran down the Beast Market, through the Market Place, and into Burton Street, painting doors red as they passed, pulling on door knockers and knocking over flower pots. At the Red Lion, they ripped down the sign and threw it into the canal. At the Old Swan Inn, in the Market Place, the Marquess was hoisted onto the shoulder of another man, to paint the carved Swan Inn sign there red. They also vandalised the Post Office and the Leicestershire banking company, before trying to overturn a caravan, in which a man was fast asleep. When he sobered up the Marquis paid for all the damage, but he and his friends were fined £100 each, a small fortune in 1837. It is from this escapade we get the term 'painting the town red'.

Some have theorized that the Spring Heeled Jack affair was nothing more than an elaborate prank, in which Henry, donning a cape and special spring-heeled boots, had a ball, frightening Londoners on dark nights. However the Marquis was killed in a riding accident in 1859, and Spring Heeled Jack was seen long after Henry's death. The idea of spring loaded boots being used on unlit lanes in early Victorian London, seems like a recipe for broken ankles, and could one, drunken aristocrat run rings round the whole of the police force for years?

Jack has also leapt out of the 19th century to make his presence felt in later years. In 1997 in the Cranbrook Hall area of Sheffield, people began to report a prowler who banged on windows and attacked people. A local reporter investigated the case. One woman said how she and her boyfriend saw a dark figure slip into a side alley as they were walking home. They both walked in the middle of the road to avoid it. But looking down the alley, they saw two red lights that they realized were red, glowing eyes. The pair fled, pursued by the weird figure, who hurled a gardening fork at the man.

On telling her father of the attack, he in turn told her of a sighting from the 1950s. Hearing mad laughter from his attic, he investigated. His attic was not separated by a wall from the others in the row of houses, they ran along as one long room. He saw a tall man, over 6 feet, dressed in black, with a black cape and glowing red eyes. He and several other house holders tried to chase the phantom, one of them almost falling through the roof. Yet the leaping figure never put a foot wrong and after half an hour's pursuit simply vanished.

Asking around, the reporter was told of a figure that ran up the side of buildings. One old woman, looking out of her window one night, saw what she thought was a burglar at the Sheffield Steel Works. But then she realized that the man was making leaps of up to 30 feet! She watched him for five minutes until he ran down the side of a pub wall and vanished into a scrap yard.

In another case, the police were called out to investigate the leaping figure. They chased him across rooftops, into the yard of the Drexel Tyre Company. The constables surrounded him and trapped him in a small room with no way out. Breaking down the door they found the room was empty. Spring Heeled Jack had vanished like a ghost.

It seems far from impossible that the world has heard the last of Spring Heeled Jack, whoever or whatever he is.

GHOULS

The word ghoul is often used interchangeably with 'ghost'. In fact, they are not the same. Ghoul is derived from the Arabic word 'gul'. It refers to a pale, desert dwelling humanoid monster, said to dig up and devour human corpses. The word 'gul' means to tear, a reference to how the creatures tore into cadavers with their teeth. In the modern day, there have been alleged encounters with humanoid creatures said to dig into graves and feed on corpses.

In 2006, researcher Barton Nunnelly, interviewed a witness called Terry about an encounter which took place in Boyle County, Kentucky, in 1970.

"Me and a friend went fishing one morning. It was in the spring about 1970. We had to walk a few miles east on the L&N Railroad tracks from Shelby City Ky. It is now Junction City Ky. We would sneak in this pond to Bass fish. I had to be home before noon one day so I started back by myself and left my friend fishing the pond. On the way back down the tracks this creature landed on the tracks in front of me. When I say landed I mean jumped, because all I saw was a blur and then there it was. It stood about 4 feet tall and stood or squatted on two legs. It was covered with real fine fur on its fat body.

It had a little round head that moved like an owl's head almost in circles. It had little round eyes and a mouth full of razor sharp teeth. Its teeth were almost transparent, kind of like a bat's. It sort of favoured a bat in the face. It had small ears and a small nose. The thing didn't have any lips just teeth. It arms looked like they came out of its chest and they were spindly. It had long fingers and a thumb on each hand with long claws. Its arms didn't have any hair or fur on them. They were slick and oily looking. The thing couldn't see good in the daylight, because it squinted a lot and it never did see me when it landed in front of me. It acted like it was watching for the trains. It would rub its tiny head on its arm every now and then. It looked right at me and couldn't see me and I was only about 30 or 40 feet from it. I could see its fur moving. The creature never made a sound. I was dumbfounded. I'd never saw anything like it before and haven't since. I wasn't scared at first, but I wasn't really sure what to expect. I had a tackle box in one hand and a fishing rod in the other, but I did have my hunting knife with me. As I tried to switch my fishing rod over to my other hand so I could get to my hunting knife, I rattled my tackle box and that's when this thing heard me. It leaned forward and squinted its eyes and when it saw me it reared back and kind of puffed up.

I thought it was going to attack me, so I jerked my knife out. When I did, this thing took off so fast it was like a blur. In one leap or jump it cleared a fence about 40 feet away and was gone. I shook my head because I couldn't believe how fast this thing could move. I took off running up to where it leaped the fence but I never saw it again. I never did see this things legs, because it squatted all the time, but they sure were powerful. This thing could have easily gotten me if it wanted to. I believe it only comes out at night and just by chance I got to see it. Maybe it was headed to its home or den after a night of hunting or whatever it does. I have thought about it and I believe it burrows in the ground like a groundhog. Because not far from where I saw this thing is an old family cemetery plot with slab graves and a rock wall around it. There are huge holes all in this cemetery under the graves. My grandmother told me of a thing the old people call a 'Graverobber' and it moves real fast and digs in graveyards, but I don't know. I know I saw this strange creature and it has to stay somewhere.

Another man I know saw it one night run across the road in front of him and it only paused for a second to look at him and then it was gone. The way he described it, it was just like what I saw. He said he's never seen anything move so fast in his life especially, on two legs. This was also on the L&N Railroad tracks about 12 miles from where I saw it. It was in Parksville Ky. where he saw it. The strange thing, it was near a cemetery as a matter of fact. That's where it came from when it ran in front of him. When I saw this creature, it was a beautiful spring morning, the sun was shining and I watched it for what seemed to be 10 minutes or more, I could have watched it longer if I hadn't made any noise. If I'd only had a camera."

Barton interviewed him again and gleaned more details about the beast.

"The creature's arms looked like they came right out of its chest," he told me. "It didn't have shoulders. I couldn't see its legs because it was squatted and its feet and legs were hidden by the rail. When it landed in front of me it just crouched down. Its legs couldn't have been very long, but they sure were powerful, to move like it did. It didn't have a tail. I couldn't make out the pupils of its eyes, its eyes were very small and round and they were very dark, maybe not completely black but almost. Its face looked mean and curious. The thing blinked a lot, it would squint its eyes as if it couldn't see very good in the daylight. Its snout or nose didn't protrude, it had a pug nose and was dark brown, no fur on it. It did favour a bat in the face. Its teeth were razor sharp and very close together. It had perfect teeth and a bunch of them. They were almost clear. I guess they looked like that because they were so sharp and thin, like a bat's teeth. It looked like they were all the same length, but the upper teeth almost covered the lower ones up. It had a fat body, its head looked too small for its body and there was no neck.

When it finally saw me it leaned forward quite a bit and squinted at me like it was trying to make out what I was, then it was gone like a speeding bullet. This things head turned like an owl's head almost in a complete circle. It had long slender arms and very long fingers with long black sharp claws, actually they looked like finger nails. It had knuckles on its fingers. There were three fingers and a thumb. Its skin on its arms looked oily and real dark brown, about the colour of used motor oil. Its fingers moved constantly. Its fur was real fine I could see it moving when the wind blew. Every now and then it would raise its arm and rub the side of its head like it was scratching, but it always moved its head towards its arm, it never would move its arm very far, it always kept them straight out in front of it. Its ears were little and short. I could barely see them above the fur. Its fur was darker on the ends and the closer it got to the body the lighter it got. Some places were greyish in colour. It looked like it didn't have any lips at all. Just mostly upper teeth in a straight line and extremely sharp. Its nose was just a small button like nose with two nose holes. It didn't have a chin. This will be hard to describe, but its thumbs were on the bottom on both hands, it almost looked like it had two right hands, its fingers were on the top and it looked like its arms came out of its chest, and they were closer together. I guess its hands looked like they were backwards, but it had long slender fingers on top of each hand and a thumb on the bottom and both thumbs were on the same side if you know what I mean."

He also reiterated what his grandparents had told him, that the older people of the area knew of the creatures, that they were mostly nocturnal and they lived in graveyards, burrowing into graves and feeding on the rotting flesh of the bodies therein.

Another witness 'Billy' spoke to Barton about seeing the same kind of creature on Caney Hill in Dorton, Pike County, Kentucky, in June of 1990.

"When I was a young teenager," he told me, "I was walking up in the hills behind my parent's house. Their house is nestled deep into the mountainside down in a valley in the eastern Kentucky hills. So, any way that you walk, you walk pretty much straight up into the mountains. After making the uphill climb, the ground levels out a bit and goes on for miles of very thick dense forest. I spotted what I thought was a very large white dog from behind about 20 yards from where I was standing. It was just a quick glimpse and I didn't know at the time that what I had just seen was much, much bigger than a dog. It was broad daylight and summer so the brush and trees were very overgrown, and I would occasionally see glimpses of it again making its way through the forest. I began to call out to it; "here puppy," and follow it up into the hillside. It must have been moving pretty fast because I then lost sight of it for about five minutes and continued to walk in the same direction. It was then that I came to a very, very large bush. The bush was directly in my path, so I had to go around it. When I did, I was in for the surprise of my life. When I peered around the corner my heart stopped.

There in front of me, less than five feet away, was a creature I had never before seen or imagined. This was not a dog. It was white and covered with hair. It was standing up on two legs hunched over with small arms very high on its body. It didn't really have shoulders. The arms were much like that of a kangaroo. It was at eye level with me but in quite a hunch which puts it at about 7 feet tall if it were standing fully erect. We stood face to face and made full eye contact for what felt like hours, neither one of us moving a muscle. The encounter probably only lasted 5 - 10 seconds before I turned and ran as fast as I could down the mountain. I was terrified. Its eyes were fire red and the way they stared into mine felt as though it was taking something from me. The terror took my breath away. I kept this story to myself for quite some time until I met another guy who began to tell me of his experience with a similar creature. Over the past 16 years I have gradually heard several other stories here in the hills. Right now, I am actually aware of 8 or 9 other people who have had personal experiences with this creature or one like it in the surrounding area where I live. It still doesn't make me less freaked out by what I saw. As far as I know, I have had the closest encounter with it.

A friend of mine used to tease me about my experience until he saw it perched in a tree at Dark Holler up near Penny road off of US-23 (he then apologized!) Another friend saw it from behind, running down the creek. A friend's mother and sister saw a white creature on two legs jump off a ledge in the cut-through and bound across the 4-lane right in front of them. (This was on US-23 at Esco). It turned and looked at them before it headed down into the ditch and out of sight.

As I am writing this, I'm aware at how crazy this all sounds! And how surreal. It's not something we talk about much around here, but these stories come out from time to time. And then you hear other people who have had, or know people who have had encounters with a mysterious white beast. Another thing to note is that all of these sightings took place within 5 -10 miles of each other. They were all near railroad tracks and/or coal mines. I was remembering this experience tonight when a friend of mine decided to look up online any stories about a white creature in the hills. Needless to say, we happened upon this site. When I saw the story about the "Gravedigger" I nearly lost it. I never had a thought about the graveyard at all. But, (its giving me chills to type this right now and to realize this tonight) my sighting was between 50 -100 yards from a very old family graveyard up in the hills. I know what I saw was real. Before I had heard the other stories, I wanted to keep it to myself, and still not many people know my story."

INVISIBLE FRIENDS

Around two-thirds of children have childhood friends visible only to themselves. These playmates, invisible to adults, are generally thought to be nothing more than the products of childhood imagination. Mike Hallowell wrote the only in-depth study of the phenomena in his excellent book *Invizikids*. Mike interviewed people from all cultures and backgrounds and found interesting patterns.

There are a number of types of invisible friends. Some look like children, others look like animals, some resemble inanimate objects brought to life. The 'sages' look like adults in culturally distinct costumes, like Indian medicine men, or Chinese mandarins. They are called sages as they give out advice to the children. Mike concluded that these invisible friends are anything but imaginary, and calls them Quasi Corporal Companions or QCCs for short. Sages are, weirdly, visible only from the waist up. One informant told Mike that their Eskimo QCC appeared waist up from the floor, as if his legs were buried.

Glen Smith's sage was in the form of a Spanish Conquistador in full armour, and holding a spear. He called himself Edwinn and insisted it was spelt with a double n. He always manifested in the living room in front of an old oak cabinet, visible only from the waist up.

One day Glen was distraught over the death of his grandma when Edwinn appeared.

"Suddenly he was there by the cabinet as usual. I can remember his breastplate and helmet were shining. I think he must have polished them constantly.

He asked me why I was unhappy, and I told him about my grandma dying. He said 'You know, the more you watch the clock, the happier you'll feel. You will never forget your grandmother, but in time you won't cry so much. Remember Glen; the more you watch the clock, the happier you'll be.

I was only eight or nine at the time and I couldn't understand what Edwinn meant. How could watching the clock stop me being upset about my grandma dying? Anyway, I tried it out. I sat watching the large clock on the living room wall for ages, and in a weird sort of way it worked. I have never forgotten my grandma but the simple act of staring at the clock on the wall helped me. Just don't ask me how."

Mike interpreted this as Edwinn meaning that time is a great healer, but putting it in a way that an 8 year old would understand.

On another occasion Glen was upset about being blamed for breaking a chair at school, something he did not do. He was punished and his mother got to hear of it. He protested his innocence, and his mother visited the school and spoke to the teacher who lied, saying he had seen Glen throwing the chair. He got detention and punishment at home. But he found out that it was the teacher's nephew who had broken the chair and that the teacher was trying to protect him by shifting the blame to Glen.

Edwinn appeared again and said *"Glen, worry not when others speak ill of you' in time their lies will pierce them like an arrow."*

Another teacher subsequently found out who had really broken the chair, and the lying teacher was forced to write a letter of apology to Glen's mum.

Jane McCloud's QCC was an animated poker!

> *"When I was a child I had an imaginary friend called Mr Poker. I think I invented that name, because that's exactly what he was: a poker.*
>
> *On the hearth in the lounge my mother and father had a small, brass stand which housed a brush, a shovel and a poker. One day, I was lying on the rug near the fire reading a comic when the poker spoke to me. I got quite a shock. It just said 'Hello' and that was that. I didn't tell my mother and father, because I knew they wouldn't have believed me.*
>
> *A few days later the poker spoke to me again. I was listening to the radio-again in the lounge-when the presenter said something about Diana Dors dying. At this point I heard the poker say 'Hello' again, and I looked over to the hearth. The poker was no longer on the stand. I couldn't believe what I was seeing. It had grown two legs and was standing on the hearth itself.*
>
> *I asked the poker what it wanted. It said 'We all have to go sometime/ and then said 'But you have other things to worry about. Your mum asked you to do the dishes, and you've forgot. Forget Diana Dors and get the dishes done.*
>
> *Mr Poker was always reminding me of things I'd forgotten, and the funny thing was that I never once questioned my sanity or thought I was going mad. I just accepted that our poker could talk, and that was that."*

Interestingly, in Japan there was a tradition that inanimate objects could gain sentience after 100 years, and how they behaved depended on how they had been treated. Known as *tsukumogami* these objects included sake bottles, sandals, paper lanterns, umbrellas, rolls of cotton, kimonos, kettles and gongs.

Kay Johnson had a celebrity QCC in the shape of Elvis!

> *"I'm a 19 year old girl now, but when I was around 4, I had an imaginary friend called Elvis.*
>
> *I don't remember a whole lot about him now but I do recall that he always wanted to dance with me, and he lived in my grandmother's garden.*
>
> *I have one very vivid memory of dancing with him in the garden-he was*

twirling me under his arm and we were laughing and eating sweets.

Okay, here comes the really strange part. As I've said my imaginary friend was called Elvis. He looked like an exaggerated version of Elvis Presley just before he died. I remember him wearing one of those white jump suits with all the gems on it and he was really overweight. This might not be strange if it weren't for the fact that, at four-years -old, I had absolutely no knowledge of the real Elvis Presley. No one in my family was an Elvis fan (not even my grandmother, who's garden my imaginary friend lived in), so I'd never even come into contact with a piece of Elvis memorabilia.

My parents didn't think anything strange about it until a few years later, when I was watching TV with them and an image of the real Elvis Presley came on the screen. I pointed to the TV and said 'That looks like Elvis.'

This probably sounds made up, or you might even think I'm crazy, but I assure you this is totally true. There are a lot of people in my family who can remember me playing and talking to Elvis, and none of them knew where I'd gotten the name from. I don't even remember that.

My grandmother told me that one day I was playing with Elvis in the garden and her neighbour asked her 'Who is Elvis? Is he a cat or something?' and she replied 'Oh no it's her imaginary fiiend! She even remembers one time when I was making him his breakfast and calling him in to eat it before it got cold.

I don't really remember what happened to Elvis, but as I got older he just stopped showing up. To this day, no one in my family can explain where I got the image of Elvis from. There was no way, at four-years-old for me to have any kind of knowledge of the real Elvis Presley, the kind of clothes he wore and the fact he was overweight."

Some QCCS seem to cross over in to the realm of fairy lore. Such was the entity that visited Barry Colson of Adelaide, Australia.

"My friend was a bit like a gnome. He'd help me out whenever I needed it; but never made life easy. Before he'd do whatever I wanted him to do, he'd ask me to bring him something. One time he asked me to bring him a matchbox filled with a mixture of tea and coffee. On another he wanted strips of newspaper. Everything he ever asked for just seemed like so much rubbish. I couldn't figure it out. The strange thing was that he would get very excited when I brought him what he wanted. He'd jump up and down as if you'd given him a handful of

dollars.

He used to have a bag attached to his belt. Everything I ever gave him went into that bag. God knows what he did with it all when he went back to wherever he came from.

But you know, he never once let me down,. If I asked him for a favour, he'd do it-providing I wasn't being greedy or asking him for something unreasonable."

Bernie McKay's QCC was called Grip Yodel. He wore a huge red hat and a tatty looking blue suit that was much patched. He also had a magick doorstop, with which he performed all kinds of tricks. The doorstop was shaped like a milk bottle and made of solid brass.

Bernie asked Grip Yodel if he wanted his mother to mend his tatty coat, but Grip said that if he tried to take anything from this world back to his world, it would vanish.

Grip would always appear by the back door. He would also always have a rope tied around his ankle. The rope vanished beneath a big stone by the back door that the family used to scrape their shoes on before coming inside. He would never venture further than the rope allowed.

On one occasion he asked Bernie why he wasn't at school, and Bernie said that the heating system had broken down. Grip Yodel waved his brass milk bottle doorstop around his head and shouted '*Wahoo!*' Bernie thought that Grip had magickally fixed the school's heating system. Instead his mother arrived saying she had just found a huge bar of chocolate that she gave to Bernie. He was convinced that Grip had made the chocolate appear by magick.

The idea of Grip being anchored to the stone is interesting, almost as if the rope was a link to his dimension. This is an idea we will return to in the next chapter.

TALKING ANIMALS
Certain animals have been known to talk. Birds, in particular corvids, parrots and mynah birds are good talkers. Some have learned 500 words or more. Apes cannot form words like humans, due to the placing of their larynx. They can, however, learn sign language. Gorillas, orang-utans and chimps have all shown that they can learn to be proficient at signing.

Yet there are strange cases of animals that are not supposed to be able to talk, speaking as fluently as a cartoon character. These can be both friendly and sinister and in some cases seem to resemble the QCCs we discussed above.

A HAIRY SCARE WITH A TALKING BEAR

Cartoon bears are a mainstay of TV, from Paddington to Yogi and Barney bear. But when a talking bear turned up in reality, it was a lot less cute.

On Friday, July 5th 1985, Greg and Stephanie McKay drove about 5 miles of Highway 10 into some woods near Greenwater in Pierce County, Washington. They set up a camp, ate supper then turned in. The next morning something attacked their camp. It was a foul smelling, eight foot tall bear.

The bear began to speak. It asked them their names and if they had permission to be on the campsite. Stephanie said that they did, but the angry bear told them to get off the site immediately. As the couple hurriedly began to pack their things away, the bear, which stood on its hind legs throughout the encounter, began to throw rocks at them. The couple were sure this was an actual bear, not a man in a bear costume or a Bigfoot. In spite of its massive size, it had a high pitched voice that was utterly inhuman. After reporting the events, Sheriff Sargent, Terry Schmitt, went to the area to investigate and found tracks that look like those of an enormous dog.

Why would the couple make up something so absurd as a talking bear? It should also be noted that lacking opposable thumbs, real bears cannot throw rocks.

TALKING DOGS

The domestication of the dog began at least 40,000 years ago. It has been suggested that the reason modern man outcompeted the strong, bitter cold adapted Neanderthals was that we had early dogs to help us hunt and to guard campsites. Today dogs assist us not just as our best friends and family members, but as guide dogs for the blind, hearing dogs for the deaf, guard dogs, police dogs, rescue dogs and medical assistance dogs that can sniff out sickness such as cancer, and even predict heart attacks, epileptic fits and diabetic attacks. With brains containing over twice as many of the neurons dealing with intelligence as the brains of cats possess, nobody can question the smartness of dogs. But they are not noted for talking! Yet there are cases where dogs have done just that.

Alan Gates told Mike Hallowell that as a boy he had a friend called Jugg, a Border Collie who could appear and disappear at will and who could talk.

> *"Jugg first appeared to me when I was about three-years old. At first I would just see him sitting there, on the rug in front of the hearth. He never made any noise-he didn't bark, or anything-and he would just look at me. He would stay for maybe five minutes, and then would just disappear.*
>
> *After the first few times, he began to talk. He would maybe just say 'Hello!" or something like that. Then he began to talk more to me, and we would have proper conversations. I can't remember what we talked about, but I seem to recall that he would encourage me to eat my dinner, tidy my*

room and such like.

Later, when I was about six-although I can't be sure-there was an incident at my school where I was beaten up by an older boy. My mother seemed to think it must have been my fault. She said I must have been cheeky to the other boy, even though I hadn't. I remember getting upset about this.

One day, around that time, Jugg came to see me. I can't recall what he said but I remember feeling much better afterwards. I supposed he must have told me that everything would turn out alright, or something. Anyway, from that day onwards I wasn't worried at all.

About two weeks after I was beaten up, the older boy then did the same thing to somebody else in my class and he got kicked out of school. I never saw him again."

Alan didn't see Jugg again until he was 8, when he turned up in the garden. They had a long conversation and Jugg said that he would not be able to see Alan again. Alan gave him a big hug and then vanished for good.

The next talkative pooch was not afraid to show those who messed with it that it could wield considerable power. On July, 1908, there was a wave of petty crime in the area of Lincoln Ave, in Pittsburgh. Two policemen were walking through a city park on the 9th on the lookout for any miscreants. The pair of coppers saw a small, black dog with curly hair that suddenly said *"Good morning."* The two police men thought it must be some kind of ventriloquist act, with some nearby person throwing their voice. They were discussing this when the dog, wagging its tail said *"I can speak for myself."* One of the officers tried to pick the dog up but it sidestepped and said *"Don't touch me."* Ignoring the request, the cops tried to grab the dog who ran away with the officers in chase. One of them managed to grab one of the dog's hind legs but fount the pooch's limb was white hot. His hand was badly burned and the talking dog escaped. As they watched it run away, it dissolved into a green mist. After telling a third officer and showing him the wound, word of the devilish dog spread like wildfire and was eventually covered in the *New York World.*

THE TALKING CAT WITH HUMAN EYES AND THE PAEDOPHILE
In November 1953, two boys aged 11 and 12 were playing Cowboys and Indians in Little Woods, Speke, a suburb of Liverpool. As it grew dark they lit a small fire and one of the boys, Alan, lit one of his arrows aflame and shot it into the air. As he did so a strange animal appeared just 20 feet away. It was a tabby cat but huge in size, around four feet tall. What's more it had blue, human eyes!

"Hello children" it said and the boys panicked, running out of the woods onto Western Avenue and almost getting hit by a bus. Back in their respective homes, the boys both told their parents their unlikely story. Unsurprisingly they were not believed.

Next day, with gathered courage they decided to return to Little Woods to see if they could find the giant, human eyed, talking cat again. As it became dark they lit another fire and sure enough the weird feline returned. Stepping out from the trees, it sat itself down on the other side of the fire from the boys. It told them that it was called 'Semeel' and that it was a 'guardian'. It warned them not to play in the woods as there was a paedophile lurking there who wanted to kill them.

The boys went straight home and told their mums and dads, who once more rejected the story. They knew the pervert in question and also knew he was away on a holiday in Wales. A week later it transpired that Semeel had been right. The man had not been in Wales but had made a camp in Little Wood in order to wait for children. The police arrested him and he confessed to trying to abduct and abuse a boy in Widnes, another suburb of Liverpool, five years earlier. Whoever or whatever Semeel was, he had saved the boys from a horrific fate.

Interestingly, the boys spelling of Semeel may be a misspelling of the name the entity gave them. 'Samael' is the name of an archangel in Hebrew lore, supposedly cast out of heaven for opposing the creation of Adam. In some legends it is he and not Satan who became the serpent that tempted Eve. He is also known as 'The Venom of God' and is a paradoxical figure, being the executioner of death sentences decreed by God. Despite being a fallen angel he is still under the command of God. He is also said to be the father of Cain and the lover of Lilith (with whom he spawns daemons). Samael seems to be both a tempter of men and a punisher of sinners. He is also thought of as the angel of death. To make things even more confusing, in the *Apocrypha of John,* a second century gnostic text, he is said to be a demiurge, a creator of the universe and takes the form of a giant serpent with a lion's head. None of this makes anything clearer.

GEF THE TALKING MONGOOSE
At last we've reached the star of our show. If there was a poster boy for Forteana, a paranormal superstar, then it's the one and only Gef, The Talking Mongoose.

Jim Irving, a former furniture salesman from Liverpool, had moved his family to the Isle of Man, a beautiful, wild location that lies between England and Ireland, He had brought a house called Cashen's Gap in a remote area of the island. The house had no electricity and no running water and it must have been quite a culture shock for the family. He brought with him his wife Margaret, his teenage daughter Vorriey, and Mona the Border Collie.

The strangeness began on Monday 14th of September 1931, the family began to hear noises coming from the attic. The sounds were animal-like growling and scratching. Soon these moved to wall cavities and they thought it may have been rats. Yet as the days went by the Irvings began to think that the noises sounded like a voice. Wrapped up in the texture of the noises were words. The voice seemed to be imitating words they used, in the way an infant mimics adults when it learns to talk.

Whatever it was, it learned fast and began to form whole sentences. The entity then began to tell the Irvings what it was. It called itself Gef and claimed to be a mongoose born in India, on June 7th, 1852. At first Gef was shy and the family only got the briefest glimpses of him but in time all of them saw him clearly. Margaret even stroked him. He was described as being a foot or more long, with yellow fur and a bushy tail. His paws looked almost like tiny human hands. As well as the Irvings, a local boy saw Gef as he bounded through a field near Cashen's Gap. His description matched that of the family.

Gef would tell jokes, sing (he was fond of *The Isle of Capri* and *Home, Sweet Home*) and proclaim to be the 'Eighth Wonder of the World'. He also made mystic proclamations, saying that nobody would ever know what he was and if they were to see him they would be '*mummified, petrified, a pillar of salt*'. Despite this he was generally friendly and even killed rabbits for the family to cook and eat, leaving them by the door.

He seemed to eat food left out for him like a mortal creature, being very fond of chocolate and biscuits When hungry he would shout '*How about some grub Jim?*'

Apparently he could travel long distances, invisibly hitching rides of buses to Douglas, the capital of the Ilse of Man, and bringing back gossip that invariably turned out to be true. He also sneaked into the flats owned by the bus drivers and described them in detail to the family. He also went on trips to the north of the Island and visited Ramsey Market.

Soon the fame of Gef spread beyond the island and found its way into newspapers around the world. A reporter from the *Manchester Daily Dispatch* who heard Gef speak wrote that the experience left him "in a state of considerable perplexity":

> "*Had I heard a weasel speak? I do not know, but I do know that I have heard today a voice that I should never have imagined could issue from a human throat. The people here at the farm who claim it is the voice of a strange animal seem sane, honest and responsible folk not likely to indulge in difficult long drawn-out practical jokes... The weasel even gave me a tip for a winner in the Grand National Horse race!*"

Gef was thought to be a weasel, before he told people he was a mongoose. The two types of animal are not closely related despite superficial resemblances.

Several investigations into Gef were carried out. Harry Price, the famous ghost hunter sent H. E Dennis, a member of the Royal Zoological Society of London to investigate. He heard Gef's high, squeaky voice and saw objects thrown, seemingly from gaps within the walls. He heard Gef moving objects, such as a ball and a chair around in his 'sanctum'. On one occasion Gef told Margaret all of the details of a conversation Jim and Dennis had during a pub lunch when only they were there. Margaret relayed the details accurately to Dennis afterwards. Dennis made three trips to Cashen's Gap in all.

Harry Price visited Cashen's Gap himself. Jim told him the Gef knew he was coming and had taken his leave saying that Price *"Put the kybosh on the spirits"* and *"Had his doubting cap on."* Price came away thinking that Gef had been created by the Irvings for their own entertainment. There were several alleged photographs of Gef taken around this time. The pictures are far from clear and in my opinion some of them look like mink stoles. Hair that Gef told the family he had plucked from himself, was analysed by the Natural History Museum and found to come from a Collie dog, probably dear old Mona. Imprints of footprints were shown to be from three distinct sources. One said was from a dog, one from a racoon, an animal not native to Europe. The third set were never identified. None of them were from a mongoose.

Other investigators came to different conclusions. Nandor Fodor (1895-1964) was a Hungarian psychoanalyst who also visited the Irvings. He didn't see Gef but he interviewed all the family and was convinced that Gef existed. He wrote:

> *"That the 'something' which is called Gef exists and talks, I hold proved. But as to what it is, opinions may differ. Once we step into the marvellous, reason and logic give us no bearings."*

Fodor stated that he found the Irvings to be *"sincere, frank and simple"* and that...

> *"deliberate deception on the part of the whole family cannot be entertained as a solution of the mystery.*
>
> *The charge of ventriloquism is best answered by the fact that Gef has been heard when each member of the family has been alternately eliminated,"* he wrote. *"It is sufficient to spend a day at Doarlish Cashen to know that, under their conditions of living, it would be impossible to carry on a ventriloquial imposition over a period of years."*

Doarlish Cashen is the name of Cashen's Gap in the native Manx language. Some people accused Vorriey of throwing her voice to trick people, but Gef was often heard when she was not at the house at all. Another theory was that Vorriey and Margaret had conspired to fake the haunting, in order to convince Jim to move back to the mainland. There is zero evidence of this and even if it were true the hoax failed miserably.

Fodor rejected the idea that Gef was a poltergeist.

> *"Poltergeists are always invisible,"* Fodor explained. *"Gef never claims to be without an animal form. He eats, drinks and sleeps...he leaves his*

teeth marks in the butter in the larder and in the fat of the bacon. He catches rabbits and performs various other services for the family. Poltergeists are an unmitigated affliction. Gef is an asset.

Poltergeist disturbances usually begin just before a young girl or boy in the house reaches the age of puberty. The disturbances die out soon after the critical period. Gef has not faded out. He grows stronger than ever, and he is visibly outgrowing the original affection which bound him to Voirrey Irving, the young daughter of the house."

Fodor thought that Gef was a real, physical mongoose, who had become psychically possessed by "a split-off part" of Jim Irving's personality.

In fact, he may have been on the right track. Probably the best explanation for Gef is that he was a tulpa or thought form. We will examine tulpas more closely in the next chapter. Suffice to say for now that they are thought to be semi-physical creatures, created by the power of the mind. Gef seemed like a gestalt of the minds of the family. He seemed to like the things Voirrey liked and disliked the things she disliked. He could speak Hindi but not well, much like Jim who had served in India. He could speak Yiddish but again, not well, much like Jim who had sold furniture to the Jewish community in Liverpool. He was, a sum of their parts with added mischief.

When the family moved in 1939, Gef did not go with them. Cashen's Gap farm was bought by a Mr Graham, who in 1947, shot an animal in the grounds that some have claimed was Gef. A picture of the animal shows that it is clearly a polecat and not a mongoose.

The house lay empty for many years and still had the reputation of being haunted right up to its demolition in 1971. In 2022 I visited the area. Nothing is left of Gef's home these days. The only clue that a house once stood there, is a well and Mrs Irving's gooseberry bushes in what was once the garden. Some friends took cuttings in the hope to grow new bushes on the mainland. One day we may be able to eat Gef gooseberries!

If Gef was a tulpa, it may be possible that he could be revitalized and return to us, if he ever truly left totally. Fortean researcher Jenny Randles says in her book *Supernatural Isle of Man*, in October of 2002, whilst driving with her mother on the outskirts of South Barrule, heading for Glen Rushen she saw a brownish yellow mongoose bound out onto the road in front of her. It stood and looked at the car before bounding off into the undergrowth.

Voirrey ended up living near her elder sister Elsie, in Cheltenham where she worked at an aviation plant. She was very reticent about talking of Gef, for fear of what people would think of her. She only took part in one interview in 1970, with Walter McGraw of *Fate Magazine*. Voirrey felt that Gef in many ways had ruined her life.

"They said I was mental or a ventriloquist. Believe me, if I was that good I'd be making money out of it now! Gef was very detrimental to my life. We were snubbed. The other children used to call me 'the spook'. I had to leave the Isle of Man and I hope that no one where I work ever knows the story. Gef has even kept me from getting married. How could I ever tell a man's family about what happened?

It was not a hoax and I wish it had never happened. If my mother and I had had our way we never would have told anybody about it. But father was sort of wrapped up in it. It was such a wonderful phenomenon that he just had to tell people about it. .

Yes there was a little animal who talked and did all those things. He said he was a mongoose and said we should call him Gef...but I do wish he had left us alone."

She passed away in 2005.

The entirety of Gef's shenanigans is far too long for this book and I would point readers to Christopher Josiffe's excellent and comprehensive work *Gef! The Strange Tale of an Extra Special Talking Mongoose* to get the full story.

WTF!?
If what has gone before just isn't weird enough for you, then we will round off this chapter with a group of cases so mind bendingly strange, they just make you go WTF!

THE DANCING COWS
In his book *Mysteries in the Mist,* WT Watson recounts a very strange story.

As a 7 year old boy, Scott Harper was visiting his grandparents on their farm in central Ohio. Scott could not understand why his grandmother did not want him to look out of the window. At the first opportunity thats what he did, as any curious child would. Scott said that the sun was setting and there was a fog. Still wondering why his grandmother didn't want him to look out of the window, he noticed a glow in the fog. He thought, at first this must be car headlights. Gradually the glow seemed to approach the house and Scott noticed that there were two sources of light. One was yellow, the other was pink. The shades of the glow were a kind of washed out pastel. As he watched the source of the lights appeared. It was two cows standing on their hind legs. One was glowing yellow the other pink. Each had a foreleg on the shoulder of the other and they appeared to be dancing. They stumbled and swayed along the road.

Scott's grandmother found him looking out of the window and became cross but then, on seeing the cows she fainted. When his grandfather got home Scott told him the whole story, despite his grandmother telling him to say nothing. Scott's grandfather took up a rifle and searched his property. He returned having found nothing but looking agitated. He grandparents never spoke of it again.

THE SHADOW STEALERS

Steve Oldale, a construction worker from Glasgow was walking in Perthshire, Scotland. He was at Schiehallion, a mountain also called 'Fairy Hill'. The time was around 5 in the evening. It was a fine day with a clear blue sky, but then suddenly a small cloud appeared, seemingly drifting in from nowhere, and a fine drizzle began causing a rainbow. As Steve watched the rainbow he began to hear the tinkling of a stream, nearby, seeming to change into something like harp music. He felt almost enraptured but suddenly he caught a movement in his eye. Looking down he saw a small man, about 4 feet tall, standing at the head of the long shadow Steve was casting. The little man was 'rolling up' Steve's shadow as somebody might roll up a sheet of cloth. Steve then noticed another figure, a little woman of the same size, armed with silver sheers who was kneeling in front of him and cutting around his feet with the shears, slicing away his shadow from his foot!

Steve shouted *"OI!"* and the little people vanished and Steve's shadow returned to normal. Steve describes them as dirty looking with leathery, wizened skin. They reminded him of the bog bodies' found in various areas in Europe. They wore caps and smocks of a Chamois leather type material stitched together.

The case resembles the fictional Peter Pan and seems to utterly defy physics, unless the events were occurring only in his head. But then again, an awful lot in this book defies physics.

THE CHARTERHOUSE GUEST AND OTHER 'FLATS'

The Charterhouse Outdoor Centre is a large bunkhouse that sleeps up to 63 people and is available for hire. It sits near the hamlet of Charterhouse, in the Mendip Hills of Somerset, England and also acts as a HQ for the local search and rescue team.

In 1982, the warden at the facility, Terry Birch, told schoolteacher Richard Gardener of something that happened to him in the building one night. Terry's duties included going up to the remote building before each holiday season, to check that the mountain rescue equipment was clean and in working order. A former Marine Commando he thought nothing of staying in the Centre overnight, rather than driving the dangerous roads in the dark.

One night in 1981, he had spent the day re-charging batteries and cleaning waterproof clothing. That night he would sleep in a bed in the centre's sick bay, as he usually did when stopping overnight. He preferred this to the draughty dormitory. He had already checked all the doors were locked. The climbed into bed and fell asleep.

Sometime later he awoke to the noise of something circling the perimeter of the building. He assumed it to be a badger or a fox. Soon after he heard the sound of claws clattering on the wooden floorboards downstairs. He was puzzled at just how the animal had found its way in but reasoned that it could also find its way back out again. He was too tired to get up and usher some animal out of the place.

However, Terry was to have no peace that night. Whatever it was came scuttling up the stairs and began to scratch and snuffle at the sick bay door. The door creaked at something pushed against it but it was firmly locked. Then there came another sound, like that of a stiff brush rubbed slowly along a polished surface. Terry realized to his horror that something was sliding *under* the door! The gap between the bottom of the door and the floor was no more than half an inch but whatever the thing was, it was managing to slide under it. It must have been utterly flat! Then the noise stopped. Whatever it was, was in the room with him. Turning over to look he caught sight of some black shape barely visible in the dark room.

Suddenly the bed rocked and he felt teeth and claws sink in to his leg. Screaming at his attacker, he kicked out connecting with some shapeless thing, that jerked and recoiled. He kicked again and again till the assailant let go, giving a high pitched shriek unlike an animal vocalization he had ever heard. Then he heard a scrabbling as the thing retreated across the room.

The visitor hit the door and Terry heard the sound of it crawling back under the gap and the scraping sound of it retreating along the corridor and down the stairs.

Jumping out of bed he switched on the light and found his bag had been opened and his clothes had been strewn everywhere. There were scratch marks in the wall and in his struggle with the creature, Terry had managed to pull some electrical wire and plaster out of the wall.

Checking downstairs Terry could find no possible way an animal could have gotten into the building.

Dermot MacManus calls these strange, compressed, slithering entities 'flats'. He records two such things in his book *Between Two Worlds*.

Mr George Hallet, a professional man from Limerick, related to him such an encounter he had when he was a boy. He was spending the summer at the Mount Temple House, near the River Shannon in Limerick in the early years of the 20th century. George was 12 at the time and had developed a habit of sleepwalking. Fortunately he didn't stray far from his bed during these attacks of somnambulism.

One night he was awoken by the chimes of the clock striking midnight, to find he had been sleepwalking again. It was pitch black and he had to wander with his hands stretched out to try and orient himself. He grasped a piece of furniture and identifying it he realized he was on the opposite side of the room to his bed. Having worked out where he was, George began to slowly walk back in the direction of the bed. Suddenly he put his foot on something flat and hairy. There were no rugs in the room. The flat, hairy thing let out a blood curdling scream as George stepped onto it. In abject terror he dived into his bed and hid beneath the covers until dawn. He was badly effected for two days afterwards. A search found nothing to account for the 'flat' and its voice, George told Dermot, was hideously human sounding.

Dermot MacManus' aunt Lottie had a run in with a 'flat' at a house called Killeaden in September of 1926. Killeaden is situated in County Mayo, Ireland. One night Lottie was awoken by the sound of her window rattling insistently. She was seized by the idea that something evil was trying to get in at her. There was a noise as if a sheet of paper were sliding between the window frame and the sash. Then there was a distinct plop as if something had dropped clumsily to the floor. This was followed by the noise of something shapeless slithering across the floor till it reached the side of her bed.

Lottie felt it roll up the bed then clasp her hands with an ice cold, inhumanly strong grip and pin her down. She felt the weight of the flat, shapeless horror shift onto her in the darkness. Using her mental will she called on all the powers of heaven to stop the thing and it did, indeed, pause in its advance. It still gripped her and squatted on the bed invisible in the darkness. Lottie entered a mental struggle to drive the thing away. Finally the thing let go, slid off the bed and crawled away to the window.

Eventually Lottie could move again and she lit a match. At first she thought the whole thing could have been an evil dream but in the match light she could see the marks on her wrists were it had gripped her and the indentation on the bed where it had lain.

OCTO-SQUATCH
In the summer of 1961, at around 11pm. Arquimides Sanchez, a truck driver, was navigating the steep roads of the Basque Mountains, in the Spanish Province of Vizcaya. Sanchez and his riding partner were nearing Puerto de Barazar, ready to unload their goods and rest for the night, when their high beams caught something at the base of the embankment on the opposite side of the road. Sanchez slammed on the brakes and he and his co-worker stared at what he descried as a 4 foot tall octopus, covered in shaggy, reddish hair. The thing had big, round, glowing eyes and it did not like the truck's headlights, as it used four of it's hairy tentacles to shield it's eyes from the light. But the creature did not move.

Fear gripped Sanchez and he ordered his friend to attack the thing with a jack-hammer. His equally scared friend refused. Sanchez threw the truck into reverse, then drove straight towards the entity, stopping mere feet in front of it. He repeated this several times to try and scare the beast away but it did not move. In the end they had to drive around it and left it simply standing in the road.

SAM THE SANDOWN CLOWN
Sinister clowns have a long and venerable history from trickster gods to the villains of horror films. And of course the greatest killer clown of them all The Joker from the Batman comics, so wonderfully portrayed by Mark Hamill in *Batman The Animated*

Series.

Coulrophobia is a recognized condition, a fear of clowns. In several countries around the world, there have been clown panics. Children report sinister clowns trying to tempt them into vans. When police investigate the clowns are never caught and the 'clown flap' is dismissed as an urban panic.

Sam the Sandown Clown is something even stranger. Neither fully clown nor fully ghost or fully alien or fully robot. Sam really is a law onto himself. The case was first reported in British UFO Research Association's Journal, Vol. 6, No. 5, January/ February 1978, in an article by Norman Oliver titled: *"Report-Extra! Ghost or Spaceman '73?."*

The Ilse of White is a popular tourist destination off the coast of Hampshire, in the south of England. In May of 1973, a 7-year-old girl 'Fay', (a pseudonym) and her friend, an un-named boy of a similar age, were walking near Lake Common on the island. The children heard a noise like an ambulance siren. They followed the noise across a golf course to a meadow near Sandown Airport.

At one point they had to cross a brook via a small wooden bridge. As they did a blue gloved hand reached up from underneath, then a whole figure emerged. The colourful being stood 7 feet tall and was dressed in a green tunic with a red collar, white trousers, and yellow pointed hood with a black knob on top and two wooden antenna protruding from either side. He had bare feet with three toes and blue gloved hands. His face was curiously stiff with triangular markings for eyes, a brown square for a nose and static yellow lips. He seemed to have no neck, the head simply sitting on the shoulders.

The clown was carrying a book that he dropped into the brook and retrieved once more. Then he leapt out of the water and moved away from the children with a high-kneed, hopping gait, not unlike that of an astronaut. He went inside a metal hut, like those seen on building sites, but soon emerged with what looked like a microphone attached via a flex to a small box.

"Hello, are you still there? The clown said in a friendly voice before continuing. *"Hello, I am all, colours, Sam"*

The kids struck up a conversation with Sam, asking if he was a man, to which he replied *"No."*

Then they asked if he was a ghost and Sam said, cryptically, *"Well not really but I am in an odd sort of way."*

They went on to ask him what he really was and he answered, *"You know."*

Sam invited the children into his hut and they entered via a small flap. Inside the hut was divided into two levels. The lower level had blue-green wallpaper and covered with a pattern of dials. It also had an electric heater and some basic wooden furniture. The upper level was not as expansive and had a metal floor.

Sam told the children that he drank water from a nearby stream, but cleaned it first. For food he collected berries. Sam then removed his hat to reveal a pair of rounded white ears and tuft of brown hair. Sam showed them how he ate berries. He placed a berry in his ear, then lunged his head forward causing the berry to disappear and reappear in one of his triangular eye sockets. After repeating the head movement the berry reached his mouth as if the whole head was hollow.

Sam said there was more of his kind and he had a secret camp on mainland England, but didn't say exactly where.

After half an hour the kids left Sam's shack and walked back across the golf course. They told an adult that they had just 'seen and spoken to a ghost' but the man laughed at them. The children also noted that there were two workmen nearby who didn't seem to notice Sam, the shack or themselves.

Fay told her father, whom Norman Oliver refers to as 'Mr Y', but not until three weeks later. Mr Y looked at the area but found nothing.

Oddly, Oliver says that Mr Y had some strange experiences of his own in the past. In 1970, whilst driving near the village of Brading, he saw a UFO. Mr Y described it as a ring of lights, cherry red and interspersed with turquoise and white. The UFO was completely silent. The object kept pace with the car and manoeuvred behind it, apparently shrinking in size. Mr Y signalled to the object with his torch for ten minutes, as it moved backwards and forwards, slowly rotating. Finally he drove to a friend's house and told him. They both went outside and watched the UFO above some trees before it finally faded out of sight.

In the following years, Mr Y claimed that he saw balls of red light in the sky that would sometimes follow him.

Stranger still, in March of 1972, he was on the cliffs at Compton bay. Looking down at the sea he saw two points of yellow light under the water. He realized these were the eyes of some huge creature watching him from under the sea. What, if anything this has to do with Sam, is unknown.

THE BIG GAY RAPIST HIPPO MAN
Yes, you did read that right. South African magistrate, Michael Osborn, dealt with a court case involving a monstrous hippo headed male rapist. Isak Niehaus records the case in his book *Witchcraft, Power and Politics: Exploring the Occult in the South African Lowveld.*

The weird story began when a sangoma was found dead in his home. He had been bludgeoned to death with a mallet he used to crush herbs, that he used in his potions. The man's son was the main suspect and he had left behind a note to say he was not trying to run away, he was just visiting his brother in Swaziland, to tell his brother what had happened and to settle matters in the traditional tribal way. He promised to return once the business was done.

True to his word, he returned a month later and turned himself in to the police. He confessed to killing his father, but said that there had been extenuating circumstances. The man was a college student and was staying up late at night reading, when a figure entered his room through a locked door. It had the body of an old man and the head of a hippo. The man became paralysed at the sight of the thing and unable to move or scream. The hippo man proceeded to sexually molest and anally rape him. The hippo man did this several times, but on the last occasion he managed to grab the mallet and breaking the big gay rapist hippo man's spell, he beat the monster around the head with the mallet repeatedly. The creature fell to the floor and grabbing a lantern the man saw that instead of the hippo-headed sex demon, he found his father, dead and naked. The old sangoma was changing himself into a hippo man, in order to rape his own son.

Michael Osborn got the defendant acquitted, due to the fact he was acting in self-defence. It seems that the jury would have done the same thing if they found themselves getting bummed off a big gay rapist hippo man!

THE KILLER SMURF PANIC
What could be less threatening than a Smurf? Tiny blue people in Phrygian caps created by the Belgian cartoonist Pierre Culliford in 1958, they have turned up in comics, books, cartoons, pop songs and films. They also turned up as homicidal fiends in an urban panic.

A panic swept through schools in Houston, Texas that schools were being invaded by murderous smurfs, who were brutally killing the head teachers. The smurfs were armed with machetes and machine guns and lurked in school toilets.

An investigation carried out by Newsweek determined that the panic had likely stemmed from a local news report, about forty members of a teenage street gang nicknamed 'The Smurfs', being arrested for petty crimes. Mike Dash of *Fortean Times* also mentions nearly fifty people being arrested around this time for *'Smurf-related goings-on'* in his book *The Borderlands.*

The fires were fanned when Jehovah's Witnesses convinced themselves that the little blue men were demonic in nature. One ex-Jehovah's Witness wrote on a forum that the first he heard of this moral panic, was when he was banned from watching the Smurfs TV show. Stories spread that Jehovah's Witnesses who had brought smurf merchandise had been plagued by demons. One story was about a little girl, who had been given Smurf-themed curtains for her bedroom, only to wake the first night and find that the demonic Smurfs had leapt out of the curtains and were dancing about laughing satanically.

A similar story ran that Jehovah's Witness family made the mistake of decorating their son's room with Smurf wallpaper, shortly after they noticed that their child was covered with tiny red marks in the morning. The mother of the family asked her son if he had noticed any bugs in his bed, but to this he replied *'No, the Smurfs come out of the wallpaper and bite me at night. In the daylight they're cute and smiling but at night they have sharp teeth'*. Terrified, the parents ripped down the wallpaper and set about redecorating the room, but the wallpaper allegedly wouldn't burn without excessive quantities of petrol being applied.

THE SLIT-MOUTHED-WOMAN.
Similar panics swept through Japanese schools in the 1970s, but instead of smurfs, a slit mouthed woman was involved. Kuchisake-onna, the slit-mouthed-woman, is a legend that states that centuries ago, during the Heian Period (794-1185), there was a beautiful woman who was the wife or concubine of a samuri. But the woman cheated on her lover and the enraged samuri slit open the sides of her mouth, leaving her with a gaping maw and telling her that *"No one will find you beautiful now."*

She was supposed to have become a yokai (a word broadly meaning ghost or monster in Japan) and the story goes that she still wanders around on foggy night till this day. Kuchisake-onna is said to cover her mouth with a surgical mask. She approaches victims (usually children) and asks *"Do you think I am beautiful?."* If they answer yes then she removes her mask and reveals the jagged tear that extends her lips all the way into her cheeks. She again asks *"Do you think I am beautiful?."* If the victim answers either yes or no the Kuchisake-onna will attack them and slash open their mouths with a knife or scalpel, so that the resulting wound resembles the mouth of the Kuchisake-onna. Reputedly there are two ways of stopping her attacking. One is to answer her question by saying *"you look normal to me"* or by offering her a piece of candy.

The Kuchisake-onna has become one of the most well-known urban legends in modern Japan. They seem to have stepped out of mythology and into the modern era in the 1970s. New stories of her origins emerged to fit in with the times. In 1979, rumors abounded in Japan that the slit mouthed woman was hunting down children. The scare stories spread via playground lore and within one year had covered most of urban Japan.

Kuchisake-onna even became the subject of a record. 'Catch Kuchisake-onna' was recorded by Pomâdo & Kometiô on King Records in August 1979, and featured a trio of girl singers wearing surgical masks. The single was not released however, due to the fact that the advertising director of King Records had received a phone call from a woman, claiming to be the Kuchisake-onna herself and accusing the company of trying to profit from her 'distinct physical characteristics'! Surely this is the only case of a ghost threatening to sue a record company. In 2005 the single was released as part of a CD from Teichiku Entertainment.

THE GIANT GLOWING DEATH MAGGOT

One of the creepiest Fortean stories, is related by Bernard J Hurwood, in his 1977 book *Ghosts, Ghouls & Other Horrors.* Hurwood gives scant details such as date and exact location, but the story is so odd it makes the cut. The creature in the story resembles the giant green maggots in the classic Doctor Who story *The Green Death.*

Many years ago, in a Yorkshire village, a postman named Mullins was walking past the graveyard on his way home when he noticed a strange light. Looking closer he saw that it was emanating from a grave. As he watched a glowing, maggot-like entity oozed up out of the grave like a giant glow-worm. Mullins said that the most horrid thing about the entity was that it had human eyes. The grave belonged to a man called Mr Peters.

The giant glowing maggot heaved itself up from the grave, and slithered off between the headstones, leaving a trail of slime behind it. Mullins bravely followed the loathsome thing and its repulsive spoor. It seemed to vanish when it reached the threshold of the vicar's house.

Mullins told his wife and his best friend, and the next night the three of them returned to the graveyard to try and spot the spectral maggot. Just as the night before, the human eyed, giant, glowing maggot heaved up from the grave and slithered away vanishing near to the vicar's house.

As it turned out the vicar and his family were taken ill and all died within the day.

The following night, the trio followed the monster maggot again and this time it crawled up to the house of the local blacksmith and vanished on the threshold. The blacksmith then took ill and died within a day.

Mullins, his wife and friend kept vigil every night afterwards, but the maggot did not reappear until the 10th night. As they followed it, to their horror they saw it vanish on the threshold of Mullins' own house! Next morning their 5-year old son was stricken by a fit of vomiting and died.

Subsequently, Mr and Mrs Mullins and their friend returned to the graveyard and dug up Peter's body and burned it, before re-filling the grave. The glowing monster maggot was never seen again, but Mullins found out later that Peters was on bad terms with both the vicar and the blacksmith. Why the killer maggot visited his house he never did find out, he had no quarrel with Peters.

The story sounds like it came from the pen of H.P Lovecraft or M.R.James. I can find

no confirmation of it elsewhere.
HAUNTED BY MUPPETS
What could be less scary than Jim Henson's loveable puppet creations? Entertaining and educating kids for well over 50 years in Sesame Street and The Muppets. Jim's work is loved all over the world.

But the great silent horror movie actor, Lon Chaney, the Man of A Thousand Faces once said *"There's nothing funny about a clown at midnight."* It's all about context. Imagine waking up to find Animal or Dr Bunsen Honeydew looming over your bed, watching you. But according to some of the correspondents to the essential Cryptonaut Podcast and other sites, phantom Muppets are an actual thing.

A lady called Nichole H contacted the Cryptonaut Podcast, or rather its sister podcast, the wonderfully named Evil Science and Magic buddies. Nicole lived in Ohio, and seemed to be the focus of a haunting in the early to mid-1980s. The haunting included knockings, sightings of 'shadow person who shoved Nicole's mother down the stairs and attempted to suffocate Nicole with her own bedclothes. Nichole claims that the thing also attacked her with her father's straight razor, cutting her arm several times.

Nichole was also examined at night by foot tall, grey skinned creatures with long pointed ears and circular black eyes. The spindly armed, goblin-like beasts crawled around her, seeming to perform a medical exam. Apparently her brother saw this too but seemed to be parallelized as the goblins crawled over his sister.

Another character was 'The Creeper Man', a tall thin entity that would dress in weird costumes like a ladybird, complete with antenna or an evil doctor in a lab coat with a mirror strapped to his head, like a mad scientist from a 1930s pulp novel.

But the Creeper Man was not the strangest phantom haunting the girl. She claims that she had nightly visits from what looked like the character Count Von Count from Sesame Street. In the show, the character is a purple skinned vampire in a cloak and dinner suit, that is obsessed with counting. The version that visited Nichole looked just the same except, she said, it had human teeth, crooked and dirty.

The Count would squat on all fours on her bed, pinching and squeezing her toes. Like his TV counterpart he would count and laugh *"One, ha, ha, ha. Two ha, ha, ha. Three, ha, ha, ha."* He would also try to get the girl to go with him and become angry when she refused. He also threatened to kill her pet cockatiel if she told her mum and dad of his visits.

A pastor who tried to bless the house, fled in fear unable to do it, and refusing to say what he had seen in the house.

When the family moved, most of the phenomena stopped, though the shadow man continued to visit Nichole into her 20s.

In January of 2014, a young man calling himself 'Clarence' wrote to the website *Your Ghost Stories.com* to tell of an experience he had in 1991, in New Orleans. He was playing truant, pretending to be sick so he could stay home from school. He was 7 at the time and resting on the sofa at his grandmother's house. He waited to hear her drive away before dropping his charade. Then he hears the door to a toy closet creak.

"Though, right now my focus is on the toy closet. The only thing that divides the den and the rec-room is a four-inch carpeted drop and so I can see clearly across to the closet, which is built into the room's far-wall. And, to a 7 year-old Clarence's wide-eyed amazement, I can see that the closet door is slowly opening.

I can't move. My eyes are fixed on the door as it casually swings open. This isn't a draft or overzealous hinges. Something is wrong here. I can feel it in my bones (just as clearly as I do now, years later as I type these words.)

"Is she gone?"

The voice is a distinct whisper, clear and yet somehow not right. The way it seems to hum on the last syllable of every word, sort of like the vibration you hear when a bass-note is played through an old amplifier. I can still hear that voice as clear as day in my memory. I wish couldn't. But I can.
The first one leans out.

He's shorter than the doorknob and looks, for lack of a less ridiculous reference, like a humanoid puppet from the American children's show Sesame Street: fuzzy light pink "skin" with a shiny black "nose", big cartoonish eyes and a tuft of troll doll-like hair atop his football-shaped head.

The only difference is this little guy isn't being puppeted by anyone and his legs are clearly visible. Seven year-old me is so fascinated by this aspect that I am suddenly overwhelmed by the desire to get closer; to get a better look at this puppet-person. My fear takes a backseat to my childish curiosity as I begin to sit up.

Then comes the GRUNT...

What seems to be an affirmative response to the first puppet's question; this new voice brings me to a halt. If there was something vaguely creepy about the first one's tone, then this second one's voice is down-right disturbing. People don't sound like that. (Make the "uh-huh" sound without opening your mouth. Did it sound sort of like "mmm-hmmm"? Now imagine that sound but louder and with a bass-like hum to it.)

The first puppet exits the closet followed by his nearly-identical companion. The two of them are only visible for a moment as they hurry behind the rec-room's wraparound sofa.

Once again my fear subsides, giving way to curiosity. Trying my best not to make a sound, I slip down off of the den's sofa and slowly crawl across to the rec-room. The first puppet is now apparently bending over something behind the rec-room sofa and his little denim rear-end (they're both wearing faded grey denim overalls) is now poking out from behind the wrap-around couch.

I make my way across the rec-room, still on my hands and knees, following the long end of the wrap-around sofa over to where I can see the first puppet's butt poking out. I'm inches away when I finally stop crawling, too afraid to go any closer. I can't even bring myself to poke my head around the bend in the sofa. So instead I simply reach out and place a hand on the puppet's back, saying in a timid whisper "Hey..."

The thing instantly whirls around, its big puppet eyes now glaring at me as I retract my hand. It leans around the corner of the sofa, bringing us face-to-face and I open my mouth to say something, though I'm not sure what it would've been.

The puppet darts out a hand and pinches my arm. HARD. Harder than I have ever been pinched before or since. Hard enough to break the skin on my forearm and inevitably cause a small scab to form.

Without a thought, I'm up and across to the side-door and then out onto the lawn where I wait for what feels like hours until my grandmother finally comes back. She starts to scold me as she exits the car, something about faking sick just so I could go outside and play... But then she sees my face.

At first, I don't want to tell her what took place. Even at seven years-old, I'm aware that these kinds of things just plain don't happen. But then she asks about the tiny smear of blood on my arm and I break down.

I tell her a condensed version of the story. I even admit to fake-sleeping earlier when she was leaving after my grandmother tries to insist that it was all nothing more than a bad dream brought on by the fever I had been faking all day. I don't admit to faking sick, of course (that would just be reckless.) So the conversation ends there, at least as far as my grandmother is concerned.

For me, though? To this day, it's still a constant debate. Am I simply crazy or did scary-ass puppet-people actually dwell in the toy-closet of my grandparent's rec-room? (I didn't say it was a logical debate.) I can remember weeks afterward sitting in class or wherever and looking at the scab on my arm and every time thinking: Oh yeah, that happened... That ACTUALLY happened"

Journalist Brian Bethel had his own run in with evil Muppets as a child in Texas.

"Long ago when I was about four or five, I stayed over at my grandparents' house for the night. My Mom stayed over with me, because Dad was out of town on business.

Granny and Granddad quite enjoyed their television programs, and had a set in each of the two bedrooms in the house, as well as in the living room. I was still stupid enough to believe TV's flickering, mind-sucking images were pretty neat, so I was in heaven.

Night came on softly, as it only can when you feel safety and warmth around you. With one of Granny's home-cooked meals in my stomach, I had begun to feel the need for slumber. So, we all piled in the monstrous king-sized bed in Granny and Granddad's master bedroom, and soon we were all snoring blissfully.

I woke up in the night and sat upright, looking around. Something had disturbed my slumber. Granddad was still snoring rhythmically, and Granny looked like she hadn't budged an inch, so I sat back and prepared to visit the realm of dream once more.

Then the television turned itself on.

Now, I'm only 22 years old (1994), but this was in the days when remote controls were the providence of the wealthy and debased. Granny and Grandad did not fit into any of those two categories.

To see a television turn itself on was an interesting thing. I sat up again to see what would come on.

At that time, the TV in my grandparents' bedroom was a black and white. I watched the white dot that had formed in the middle expand to full screen, but only the static of a dead channel appeared.

Then, images began to appear.

I couldn't really describe them. They were sort of shadowy things at first, but they seemed to be -- for lack of a better term -- "scoping me out." Slowly, an actual image began to appear.
Muppets! The Muppets were on! I was exalted, elated. I wanted to wake my grandparents up, but I then started to feel a bit uncomfortable about what was happening on the television.

Muppets did not usually have fangs as I recall. At least, not ones that looked so ... real and out of place in an otherwise standard Muppet-style mouth.

I realize this is sounding goofy. You were warned.

Well, uncertain about what to do, I decided to keep a close watch on the television. Herry Monster, as a refrigerator magnet.

The "Muppets" looked at me. It was common of course on Sesame Street and the Muppet Show for them to acknowledge the audience, so I wasn't alarmed so much by that.

I describe these things as Muppets because that's primarily what they looked like. Other than the rows and rows of unusual teeth, one looked vaguely Groveresque and the other sort of reminded me of Herry the Monster (don't know if he's even ON Sesame Street anymore).

The Groveresque Muppet leaned over and pointed at me, while whispering something to his companion. They looked at me in unison, whispering all the while in a strange, unusually guttural tongue. It sounded completely random, although it did somehow seem to follow the meter and pattern of a language.

I noticed when the Groveresque Muppet pointed, he had very long, distinct talons on his furry hands. This, too, was quite disturbing.

The Muppets began to dance, sing and cavort about in that strange language of theirs. It was sort of amusing, I recall. I began to feel a bit more at ease.

The Muppets motioned for me to come forward.

I shook my head.

The Muppets tried again.

I shook my head. I was beginning to feel frightened. If there ever was a way Muppets could look p***ed, these guys were doing it. With all those rows of fangs, it wasn't pretty.

I should mention that all of this singing, dancing, cavorting horror that was going on seemed to in no way disturb my grandparents. This disturbed me as well, because these guys were LOUD.

I got out of bed and crept into the living room, being careful to avoid approaching the screen. The Muppet-things wore visages of absolute anger now, motioning violently for me to approach the screen. My attempts rouse my grandmother and grandfather were in vain; they would not stir.

I ran into the living room, crying. I collapsed in my Granddad's chair, buried my face into the fabric, and began to weep, certain doom had come for me.

I looked over at the television in the corner of the living room. It had already begun to turn itself on, the shadow-forms that had dissolved into the Henson-styled horrors already beginning to flicker across its surface.

I screamed, rooted to the spot. But as the scream left my lungs and two grinning, fanged faces burst into being on the television screen, faintly and then with increasing tempo I heard footsteps.

The things in the television looked worried, swirled into their shadow-forms and were gone.
The television winked out just as my Mother ran into the room.

After consoling me, we went to check on Granny and Granddad. They were both awake, and had heard me scream.

We all sat up for a while talking, and eventually the warmth and love returned to fill the chill in my soul.

I went to bed and nothing more happened that night. Or any other night I stayed with my grandparents.

A waking dream? Probably so. But one that still fills me with terror.

I was awake of course, Mom will still assert, when I was in the chair. So, I did move somehow from the bed to the chair.

Still, it gives one pause. What better way to snare a young boy than to show him something he loves, then pull him in unawares?

Whatever those two things were, I'm sure the Muppet-forms were not their natural shape. I'm sure the fangs and talons were part of it, though.

If they're still out there, I hope they haven't had much time to practice those forms. If they could get them just right ...

I still wonder what would have happened -- dream or not -- if I had put my then-small hands up to the screen there in my grandparents' bedroom.

Perhaps nothing.

And then again, perhaps it's better not to know."

THE WALL DOUGHNUTS

In March 1966, 15 year old Kathy Reeves, of Pioneer Mountain, Oregon, was walking home with a friend along Pioneer Road. Seeing a reddish light in a field, the friends thought a fire had broken out and investigated. The friends found not a fire but a glowing, see through dome that seemed to contain 'boiling smoke'. The confused pair decided to press on home but saw another light. Kathy likened it to a torch with a cover on the end preventing the beam shining out. Hence there was only a circle of light. Thinking that somebody was playing a trick on her, Kathy threw a rock at the light. Suddenly a ring of lights appeared all around the first light. Getting scared the girls ran home. Little did they know that these events heralded a cavalcade of weirdness, that would descend on Kathy's home.

Kathy's home began to be haunted by what the family called 'wall doughnuts', doughnut shaped lights that would crawl up the walls. These resemble the Mysterons from the TV show *Captain Scarlet*.

The family contacted Deputy Sheriff Thomson Wayne Price, who called on the house many times and saw the wall doughnuts and heard a weird noise that accompanied them, that he likened to a high speed saw or a giant spinning top.

On March 30th, Max Taylor, a family friend, camped outside the house to observe the phenomena from the outside. He reported two pulsating blue lights on either side of the house. Deputy Price was called again and when he arrived there at 1.30am there were a group of 12 people watching the lights. Price also saw a large orange object in the sky that was accompanied by the same noise as the wall doughnuts.

One night Mrs Reeves woke up to find her bedroom filled with a rosy glow. Hanging in the doorway, was a watermelon coloured, translucent cloud that was visible for several seconds before vanishing.

Kathy saw three walking tree stumps, crawling across the pasture outside of the house.

They seemed to change colour from orange to blue, to white to yellow to 'watermelon' coloured. These resemble the ghastly entity seen by a young Jon Pertwee, in the haunted room of his friend's house.

Other people in the area reported a craft with a dome and a searchlight, something that looked like revolving Christmas tree lights and a house sized glowing sphere.

The family finally left the house, unable to cope with the weird events. The next owner, the wonderfully named Delbert Mapes, said that he tacked sheets over the windows to keep out the lights.

My good friend, Mohammed Bhula, encountered light phenomena very like that seen by Kathy and her friend. It occurred in 2022 in Nuneaton, Warwickshire, England. He was walking along the Coventry canal between the Cat Gallows Bridge and the Cock and Bear Bridge at about 3. AM. He saw a circular light up ahead that resembled a torch without a beam, a small circle of light. As he got closer the light seemed to spilt into a larger circle of smaller lights. At this point he became afraid and swiftly left the canal path, walking home by road, a route that took him a far longer time.

These highest of high strangeness cases make no sense. Maybe one just can't understand them, in the way a greenfly couldn't understand an electron microscope. It seems that the phenomena interacts with human minds on some level. Some of these cases sound almost like 'floor shows' put on for the benefit of the witness. But why? This is what we will be looking at in the final chapter.

CHAPTER FIVE: WHAT THE HELL IS GOING ON?

"Looking so hard for a cause
And it don't care what it is
And never really ever seeing eye to eye
Though it doesn't really mind, perhaps it's why
It never really saw"

The Cardiacs, *Is This The Life?*

W ell that, as Charles Fort would have said, was a 'parade of the dammed'. If some of the cases we have looked at were submitted as scripts for films and TV shows, they would have been rejected for being too strange or incomprehensible. Yet these things occur, year in, year out, to ordinary people from every walk of life. These nonsensical events seem to exist purely to frighten and baffle. Many have a dream like feel to them but have real effects on the world around them. Others seem like surrealist art. And the cases I have looked at are just the tip of the high strangeness iceberg. If I had tried to include everything, we would have ended up with something the size of Richard Corliss' Source Book Project.

In this final chapter I will try to make some kind of sense of it all. My friend Jon Downes once said, that studying the Fortean was like trying to do a jigsaw that has some pieces missing, pieces from other jigsaws inserted into the box and the picture on the lid of the box has nothing to do with the picture on the jigsaw. Maybe we can make a small part of the picture, that will give us some clue to the larger image.

MISTAKES
People mistake the mundane for the strange, particularly in the age of Youtube. How many lousy clips have you seen of out of focus objects, that the poster thinks is a UFO / Bigfoot/ ghost / lake monster? The paranormal is the last place you should go but for many it's the first.

As a child at primary school in England, back in the mid-1970s, I glanced out of the window one summer morning and was convinced I saw a spaceship. It looked like a battleship grey object of huge size, that seemed to have fins and landing gear. The other children saw it and became excited and scared. My seven-year-old self was convinced it was about to land and life would change forever. I wondered how life would be under our new alien rulers. Despite the teacher's protestations, we kids all ran outside to look at the huge UFO. But then we realized what it was, a Goodyear blimp, seen at a weird angle. Once that angle changed, we all saw the object for what it really was and felt a bit silly. But I can still remember the feeling of relief that we were not getting invaded!

There is a big difference from a class full of 7-year-old children to adults, or so you would think.

A former friend of mine once told me a story he had heard from a policeman. The copper had been on patrol with a colleague one night, driving around the outskirts of Plymouth in Devon, England. They came across a car, upside down in a ditch. They stopped to see if the driver was ok. Luckily he was and he managed to clamber out of the car unhurt but shaken. When the police questioned him he told them that he had crashed into the ditch whilst swerving to avoid Bigfoot!

The police naturally assumed the driver was drunk and breathalysed him. Imagine there surprise when they found out he was dead cold sober. They arrange for the car to be moved and drove back to the station scratching their heads.

A few weeks later the mystery was solved. Apparently there had been a Star Wars convention in a nearby hotel. One attendee, dressed as Chewbacca had gotten very drunk and wandered off round the country lanes and was the errant 'bigfoot' spotted by the hapless driver!

In 1948, a set of huge, three-toed tracks were found running along the beach at Clearwater, Florida. A few days later, more of the same tracks turned up along the banks of the Suwannee River. Around about the same time there were vague reports of a strange, bipedal creature seen wandering the swamplands of the area. Cryptozoologist Ivan T. Sanderson, linked the tracks with the sightings and concluded that the tracks were left by some unknown species of giant penguin! Sanderson believed that the monster bird had wandered away from its habitat and was now stalking the swamps of Florida. Nobody did explain the creature seen, but in 1988, *St Petersburg Times* reporter Jan Kirby revealed that the tracks were made by Tony Signorini and Al Williams, who, having seen fossilized dinosaur tracks, decided to make their own using cast iron feet that they strapped onto their own boots to emulate the fossil tracks. Sanderson had put 2 and 2 together and come up with five. Surely one of the greatest cases in history of someone's imagination running away with them.

Even professional people make mistakes. British astronaut Tim Peak was aboard the International Space Station in 2015, he saw a formation of four twinkling lights outside of the space station and thought he was looking at a UFO. The crew figured out that the lights were in fact crystallized droplets of urine, from a nearby Russian space probe floating past the window!

Vic Tandy (1955 – 2005) was a British lecturer for information technology at Coventry University. In the early 1980s, he was working in a laboratory designing medical equipment. There were stories of the lab being haunted, but Tandy gave no notice to these. Then one evening, whilst working in the lab, he broke out in cold sweat. Tandy felt the distinct impression that he was being watched. Then he saw a grey shape in the corner of his eye. Spinning round to face the ghost, he found that it dissolved when seen face on. Tandy hurried home, frightened by what he had seen.

The next day he was back in the lab. Tandy was a keen fencer in his spare time and had a foil he was working on, in a vice in the lab. He noticed that the sword blade was vibrating up and down very fast. He found that the vibrations were caused by a standing sound wave that was bouncing between the end walls of the laboratory and reached a peak of intensity in the centre of the room. He tracked the sound wave to a newly installed extractor fan and calculated that the frequency of the standing wave was about 19hz cycles per second. 19Hz is infra sound, below human hearing. Infra sound is known to cause unease, dizziness, blurred vision, hyperventilation and fear, possibly leading to panic attacks.

Tandy realized his 'ghost' was nothing more than an effect of a sound wave made by the extractor fan. Investigating a supposedly haunted pub cellar in Coventry, where people had seen a grey lady and experienced feelings of unease. He discovered a 19hz standing wave here as well. One wonders how many other ghosts are products of infra-sound?

But can we really write off every single case in this book as a mistake, including multiple witness cases.

HOAXES
Deliberate hoaxes are the explanation touted for most Fortean phenomena by sceptics. Hoaxes have indeed occurred in all areas of the field. The best know was the case of the surgeons photograph.

In the 1930s, monster fever was sweeping the nation with many people claiming to have seen a massive animal in Loch Ness. A big game hunter by the name of Marmaduke Wetherell, had been hired by the *Daily Mail* to track down the beast in 1933. Wetherell soon came across a set of tracks in the mud, on the Loch's shore. Whatever made them seemed to have been huge. Wetherell announced that he had found the spoor of the monster and made plaster casts of the tracks and sent them to the Natural History Museum in London. They were swiftly identified as being made by a preserved hippo's foot, quite possibly one used as an umbrella stand. Wetherell had been pranked and was utterly humiliated. He vanished from public view. But he was plotting his revenge.

On April 21, 1934, the *Daily Mail* published the now iconic photograph of a long necked object in Loch Ness. It was taken by a London surgeon Robert Kenneth Wilson. The picture was the most famous claimed image of the monster for 60 years. In 1994 Christopher Spurling confessed, at the age of 90, that the picture was a hoax. Spurling was the stepson of Marmaduke Wetherell, and the shamed big game hunter had approached Spurling, who was a sculptor, to create a fake picture to get his own back on those who had tricked him. *"We'll give them their monster,"* Wetherell said.

Spurling created a serpentine head and neck from plastic wood (saw dust mixed with a binder), and attached it to a toy submarine brought from the now defunct supermarket, Woolworths. The model was floated in the loch and photographed. Spurling's friend Robert Kenneth Wilson, was called in as a respectable front man who claimed to have seen and photographed the beast. It goes to show just how long a hoax can last and how far it can travel. The surgeon's photograph was known throughout the world.

The most prolific Loch Ness hoaxer however, was Frank Searle. Searle was a photographer, who in 1969 set up a tent on the south shore of the loch in order to photograph the monster. Frank spent the next 14 years there, and produced a string of photographs that he sold to local and national newspapers. Each and every one of them was a fake. They included shots of logs, posts sticking out of the water that he altered to look like a monster's head and neck, and a painting of a *Brontosaurus excelsus* cut out of a postcard, pasted onto a photograph of the loch and re-photographed. When his chicanery was exposed, Searle turned nasty, even attacking another research group's boat with a fire bomb. He finally left the loch to live in obscurity in Fleetwood, where he died in 2005.

Sea serpents too, have seen their share of hoaxes. In 1845, Dr Albert Koch exhibited a remarkable beast at the Apollo Saloon, on New York's Broadway. His specimen was a fossilized sea dragon fully 35 metres (117 feet) long. The exhibit consisted of a long, toothy skull raised high on a snake-like neck that continued into a sinuous, elongate vertebral-column, gigantic ribs and the remains of paddles. Koch called his monster *Hydragos sillmanii,* in honour of the wonderfully named Benjamin Silliman, a scientist who had recognised the existence of sea serpents in 1827. The good doctor happily charged 25 cents per visitor, for the privilege of seeing the dragon. His charge was immensely popular and he did a roaring trade - until one un-fortuitous day when a member of the public with zoological knowledge, happened to visit the Apollo.

Professor Jeffries Wyman, who was a competent anatomist, instantly saw Dr Koch's trickery. He noted that the "dragon" had teeth with double roots, which is a characteristic found only in mammals. He also exposed the fact that the specimen was made up of several different skeletons, giving it an artificially long body. He published a damning account, concluding that the monster was nothing more than the skeletons of several fossil whales, strung together. Koch had indeed made a spectacular hoax. The bones belonged to several specimens of the extinct, archaic whale, Basilosaurus (so termed because its original discoverer, Dr Richard Harlan, wrongly thought it to be a reptile). Koch was nothing if not resourceful. Having been exposed as a sham-artiste in the States, he merely packed up his serpent and crossed the Atlantic.

Re-naming his brute *Hydragos harlani* (after Dr Richard Harlan as Professor Silliman now refused to have anything to do with it), he exhibited it in Dresden. However the jig was up. The eminent English palaeontologist, Gideon Mantell (the discoverer of the dinosaur *Iguanodon mantelli*), wrote to the editor of the *Illustrated London News* with a damning letter. He pointed out that as well as his current scam, some years earlier he had collected the bones of mastodons and mammoths, and reconstructed them into a species of his own invention called Missourium. This fake-monster had been displayed in the Egyptian Hall, Piccadilly. The British Museum actually brought Koch's collection and reconstructed a fine Mastodon out of them.

Mantell goes on to say:

> *"Not content with the interest which the fossils which he collected in various parts of the United States really possess, Mr Koch, with the view of exciting the curiosity of the ignorant multitude, strung together all the vertebrae he could obtain of the Basilosaurus, arranged them in a serpentine form; manufactured a skull and claws, and exhibited the monster as a fossil sea serpent, under the name above mentioned Hydragos."*

Amazingly, there were those who still believed Koch. In Dresden the chimera was examined by Carl Gustave Carus, a student of comparative anatomy. He attempted to restore the cranium. Only a portion of this had been found by Koch, forcing him to fake the rest. Carus restored it as the skull of a giant reptile, completely in error.

Sometimes hoaxes can turn deadly. In Montana in 2012 Randy Lee Tenley decided to dress up in a ghillie suit, a kind of shaggy outfit favoured by military snipers and game hunters, in order to scare people, by pretending to be Bigfoot. The 44 year old visited southbound lanes on U.S. Highway 93 near Kalispell, attempting to spook drivers. He was struck twice by cars, both driven by teenaged girls. Highway Patrol spokesman Sergeant Steve Lavin said.

> *"From what I understand, at least one of his friends said that he was trying to induce a sasquatch sighting by using the suit along the highway. This is a first for me after 20 years on the highway patrol. It's strange."*

A major police and armed forces response occurred when no less than six spaceships apparently landed across the south of England, on the 4th of September 1967. The silver craft were about 1.4m long, silver and shaped somewhat like blood cells. The six craft were located along the 51st line of latitude, from the Thames Estuary to the Bristol Channel, roughly equidistant from each other. One each at a new housing estate near Queenborough on the Isle of Sheppey, Bromley golf course in south London, a horse paddock in Winkfield village, the village of Welford, in Berkshire, Chippenham in Wiltshire and on Dial Hill in Clevedon in Somerset.

The army, police force and bomb disposal units were mobilized and the craft were cordoned off. Royal Air Force helicopters were marshalled. The Ministry of Defence were informed and the army blew up the saucer found at Chippenham. Another of the ships was sent to the Atomic Weapons Establishment at Aldermaston and one to the guided weapons division of the British Aircraft Corporation. When one saucer was drilled into by the police a foul smelling slime exploded from the craft coating them.

The 'spaceships' had been built by students, Christopher Southall and Roger Palmer, at Farnborough College. They had created the spaceships to raise money for the college Rag Week. The pair had no idea their hoax would go quite so far. The ships were rigged to make a noise when moved and the foul alien slime had been made from flour and water. No action was taken against the pair, and they raised £2000 for charity, a huge amount in 1967.

In 1992, whilst in Cleveland, Ohio, musician and film maker Ray Santilli was searching for archive film of old rock and roll stars. He claimed he met a former US military cameraman, Jack Barnett, who passed on a 17 minute piece of film to him. Supposedly, the footage was filmed in July of 1947, around the time of the famous Roswell UFO crash. The film shows what appears to be two doctors in hazmat suits dissecting a classic 'grey alien', with an enlarged cranium and black eyes. The surgeons remove the creatures innards and brain.

The film and story, released when the TV show *The X-Files* was at its most popular, caused a sensation and a whole documentary was made about it.

The autopsy was, in fact, filmed in a flat in Rochester. The 'alien' was a manikin filled with the body parts of animals brought from a butcher in Smithfield market.

The most outrageous case of a money making hoax, must be that of the Amityville Horror. On the night of November 13th, 1974, Ronald DeFeo Jr. shot dead his mother, father, two brothers and two sisters, as they slept in their beds at 112 Ocean Avenue in Amityville, New York. Later Ronald claimed that a hooded, black handed female demon handed him the .35 calibre rifle and told him to murder his family. He was later imprisoned for life, the jury being unimpressed by his story.

In December of 1975, George and Kathleen Lutz and their three children moved into the house. They left 28 days later claiming to have been driven out by vile smells, swarms of flies, encounters with hooded entities, scratch marks appearing on their bodies and the manifestation of a demonic flying pig called 'Jodie'.

No subsequent owners of the house ever reported a single odd thing occurring there. The Lutz family also acted in a very odd way. George Lutz contacted Steven Kaplan - who claimed to be a parapsychologist, to do an investigation on the house, Kaplan in return told the family that he wanted whatever they found, whether it supported their story or not to be made public. George Lutz cancelled the investigation, claiming not to have wanted the publicity. But on the very morning that he called off Kaplan, the family had attended a large press conference about the case. Even more suspiciously the press conference was organized by none other than William Weber, the defence attorney for Ronald DeFeo!

William Weber, in an interview for a local radio station in 1979, says that he had been contacted by Paul Hoffman, a journalist, indicating that the Lutz family wanted to meet with him. At this point, Weber had already been given an oral commitment that the Lutz family would get a large advance for both a book and a movie. He said...

> *"We were there from about 9 or 10 in the evening to 3 in the morning drinking wine and I can't tell you how many bottles of wine we had, but it was many. I had photographs of the crime scene and based upon the photographs and other facts that had related, we developed what ultimately turned out to be part of the Lutz's version of the possession aspect of the case."*

Ronald DeFeo was a known drug user and was also known to be violent. He often fought with his father, who was equally violent. A mixture of drugs and rage was the cause of the murders, not 'demonic entities'.

The Lutz story went on to be a bestselling book and a popular film franchise. They made a bomb out of book sales and movie rights. The Amityville Horror was a cynical money making scheme, one based on the deaths of 6 people. The whole affair leaves a very bad taste in the mouth. The real horror at Amityville was human violence, followed by human greed.

So hoaxes do occur for all sorts of reasons. Mischief, revenge, money making or simply an urge to scare people silly. But can *all* the cases we have examined be hoaxes? Some are so spectacular that they would have cost hundreds of thousands of pounds to carry out, even if the technology was available. In order for all high strangeness cases to be hoaxes, there must be an international cabal of millionaire pranksters, who employ hundreds of people to carry out nonsensical rituals and acts all over the world, with no other goal but to scare and baffle the public.

HALLUCINATIONS

Hallucinations can be caused by all kinds of factors, from drug use to illness and mental disease. Some reported Fortean phenomena can surely be put down to some form of hallucination.

Charles Bonnet syndrome causes the deterioration of sight and can lead to hallucinations. When a person starts to lose their sight, their brain doesn't receive as much information as it used to. It's thought the brain sometimes responds by filling in the gaps with fantasy patterns or images that it's stored. Records show that these hallucinations can continue for years. My friend and fellow Fortean Jackie Tonks, a social worker, once had a client with Charles Bonnet syndrome who saw rabbits armed with AK 47s attacking him!

Schizophrenia can also cause hallucinations. These can include hearing phantom voices or smelling phantom odours. In 1799, a Prussian bookseller named Christoph Friedrich Nicolai, read a paper to the Royal Society of Berlin, entitled *"A Memoir on the Appearance of Spectres or Phantoms occasioned by Disease."* In this, he described how one morning in February 1791, during a period of considerable stress and melancholy in his personal life, he saw the apparition of a deceased person in the presence of his wife, who, however, could see nothing. This apparition haunted him for the duration of the day and, in the subsequent weeks, the number of these figures began to increase. Nicolai, a sceptic by nature realized that these ghosts were caused by mental imbalance.

Sarah Myers, a behavioural neuroscientist, who suffers from the condition herself, was once lecturing for the National Alliance on Mental Illness, when one of the audience said *"When my daughter gets angry, she sees black creatures crawling on the floor."* This instantly brings to mind some of the creatures seen crawling on the floor in haunting cases.

The artist, social reformer and visionary, William Blake, (28 November 1757 – 12 August 1827) had strange visions throughout his life. At the age of 4 Blake claimed to have seen 'God' putting his head through the window, causing the boy to scream. Later, aged around 10 at Peckham Rye he saw "a *tree filled with angels, bright angelic wings bespangling every bough like stars."*

The visions continued throughout his life. One of his most famous paintings *The Ghost of a Flea,* painted between 1819 and 1820, was based on a vision. Blake's friend John Varley, had commissioned him to paint a number of portraits of spirits. Varley was a student of astrology and physiognomy (the practice of assessing a person's character or personality from their outer appearance). He was a believer in spirits, but was frustrated by his inability to see them, unlike Blake who could see them clearly. Varley would attempt to summon spirits and Blake would draw and paint them. Blake had the image of the phantom flea appear to him in 1819 during a séance. Varley wrote...

> *"As I was anxious to make the most correct investigation in my power, of the truth of these visions, on hearing of this spiritual apparition of a Flea, I asked him if he could draw for me the resemblance of what he saw: he instantly said, 'I see him now before me.' I therefore gave him paper and a pencil with which he drew the portrait... I felt convinced by his mode of proceeding, that he had a real image before him, for he left off, and began on another part of the paper, to make a separate drawing of the mouth of the Flea, which the spirit having opened, he was prevented from proceeding with the first sketch, till he had closed it."*

Blake said that the flea had told him that…

> *'It was first intended to make me as big as a bullock; but then when it was considered from my construction, so armed—and so powerful withal, that*

in proportion to my bulk, (mischievous as I now am) that I should have been a too mighty destroyer; it was determined to make me—no bigger than I am."

Blake's painting does not resemble a flea but looks more like a massively muscular, reptile / human hybrid.

Author John Higgs, in his Blake biography, *William Blake vs the World* cites the work of neuroscientist Dr Adam Zeman, who has studied the imagination for decades. He first described in 2015 the condition of aphantasia, where some people were found to be unable to visualise mental images. In other words, they had no mind's eye.

Blake seemed to be the polar opposite end of the spectrum having hyperphantasia. Dr Zeman says.

> *"Some people with hyperphantasia say that it is hard for them to be sure whether they've imagined something, or it actually happened because their imagining is very vivid."*

Higgs says:

> *"The limited, rational or logical part of our brains, which Blake characterised as Urizen, is actually only a model of how we understand the world. We think it's real, we think it's true. When it feels under threat, it lashes out and tries to defend itself."*

Blake himself said: *"What is now proved was once only imagined."*

Myers writes also of her own experiences. *"I feel like I am two feet above, apart from my body."* It is impossible to read this without thinking of alien abduction cases and bedroom invaders. She goes on …

> *"I see spiders crawling everywhere in my field of vision. I see people chanting, and I am paranoid that there is a mob wanting to overthrow the government. I draw these things. I send the drawings to a friend, and he says they are cute. But, to me, they are really scary."*

Epilepsy may be the cause of some experiences. Joan of Arc claimed to hear the voice of God and see visions of angels and saints during France's Hundred Year War with England. The claims led to her becoming a military leader and winning several battles, before being captured and burned at the stake aged just 19.

Dr. Guiseppe d'Orsi, a neurologist at the University of Foggia in Italy, and Paola Tinuper, an associate professor of biomedical and neuromotor sciences at the University of Bologna, have theorized that Joan had a type of epilepsy that affects the part of the brain responsible for hearing, or idiopathic partial epilepsy with auditory features. Visual hallucinations are also experienced in this kind of epilepsy. Joan is reported to have said, that the "sound of bells" sometimes triggered the voices. Hearing certain sounds can be a trigger for seizures, d'Orsi and Tinuper said.

However, there is problem with passing off all cases of high strangeness experience as hallucinations. Many are multiple witness cases, wherein reports the same phenomena.

Hallucinations cannot leave behind physical traces, as many of our cases do. Also, many sufferers realize full well, that what they are seeing is not real. As Sarah Myers says...

> *"When a hallucination scene is unfolding before my eyes, the scene does not interact with anything of the environment, rather, they behave within the reality, as if I am accessing two dimensions at once. My hallucinations never pick up objects or pay regards to any law of physics. They do not walk with gravity."*

MASS HYSTERIA

Mass hysteria is a collective illusion of some kind of threat, that passes through a community via rumour. The illusion acts almost like a disease sweeping through a population causing panic. A fine example is that of the Halifax Slasher. In 1938, a series of attacks on women were reported in the Yorkshire town of Halifax. Mary Gledhill and Gertrude Watts claimed to have been attacked by a mysterious man with a mallet and bright buckles on his shoes. Soon after, more reports of a man attacking women with a razor occurred and the attacker was dubbed the Halifax Slasher by the newspapers. Businesses shuttered their windows and a vigilante group was formed. Several innocent men were beaten by the mob. A reward was offered for the Slashers capture. Clifford Edwards, a local man who had gone to help, was later accused of being the slasher himself. Soon a crowd had gathered and after they had started to chant for his death, police had to escort him home. A call to Scotland Yard for help was put in and two detectives arrived on November 29th.

As the detectives began questioning the victims, their stories collapsed. Suddenly, everyone began confessing that they had actually cut or injured themselves. One woman said she fought with her boyfriend and, upset, sliced her arm because she had heard about the Halifax Slasher. After nine of the 12 victims confessed to self-harm, police closed the investigation. The nine were all criminally charged, with four going to jail.

On 2 December, the *Halifax Courier* reported ...

> *"Carry on Halifax! The Slasher scare is over... The theory that a half-crazed, wild-eyed man has been wandering around, attacking helpless women in dark streets, is exploded... There never was, nor is there*

likely to be, any real danger to the general public. There is no doubt that following certain happenings public feeling has grown, and that many small incidents have been magnified in the public mind until a real state of alarm was caused. This assurance that there is no real cause for alarm, in short, no properly authenticated wholesale attacks by such a person as the bogey man known as the 'Slasher', should allay the public fear..."

In 1956, a similar case occurred in Taipei, Taiwan. A mysterious man was said to slip past people in crowds, slashing them as he passed. About 24 people reported being attacked. But when the police interviewed victims, it became apparent that, like in Halifax, there was no slasher. Spurred by hysterical news reports, people thought that regular minor cuts on their bodies were the work of a crazed killer. In one case, the "slash" was an old injury which the person had scratched and re-opened.

In May of 2001, stories began to circulate in New Delhi, of an ape like monster called the Monkey Man (not to be confused with the Man Monkey of Ranton!). This hairy, 4 to 6 foot tall beast, was said to slash at its victims with metallic claws not unlike Spring Heeled Jack. The Monkey Man was said to make huge leaps from building to building. People sleeping outside on rooftops to escape the heat were supposed to be its main victims. As panic swept the city, two men and one pregnant woman were killed when they were pushed from high buildings by crowds of people fleeing the monster.

Sanal Edamaruku of the Indian Rationalists Society wrote...

"We went out to have a close look at the victims' injuries, which had become something like the last bastion of the spook. We succeeded in tracing most of the known cases and were rather "disappointed": There was not a single serious wound, only little scratches, cuts and rubbings, which under normal circumstances would not get any attention. Red marks, which could well be caused by rubbing of naked skin on the hard jute strings of the traditional beds, small scratch marks looking like opened mosquito bites, four line scratch designs in fork size and other common little injuries. Interestingly there was no uniformity in them, though they were claimed to come from the same source. Most of the injuries occurred on the legs, which is quite unusual if there was an upright attacker of six feet height. With every single case we were more convinced that all these injuries were self-inflicted, deliberately or unknowingly."

As time went by the ape-like Monkey Man morphed into a kind of cyborg with a helmet and lights on its chest. Stories then began about how the attacker had been created in a lab and was being controlled by a mad scientist. The local police sent out 3,000 men to try and catch the villain, which poured petrol onto the flames of the panic. A reward of 50,000 rupees was offered for the capture of the beast.

By the end of May the reports began to peter out until they died off utterly. No trace of the Monkey Man was ever found and it is likely that he had no existence outside of the minds of the poor of New Delhi.

The following year, a new scare started in Southern India in the state of Uttar Pradesh. Known as the Muhnochwa or 'face scratcher' it was a flying, rugby ball shaped object with lights on it and mechanical claws. It was said to emerge from the dark and scratch the faces of its victims. It haunted the state capital Lucknow.

Local journalists said that people are staying awake through the night, taking turn to guard their homes. Many people have removed their TV antennas, fearing that these might attract the "face scratcher."

Police shot dead one man and injured 12 others, when a mob of hundreds stormed the police station in Barabanki, demanding protection against the Muhnochwa. One theory was that the assailant was a genetically-engineered insect, introduced by "anti-national elements."

Like the Monkey Man before it, the face scratcher simply melted away.

Mass hysteria could well be behind some of the phenomena in this book. But other phenomena do not fit the mould very well. Some of our high strangeness events are long lasting, unlike the fleeting mass hysteria. Mass hysteria fits better than the other hypothesis so far but it cannot explain all Fortean phenomena.

CHEMICAL CAUSES

Drugs and alcohol can cause hallucinations, but to suggest all witnesses since the year dot have been drunk or tripping, stretches credulity to breaking point. There are naturally occurring chemicals and substances that can cause strange visions though.

In 1921, an odd story appeared in the *American Journal of Ophthalmology*. William Wilmer wrote an account of a patient of his Mrs. H, describing what happened after her family and servants moved on November 15, 1912, into a "large, rambling, high-studded house, built around 1870. The house had been empty for 10 years and did not have electricity, being lit by gas. Within days the family began to feel depressed in the house.

> *"One morning I heard footsteps in the room over my head. I hurried up the stairs. To my surprise the room was empty. I passed into the next room, and then into all the rooms on that floor, and then to the floor above, to find that I was the only person in that part of the house.*
>
> *I had not been in the house more than a couple of weeks when I began to have severe headaches and to feel weak and tired. I took iron pills*

three times a day and spent a couple of hours each afternoon in my room, lying down and resting, a rather discouraging process, as after resting my headache was always worse than it had been before."

Her husband 'G' had feelings of being watched from behind and her children abandoned their playroom, despite their toys being stored there. The kids became pale and listless, losing their appetites. She took them away for Christmas but her husband stayed behind.

"While we were away, G was frequently disturbed at night. Several times he was awakened by a bell ringing, but on going to the front and back doors, he could find no one at either. Also several times he was awakened by what he thought was the telephone bell. One night he was roused by hearing the fire department dashing up the street and coming to a stop nearby. He hurried to the window and found the street quiet and deserted."

Mrs H and the children felt better whilst they were away, but on returning the gloom of the house soon overtook them. The children fell ill but oddly seemed better out of doors.

"My headaches returned, and I frequently felt as if a string had been tied tightly around my left arm. One night I was awakened by a heavy door slamming quite near me. It woke G too, and he said to me, 'What was that?' 'Only the door of the room,' I replied; but as I grew more wide awake I realized that it could not be any one of the doors of the room as they were tightly closed.

Another time, a little before daylight, I was awakened by heavy footsteps going down a staircase behind the wall at the head of my bed. Then a number of crashes downstairs, as if several pots and pans had been hit together or against the kitchen stove. Soon I realized that there was no staircase behind the wall, only the thickly carpeted front stairs on which no footsteps could be heard. Also that it would be impossible in my room to hear any sounds from the kitchen, no matter how loud.

On one occasion, in the middle of the morning, as I passed from the drawing room into the dining room, I was surprised to see at the further end of the dining room, coming towards me, a strange woman, dark haired and dressed in black. As I walked steadily on into the dining room to meet her, she disappeared, and in her place I saw a reflection of myself in the mirror, dressed in a light silk waist. I laughed at myself, and wondered how the lights and mirrors could have played me such a trick. This happened three different times, always with the same surprise to me and the same relief when the vision turned into myself.

As I was dressing for breakfast one morning B (four years old) came to my room and asked me why I had called him. I told him that I had not called him; that I had not been in his room. With big and startled eyes, he said, 'Who was it then that called me? Who made that pounding noise?' I told him it was undoubtedly the wind rattling his window. 'No,' he said, 'it was not that, it was somebody that called me. Who was it?' And so on he talked, insisting that he had been called, and for me to explain who it had been."

The children lost their appetites and the house plants began to die. The husband had a pain in his head and even the servants looked pale and ill.

"On the night of January 15 we went to the opera. That night I had vague and strange dreams, which appeared to last for hours. When the morning came, I felt too tired and ill to get up. G told me that in the middle of the night he woke up, feeling as if someone had grabbed him by the throat and was trying to strangle him. He sat up in bed and had a violent fit of coughing, which lasted about five minutes. His first thought had been that burglars were in the house, but as everything was quiet he instantly dismissed that idea. It then flashed across his mind that I had been playing a joke on him, but upon looking at me, he saw that I was in a heavy sleep, very much as if I had been drugged. Until we lived in this house, I had always been a light sleeper, waking at the slightest sound. In this house, however, nothing seemed to wake or disturb me. Quite the contrary with G, for in the past he had always slept heavily, never hearing a sound and nothing disturbed him. Now he was continually waking, answering the telephone and the doorbell, which had never rung, and looking for burglars, who never materialized.

That morning after breakfast, as was my usual custom, I sent for the children's nurse, a Scotch woman who had lived with me for several years. She looked worn out, and when I asked how the children had slept she burst out with, 'It has been a most terrible night. This house is haunted."

Mrs H dismissed the idea as nonsense, but the nurse continued.

"While you were at the opera, about half past eight, B woke up and ran screaming through the hall to my room, "Don't let that big fat man touch me." He was terrified. It took Fraulein and me until ten o'clock

to calm him. He slept the rest of the night with me, in my room. Fraulein slept in B's bed, besides G Jr., to protect him.

G Jr. did not wake up all night but the muscles of his face kept twitching, as if someone was continually pinching him. In the morning when he woke, he said indignantly to Fraulein, "Why have you been sitting on top of me?" And when she told him that she had not been sitting upon him, but had been in the bed next to him, he said, "No, you have been sitting on top of me, and you were awfully heavy, too."

The nurse also told Mrs H that she and her German friend would often hear heavy footsteps from empty rooms, on the top floor of the house at night. On searching the rooms they found them empty. She continued.

"Some nights after I have been in bed for a while, I have felt as if the bed clothes were jerked off me, and I have also felt as if I had been struck on the shoulder. One night I woke up and saw sitting on the foot of my bed a man and a woman. The woman was young, dark and slight, and wore a large picture hat. The man was older, smooth shaven and a little bald. I was paralysed and could not move, when suddenly I felt a tap on my shoulder and I was able to sit up, and the man and the woman faded away. Sometimes, after I have gone to bed, the noises from the storeroom are tremendous. It does not happen every night; perhaps a week or ten days will pass, and then again it may be several nights in succession. Sometimes it sounds as if furniture was being piled against the door, as if china was being moved about, and occasionally a long and fearful sigh or wail."

The servants told Mrs H of seeing figures in white and purple flitting around their beds and the governess, the nurse's German friend, said how her four poster bed had been shaken.

"Saturday morning, the eighteenth of January, G's brother told us that he thought we were all being poisoned; that several years before he had read an article which told how a whole family had been poisoned by gas and had had the most curious delusions and experiences. He advised us to see Professor S at once. As he was out of town, his assistant, Mr. S, came at once to our house.

We told him how listless and ill the children appeared. He found one of them lying on the floor, and the other two in bed. We related the experiences of the children and servants, and told him about the plants. He examined the house thoroughly from top to bottom and interviewed the servants. He found the furnace in a very bad condition, the combustion being imperfect, the fumes, instead of going up the chimney, were pouring gases of carbon monoxide into our rooms. He advised us not to let the children sleep in the house another night. If they did, he said we might find in the morning that some one of them would never wake again.

Early in the afternoon our physician arrived and examined the children and agreed with Mr. S that they were being poisoned. ... He also stated that none of us ought to stay in the house another night."

The medieval affliction know as Saint Anthony's fire was named after a 3rd century hermit, who lived near the Nile in Egypt. In his solitude he was supposed to have been tormented by deamons taking the forms of women, wild beasts, creeping things, gigantic bodies, and troops of soldiers. He became the patron saint of butchers, grave diggers and those with skin disease. One of the symptoms of Saint Anthony's fire was vivid hallucinations of deamons trying to drag the sufferers to hell, or grabbing them and flying off. The arms and legs of the victims would turn bright red, hence the name of the disease, and gangrene would set in due to constriction of the blood vessels, causing fingers and toes to drop off. Mental effects included mania and psychosis. Very few survived once Saint Anthony's fire took hold of them.

The cause of both the burning limbs and visions of deamons was ergot poisoning. The ergot fungus *Claviceps purpurea* grows on cereal crops. Ergotamine, the chemical the fungus produces is bio-chemically very similar to Lysergic acid or LSD that is known to cause hallucinations. The fungus grows well in cool, wet years.

In 944, an outbreak of Saint Anthony's fire killed no less than 40,000 people in Southern France when infected rye was ground into flour.

The Order of Hospitallers of St. Anthony, dating back to 1095, was a religious order set up to fight the disease. They built more than 370 hospitals. The hospitals were painted red, so that even illiterate peasants would know what disorder they treated. They were a success and historians think that the lack of rye bread being fed to patients in the hospitals, may have been the key.

But outbreaks of Saint Anthony's fire are not restricted to medieval times. In 1951, in the village of Pont-Saint-Esprit, southern France an outbreak effected 250 people, killing 7 and putting 20 in an asylum. The cause was tainted flour from the mill of Maurice Maillet.

Victims reported nausea, vomiting, cold chills, heat waves. These symptoms eventually worsened, with added hallucinatory crises and convulsions. The situation in the town deteriorated in the following days. On the night of 24 August, a man believed himself to be an aeroplane and died by jumping from a second-story window, an 11-year-old boy tried to strangle his mother and another villager climbed up a dung hill naked. Others thought their limbs had swollen or shrunk, some believed snakes were wrapped around their bodies or that they were being chased by lions and tigers. Strangest of all some thought that they were being pursued by bandits with the ears of donkeys! One man was so badly effected he chewed through the straps of his straight jacket, breaking his teeth.

That was 20th century France, now imagine the same scenario hundreds of years before. It would have been a demonic infestation!

Both of these cases are highly persuasive that chemical causes may be behind some Fortean phenomena. But once more we cannot extend this to all cases. What of the sightings in wilderness areas, away from mass produced foods, or gas poisoning in enclosed buildings? What of the phenomena that leaves traces? In both of the above cases the offending substances were easily identified. In most high strangeness scenarios no such culprit is uncovered.

SLEEP PARALYSIS

Sleep paralysis is a glitch of the brain between wakefulness and REM (Rapid Eye Movement) sleep. In REM the sleeper experiences life-like dreams. In order to stop the body acting out these dreams the brain temporarily parallelizes the body. A small amount of neurochemicals tilt the body between sleep and wakefulness. Sometimes this 'switch' fails and the brain awakes whilst the body is still under REM paralysis. The effects of this are twofold. Firstly there is an inability to move that often feels like something is holding you down or sitting on your chest. Secondly the dreams of the REM state can intrude on waking consciousness as sinister figures lurking in the bedroom.

There are a number of culture bound explanations for sleep paralysis. In Newfoundland it is called the 'old hag' and takes the form of a witch that sits on the sleeper's chest, making it hard for them to breath. David J. Hufford of the Memorial University of Newfoundland Folklore and Language Archive, wrote a book on the old hag entitled *The Terror that Comes by Night.* The locals referred to such attacks as being hagged or hag ridden. Hufford reports that people thought they could summon the hag by reciting the Lord's Prayer backwards and kill her by sleeping on their backs with knifes or boards with upright nails against their chests designed to stab her when she sits on them. One man even accused another of summoning the hag to attack him.

A group of horror film makers, Called Grindmind, consisting of John Carter, Justin Wiseman, Francois Van Zyl and Shane Mills, put out a call on social media, in 2019 for anybody who had experienced the old hag, as they were making a film on the subject. Their inboxes were inundated with accounts. One woman wrote...

> *"I was sound asleep, cosy and warm under the covers.*
> *Suddenly I was awoken by a shadowy figure that was*
> *hunched, almost perched, like a human-sized bird, at the*
> *bottom of my bed. She had long, pointy fingers. Glowing*

eyes. A crimson mouth with black teeth. Quickly she jumped from the footboard onto my legs. I tried to scream for help but I couldn't make a sound.

The figure was slithering up towards my face and I knew if it got too close it would somehow devour me. I tried to move, but now it was on my chest and I couldn't escape. Again, I cried for help but the most I could muster was a whisper. Over and over I repeated that quiet scream, hoping someone might hear me. Finally, I jolted upright. I looked around but the figure had vanished. I comforted myself with the idea it was a nightmare, but an eerie feeling crept over me. I clung to my husband and prayed the Old Hag wouldn't return when I closed my eyes."

Journalist Pasha Malla once interviewed a woman called Carolyn Gill, who gave him the following account.

"A woman was crawling across the floor and up onto the bed. I could actually feel the mattress shifting under her weight, but I couldn't move, I couldn't do anything. She got on top of my chest and started strangling me. That's when I got a good look at her face: it was me."

Other cultures visualize the nocturnal attacker as something else. For example in Catalan folklore, the pesanta is a great black dog that comes to sleepers, and places it's paws on their chest to prevent them from breathing.

In Greece a daemon called the mora, vrahnas or varypnas sits on the sleeper's chest and tries to steel their speech.

In Sardinia, the ammuttadori is a deamon that sits on the chest and rakes the sleeper with its claws. It has 7 red caps on its head. If the victim can remove one, they will subsequently find hidden treasure. In Persia there is a very similar tradition. The bakhtak is a ghost that sits on the sleeper making it hard to breath. If the victim can grab its nose, the ghost must tell them were hidden treasure is. Its name even means 'small fortune'.

In the Philippines, the batibat or bangungot is a spirit that takes the form of a grotesquely fat woman who invades dreams in an attempt to suffocate the sleeper. The victim must wiggle their toes to break its spell.

In the traditions of the Sami people of northern Finland, the deattán is an evil fairy that transforms into a bird that perches on the sleeper's chest giving them bad dreams.

A PERSONAL ACCOUNT: FOUR VISITS FROM THE HAG

I have experienced sleep paralysis on three occasions. On the first I awoke and found I was unable to move. It was already light and I had the feeling of some entity looming over me, though I could see nothing visually. Then the 'thing' seemed to withdraw. I had the feeling that whatever it was, it was simply the 'fingertip' of something much bigger. It withdrew, backwards, away from my bed and out of the door. I reiterate that I actually saw nothing, the thing was just a feeling. Suddenly movement came back to my body again.

The second time started much as the first, but this time I realized that what I was experiencing was sleep paralysis. I simply relaxed and observed. Again there was a feeling of something in the room. Again, it seemed to withdraw by being pulled backwards. Then I was in control of my body once more.

The third time I had sleep paralysis was in an old shack in Russia. I was on an expedition searching for the almasty, a hairy Russian wildman. It was our second night in the old farmhouse, abandoned since 1973 (naturally). On the first night we may have got close to an almasty, but that is a story for another time and another book. On the second night I fell asleep on a filthy old mattress. I had a vivid dream. In the dream a tall, blue black hooded figure with large shoulder pads and blue glowing eyes under the hood was attacking me. The black clad entity was strangling me with very long, multi jointed fingers. I could not move or scream. I knew I had to force myself to wake up or I would die. I vividly recall forcing myself to wake up. When I finally did the hooded entity was gone.

The fourth time was just as harrowing. I had been on the hunt for a road ghost in Newport Pagnell. I was staying in a hotel that was once a mansion house. I awoke to feel myself unable to move, with the feeling of something vaguely humanoid and rubbery on top of me. Once more there was no visual element but in my mind's eye I saw a white thing, about 5 feet tall. It looked like an outsized, white jelly baby. Its head was featureless and spherical, the stumpy arms and legs lacked hands and feet. It snuggled its round head into my neck. I was finding it hard to breathe and recall trying to turn my head and bite the rubbery white flesh of the thing. I managed feeble cries for help and my friend, whom I was sharing a room with called out to me. This seemed to break the sleep paralysis and I awoke properly. It was a much more vivid and personal experience than before, and I can readily see how some could interpret it as an attack from a ghost, monster or alien.

WAKING DREAMS

Waking dreams or hypnagogic hallucinations, occur as you're falling asleep. They differ from schizophrenic hallucinations in that they are more often visual, involving animals or figures. Those with narcolepsy, a chronic neurological disorder that affects the brain's ability to control sleep-wake cycles, are prone to hypnagogic hallucinations. Other prone groups include with insomnia, excessive daytime sleepiness, and mental health disorders.

The reverse of this is hallucinations when one is waking up. These are called hypnopompic hallucinations. These types of hallucinations are similar to hypnagogic hallucinations, but they may feel more like an extension of a dream. Sleep paralysis is often associated with hypnopompic hallucinations.

Both waking dreams and sleep paralysis have limited use in explaining high strangeness cases, outside of bedroom invaders. They could not be invoked to explain sea monster, hairy giants, poltergeists or UFOs seen when the observer is outside or in daylight.

THE GLOBAL MONSTER TEMPLATE

I can readily recall where I was when the thought first struck me. One could hardly forget, gigantic statues of Hindu gods such as Kali, Hanuman, and Ganesh, swathed in jungle creepers rising out of the thick tropical vegetation of the Indo-Chinese forest. I was in Thailand, an ostensibly Buddhist country, but as with most ancient cultures there was a cross pollination from neighbouring religions and countries, hence the Hindu statues.

I was in Thailand in the year 2000, year of the dragon no less, to investigate reports of the naga, a gigantic crested serpent reported in and around the Mekong River. Indo-Chinese legends also speak of the singa, a huge golden lion and the garuda, a creature half bird and half man. I remember thinking to myself 'I could be in Cornwall'. In Cornwall there are stories of a sea serpent called Morgawr, mysterious big cats and the grotesque, surreal Owlman of Mawnan, a thing seemingly half man half owl.

It was then I began to formulate my Global Monster Template theory. It seems that in every culture you find the same monsters repeated again and again in both folklore and alleged modern sightings. Certain categories or combinations thereof are always reported. My template ran thus...

- Dragons: The most powerful, ancient and widespread of all monsters. Giant reptiles, under many names, they are more associated with water than fire worldwide. Often venerated or feared.

- Hairy Giants: Shambling, ape-like man beasts ranging from dwarfs to giants. Sometimes foul smelling, they have primitive tool use but no fire.

- Little People: Goblins, elves and analogues of such are found worldwide. Modern day reports are surprisingly common. They are totally unlike the 'Disney' idea of a fairy.

- Monster Dogs: From demonic black dogs such as the barguest and the shuck to the shape shifting tanukies and kitsunes of Japan and the infamous werewolves.

- Monster Cats: Phantom panthers, mystery pumas, livestock killers that can't be caught and corpse animating, two tailed Asian horrors, and weretigers in Sumatra.

- Giant birds: Rocs, thunderbirds and outsized avians, as well as bird human hybrids like tengu and garuda as well as modern day horrors.

However, there are also cases of 'high strangeness' associated with the categories too. In these cases it seems that we are not dealing with flesh and blood animals in the accepted sense, but something altogether weirder. The creatures in these reports seem like the inhabitants of a nightmare menagerie. I christened this 'The Global Monster Template'.

The creatures from The Global Monster Template all sounded oddly familiar and on pondering this, another thought hit me. This Global Monster Template may have its genesis back in the mists of time, in our own evolutionary past. Several million years ago, our Australopithecine ancestors on the plains of East Africa had a struggle to survive. Our ancestors were being preyed upon by, and were in competition with, various formidable creatures. Crocodiles and pythons ate them, as did big cats, African hunting dogs, and large birds of prey such as eagles. They competed against other, primates such as giant baboons and other races of hominids, some larger than them and some smaller.

All of these creatures can be slotted nicely into the universal monster template. There seems to be groups of monsters reported all round the world, in every culture. These archetypes include dragons, giants, little people, monster birds, mystery big cats, and monstrous dogs. All of them have a direct link back to our ancestral horrors. Coincidence, I think not. Also, volcanic activity may have caused strange lights in the sky and moving out of the jungle, with its canopy of vegetation, the Australopithecine could have observed things like lightning, comets, shooting starts and other natural phenomena, that would have baffled and frightened them. Maybe at the same time they began to grasp their own mortality and have a concept of death and the fact that they too would one day die, something that could happen at the drop of a hat in such a deadly, predator filled environment.

The concept of 'fossil memories' runs that modern day organisms, including humans, retain behaviour developed for dealing with predators and competitors from our ancient past. These behavioural quirks are passed down in our genes. An example is how, in homes with more than one story, the bedroom is almost always upstairs. This is a throwback to sleeping in trees to avoid predators. As well as behaviour, latent fears are passed down too. We associate black with evil, villains dress in black, this is due to a primordial fear of the dark, when human vision is poor but the vision of primate eating predators like leopards, is acute. Our ancestors were at a disadvantage. The Global Monster Template is formed from the ancient terrors of our early African ancestors, passed down in our genes through millions of years, from generation to generation through species to species. Even after three million years of more of evolution, those fears remain with us. But how does all this link into high strangeness Fortean cases reported today?

THOUGHT FORMS

The Buddhist monks of Tibet, Nepal, and other parts of the Orient have long claimed to be able to create tangible objects with the power of their minds alone. Through deep concentration, and extreme mental discipline, it is said they can create a kind of spirit being - an artificial ghost if you will - that is so convincing that it is often mistaken for a real person or animal. These mind beings are called "tulpas." Westerners have experimented with them to varying degrees of success.

Perhaps the most renowned of these was a remarkable Frenchwoman called Dame Alexandra David-Neel. Born in 1868, she lived during a period where women were considered very much as second class citizens and were expected to live their lives as dutiful obedient wives. This makes her 100 year life as an explorer and mystic, even more incredible. She travelled extensively in the Himalayas and eventually became a Lama, (the highest ranking Tibetan Buddhist – or Lamaist - priest) in Tibet.

She created a thought form or tulpa herself, with remarkable consequences. She relates the happenings in her book *Magic and Mystery in Tibet*.

> "I could hardly deny the possibility of visualising and animating a tulpa. Besides having had a few opportunities of seeing thought forms, my habitual incredulity led me to make experiments for myself, and my efforts were attended by some success. In order to avoid being influenced by the forms of the lamaist deities, which I saw daily around me in paintings and images, I chose for my experiment a most insignificant character: a monk, short and fat, of an innocent and jolly type. I shut myself in doors and proceeded to perform the prescribed concentration of thought and other rites. After a few months the phantom monk was formed. His form grew gradually fixed and lifelike looking. He became a kind of guest, living in my apartment. I then broke my seclusion and started for a tour, with servants and tenants. The monk included himself in the party. Though I lived in the open, riding on horseback for miles each day, the illusion persisted. I saw the far trapa, now and then. It was not necessary for me to think of him to make him appear. The phantom performed various actions of the kind that are natural to travellers and that I had not commanded. For instance, he walked, stopped, looked round him. The illusion was mostly visual, but sometimes I felt as if a robe was lightly rubbing against me, and once a hand seemed to touch my shoulder. The features which I had imagined, when building my phantom, gradually underwent a

change. The fat, chubby-cheeked fellow grew leaner, his face assumed a vaguely mocking, sly, malignant look. He became more troublesome and bold. In brief, he escaped my control. Once, a herdsman who brought me a present of butter saw the tulpa in my tent and took it for a live lama. I ought to have let the phenomenon follow its course, but the presence of that unwanted companion began to prove trying on my nerves; it turned into a "daymare." Moreover I was beginning to plan my journey to Lhasa, and needed a quiet brain devoid of any other preoccupations, so I decided to dissolve the phantom. I succeeded, but only after six months of hard struggle. My mind creature was tenacious of life."

Another remarkable woman who had experience with tulpas, was Violet Mary Firth - better known by her pen name of Dion Fortune. In perhaps her best-known work, *Psychic Self - Defence*, she describes how such creatures can be created inadvertently. After being severely wronged, she lay brooding on her bed one night. In a state between sleep and wakefulness, she began to think of Fenris - the demonic wolf who devours Odin in Ragnarok, the death of the gods in Norse myth. After a strange drawing-out feeling from her solar-plexus, she was horrified to see a huge, snarling, wolf materialize on the bed next to her.

"I knew nothing about the art of making elementals at that time, but had accidentally stumbled upon the right method - the brooding highly charged with emotion, the invocation of the appropriate natural force, the condition between sleeping and waking in which the etheric double readily extrudes.

I was horrified at what I had done, and knew I was in a tight corner and that everything depended on me keeping my head. I had enough experience of practical occultism to know the thing I had called into visible manifestation could be controlled by my will provided I did not panic; but if I lost my nerve and it got the upper hand, I had a Frankenstein monster to cope with."

She got an idea of the phantom's power the next day, when several of her housemates complained of nightmares about wolves. She came to the conclusion that the wolf was really part of herself, extruded and after revenge. She made the decision to forgo her revenge, and attempt to reabsorb the beast. She called it forth into her room once more.

"I obtained an excellent materialization in the half-light, and could have sworn a big Alsatian was standing there looking at me. It was tangible, even to the dog like odour. From it to me stretched a shadowy line of ectoplasm, one end was attached to my solar plexus, and the other disappeared in the shaggy fur of its belly, but I could not see the actual point of attachment. I began by an effort of the will

and imagination to draw the life out of it along this silver cord, as if sucking lemonade up a straw. The wolf form began to fade, the cord thickened and grew more substantial. A violent emotional upheaval started in myself; I felt the most furious impulses to go berserk and rend and tear anything and anybody that came to hand, like the Malay running amok. I conquered this impulse with an effort, and the upheaval subsided. The wolf form had now faded into a shapeless grey mist. This too was absorbed along the silver cord. The tension relaxed and I found myself bathed in perspiration. That, as far as I know, was the end of the incident."

Fate magazine ran a story in 1960, written by Nicholas Mamontoff, the son of one of a group of Russian occultists who studied under a Tibetan guru. The mystic had told the 'Brotherhood of the Rising Sun', that western scientists had never known how powerful the human mind is, or what miracles it could work. In 1912, he led the group in an experiment to create an egrigor - another term for a tulpa. One of The Brotherhood had suggested they create a dragon, but the guru suggested that they create something harmless. They decided on a 'Puss in Boots' character, and concentrated on the image for about half an hour. Gradually, a cloud began to form, that condensed into a red haired cat. Its clothes, however, were ill-formed. The guru suggested that they gave it only boots, eschewing the hat, coat and other items. This improved the creature's clarity.

> *"Within a few moments the features of the cat stabilized and on its hind feet were a pair of Russian boots. The egrigor was motionless and looked like a poorly developed photograph."*

Helen Duncan, the last woman to be prosecuted for witchcraft in 1944, once claimed she had seen a fairy materialize on a medium's had during a séance.

Robin Furman the parapsychologists of' Ghostbusters U.K' – whom we met in chapter two, may once have brought forth a tulpa of his own. He suffered an unpleasant childhood at the hands of his violent father. Once - after being sent to bed with a thrashing - he had a very strange visitor.

> *"I looked up to see an enormous dragon walking straight through my bedroom wall. It was green with huge wings. The thing I remember most about it was its tail that it held up in the air in long coils. I wasn't afraid of it. It just walked through my room and out down the corridor. I ran down stairs shouting for my parents to come and look at the dragon but they could not see it."*

Perhaps Robin's dragon was an 'unconscious tulpa'. Maybe his subconscious had created a 'guardian' for him in the shape of the biggest, fiercest, creature it knew - a dragon.

Another example of involuntary thought-form creation, is mentioned in W. Y. Evens-Wentz in the book *The Tibetan Book of the Great Liberation.*

> *"Mediums in the Occident can, while entranced, automatically and unconsciously create materialisations which are much less palpable than the consciously produced tulpas by exuding "ectoplasm" from their own bodies. Similarly, as is suggested by instances of phantasms of the living reported by psychic research, a thought form may be made to emanate from one human mind and by hallucinatory perceived by another, although possessed of little or no palpableness."*

In 1972, an experiment was carried out in Toronto, Canada by a group called the Toronto Society for Psychical Research. It was led by parapsychologist Dr. A.R. George Owen and psychologist Dr. Joel Whitton, its aim was to create an artificial ghost using the mind power of participants. The team consisted of Dr. Owen's wife, Iris, an industrial designer and his wife, a heating engineer, an accountant, a bookkeeper, and a sociology student. They created a fictional character, Philip Aylesford who was born in 1624 in England. Philip served in the military and was knighted at the young age of 16. He fought on the Royalist side in the English Civil War. He was married to a frigid wife, Dorothea, but fell in love with a gypsy girl called Margo, and had an affair with her. He brought Margo back in secret to live in the gatehouse, near the stables of Diddington Manor - his family home. Later Dorothea discovered her and accused her of using witchcraft in order to steal her husband. She was later burned at the stake. Philip, scared for his reputation, did nothing to stop the events, he was filled with guilt and committed suicide in 1654. Philip never existed and his whole back story was made up by the participants.

The group visualized and concentrated on Philip, willing him into existence. Dr Owen changed the experiment conditions, altering several key environmental variables– dimming the lights, for instance– to more closely resemble a conventional séance. The group recorded rappings, voices, strange breezes and a table that moved about on its own. The table would even slide along and trap members in the corner of the room. Philip seemed to gain a personality of his own.

Similar experiments were conducted, to see if similar results could be replicated, with ghosts named Lilith, a French Canadian spy, Sebastian, a medieval alchemist and Axel, a man from the future.

And all the while participants reported similar supernatural phenomena in the room.

It seems that these artificial ghosts where created by the power of the group's minds working together.

Sometimes it seems that even fictional characters can leave the page and be seen by their creators. John Constantine is a character created by the legendary comic book writer, Alan Moore, and debuting in issue 37 of the DC comic *Swamp Thing* in 1985. Constantine is a working class, British occultist from Liverpool. He bears a strong resemblance to Sting, singer in the band *The Police.* Unshaven, wearing a long raincoat and depicted as constantly having a fag hanging from his lips, Constantine is like an occult cross between Gene Hunt from the tv series *Life on Mars* and Steerpike from Mervyn Peak's gothic fantasy *Titus Groan.* The character proved such a hit, that he was soon staring in his own comic *Hellblazer.*

Several writers working on the comic swear they have seen Constantine in real life, despite the fact that he is a fictional character in a comic. Alan Moore once met him in a cafe.

"One interesting anecdote that I should point out is that one day, I was in Westminster in London -- this was after we had introduced the character -- and I was sitting in a sandwich bar. All of a sudden, up the stairs came John Constantine. He was wearing the trenchcoat, a short cut -- he looked -- no, he didn't even look exactly like Sting. He looked exactly like John Constantine. He looked at me, stared me straight in the eyes, smiled, nodded almost conspiratorially, and then just walked off around the corner to the other part of the snack bar. I sat there and thought, should I go around that corner and see if he is really there, or should I just eat my sandwich and leave? I opted for the latter; I thought it was the safest. I'm not making any claims to anything. I'm just saying that it happened. Strange little story."

Another time he supposedly ran into Alan and whispered in his ear. *"I'll tell you the ultimate secret of magic. Any cunt could do it."*

Jamie Delano who took over writing after Alan Moore had his own encounter.

"When I was writing him I walked past him outside the British Museum in Bloomsbury. I didn't realise I'd walked past him until I'd gone fifty yards down the road, I looked round, and he was just vanishing round the corner. So yeah, it was all very real and immediate, and the stuff I was trying to make stories out of was real and immediate."

Others who worked on *Hellblazer* also caught glimpses. Peter Milligan saw Constantine at a party around 2009 and rushed after him, only to find he'd disappeared. Brian Azzarello saw him at a Chicago bar in the early 2000's but avoided him.

Further clues about tulpas can be gleaned from the case of one Franek Kluski (1873-1943). His real name was Teofil Modrzejewski, and from an early age he knew that he was "different." As a boy, he claimed to be able to see dead relatives and animals. He also had out-of-body experiences. Importantly, other children with him at the time, claimed to also be able to see 'the dead', as if he were passing on his power to those around him.

When Kluski grew up, he worked partly as an engineer, and partly as a professional medium. His speciality was the 'materialization' of spirit-animals, and he seemed to have a whole phantom menagerie at his beck and call. Sitters at his séances saw a big cat like a lioness that would stalk around them, lashing its tail, and leaving behind a strong, acrid smell that lingered for some time. Another beast was christened Pithecanthropus by witnesses after the now defunct name for the primitive human ancestor *Homo erectus*. The creature seemed part-ape part-man. The brute seemed benevolent in nature, but possessed vast strength. One witness, Colonel Norbert Ocholowicz noted:

> *"It could easily move a heavy bookcase filled with books through the room, carry a sofa over the heads of the sitters, or lift the heaviest persons, in their chairs, to the height of a tall person."*

Other sitters felt the 'ape-man' rub its furry hide against their cheeks, and lick their hands, revealing that the creature could be felt as well as seen. It too left behind a foul odour. A massive black dog also appeared during the sessions, much like the ones reported in eyewitness cases for hundreds of years.

Another member of Kluski's 'Zoo' was an owl-like bird, that would apparently materialize in mid-air and fly noisily around the room. On August 30th 1919, at a séance in Warsaw, the bird was photographed perching on Kluski's head. The shot revealed the bird to be remarkably similar to *Caprimulgus europaeus*, the European nightjar. Observant readers will have noticed something. Kluski appeared to have been manifesting a number of the great monster archetypes.

Kluski's seances were not the only ones where an ape-man turned up. Stan Gooch was a senior research fellow at the National Children's Bureau and the author of several works on child psychology, but in 1958 he attended a séance at the home of a friend in Coventry, England. He sat facing the medium, he experienced a sense of light-headedness, then a rushing sensation, as if the room was full of a great wind, and he heard a noise like roaring waters. He felt as if a barrier had collapsed and became unconscious. When he came to, he was told that he had entered a trance state, and that several 'entities' had spoken through his mouth, including a cousin who had died in the war."

Afterwards he began to attend the séances on a regular basis. During one of these a strange creature manifested...

> *"This was a crouching ape-like shape, which became clearer as the*

moments passed. I guess it approximated to most people's idea of what an ancient cave man would look like. Yet one could not make out too much detail - the eyes were hidden, for example. It stood in half shadow, watching us, breathing heavily as if nervous. I must say, though, that I sensed rather than heard the breathing. I could not decide whether our visitor was wearing the skin of some animal, or whether it had a rough coat of hair of its own."

The thing did not respond to attempts to communicate with it.

Maybe Kluski was unconsciously manifesting the fears of the collective mind of the human race. This may sound absurd, but there is evidence that such a thing actually exists.

M-FIELDS
British biologist, Rupert Sheldrake, infuriated adherents to academic dogma in 1981 when he published his revolutionary theories in a book entitled *A New Science of Life*. In this book Sheldrake raised the question of how, if every DNA molecule contained the coded information to make a specific creature, did the body know just what went where. For example, how did it know to grow skin-cells, and not, say, muscle-cells in the right areas. Also, many animals (like some lizards), can regrow lost-limbs, whilst others, such as echinoderms, can be totally destroyed, (for example by putting them in a liquidizer), but each piece will regrow into a fully-formed adult. Sheldrake realised that, contained within the DNA, must be something akin to a 'blueprint' for each species, a life-shaping field unique to each life-form that orders the DNA. He called his hypothetical 'blueprints' morphogenic-fields, or m-fields for short. The m-field theory might also explain how subjective information like emotions and memories are retained. The cells in our bodies are constantly dying and being replaced, and this includes brain-tissue. Yet we retain our memories and personalities, except under conditions of severe or maximum brain-damage, (even minimal to moderate brain-damage is self-repairable) ergo something must be making the new atoms follow the exact patterns of their forbears.

This m-field template may be the key to understanding other biological mysteries such as migration. Darwin believed that this kind of information was passed on in the genetic-characteristics of the parents, but some startling experiments have challenged this view.

In the USA a series of experiments were carried out on rats. The rats had to learn how to escape from a pool of water without following the most logical course, as this had been rigged to give them an electric-shock. The first generation took a number of attempts to learn this. The young of these rats took less time to work out the problem. This seemed to be supporting the Darwinian idea, but identical experiments were being carried out in another country with rats that had no genetic relationship to the ones in America. These rats took even less time to solve the puzzle than the second generation of rats in the American labs.

Sheldrake believed that this was because of a shared m-field. He hypothesised the m-fields of all individuals of a species, were linked to a huge gestalt m-field. He proposed that evolutionary changes, behavioural patterns, and information were shared at a subconscious level between the whole species. When individuals pick up advantageous new behavioural traits, it is incorporated into the gestalt. He believed this was passed on by resonance, rather like the way that the energy wave from a plucked string on an instrument, can resonate onto another string on the same instrument that has not been plucked. This works because part of the unplucked string has the potential to resonate at frequencies in common with the vibrating string, and thus can resonate in harmony. In music this is called harmonic resonance. Sheldrake called his biological analogue, morphic resonance.

Of course the inverse of this also occurred, wherein the individual's behaviour is altered by the m-field of the species. Animals with fewer turnovers of generations - those with longer life spans, would have m-fields that work more slowly. But they work nonetheless. Some Einstein of the sheep-world worked out how to cross cattle-grids in Britain. The sheep curled up in a ball, and rolled across the grid! Initially only a few did this trick, but within weeks sheep all over the world were making P.O.W style escapes from farms.

This would seem to be the ideal way for gigantic "racial thought forms" to occur. Perhaps we should seek the origin of dragons and other monsters, in the jungles of our own minds, and in the fossil memories handed down to us in our genes from our remote ancestors. UFOs and ghosts may stem from the confusing new lights seen in the night sky, by our once jungle dwelling primate ancestors and their own awakenings to the nature of death.

Sheldrake himself seems to support this notion.

> "In the early stages of a form's history, the morphogenetic field will be relatively ill-defined and significantly influenced by individual variants. But as time goes on, the cumulative influence of countless previous systems will confer an ever-increasing stability on the field; the more probable the average type becomes, the more likely that it will be repeated in the future."

Perhaps our fossil memories can be triggered by certain things in our surroundings. Maybe some kinds of electromagnetic-interference coupled with the right person, with the right brain chemistry, in the right place, at the right time, can create a monster. If the brain, an electro-chemical computer, is "shorted" it "re-boots" like an mechanical computer, and for a while switches to its most primitive "operating-system." In this condition, our m-field kicks in, and together with our fossil-memories, creates a defence mechanism, the primal fear, 'flight or fight', taken to its extreme in the creation of something visible and (for a time at least), tangible.

These horrors have been with us since time immemorial and crawl, flap and stalk through the folklore of all cultures. In his book, *Appearances of the Dead: A Cultural History of Ghosts* the historian, R.C Finucane, quotes the Capuchin monk Nole Taillepied's *A Treatise of Ghosts* (1588), to show both that the Devil was commonly believed to disguise himself as a ghost, and that 'evil spirits may appear as a lion, bear, black dog, toad, serpent or cat'. The Global Monster Template strikes again!

A PERSONAL ACCOUNT: THE WEB OF FEAR

The author himself has experimented with tulpa creation. The story of 'The Great Leeds Spider Plague' is blackly comical, and may seem in parts like an irrelevance, or a grand prank in appalling-taste. However, it illustrates how relatively easy it is to create such a being, and how its continued persistence can be ensured.

The story begins in February, 1997, whilst I was at University in Leeds. My student-digs were in a three-story house built in suburbia at the beginning of the 20th Century. It had a large and rambling cellar that was unused due to its dank and musty nature. I was 27 at the time, and was throwing a party to celebrate ten years of being a "Goth." To those humble souls not in the know, Goth is a youth sub-culture. The illegitimate offspring of punk and new-romantic. Its adherents dress like members of the Munsters or the Adams Family, and revel in the dulcet tones of such bands as *Bauhaus, Joy Division, Fad Gadget,* and *The Cure.* An attraction to all things dark, strange, and uncanny is also a bonus. The whole movement began in Leeds in the late 1970s and had its heyday in the early 1980s. I felt that ten years of dressing like an undertaker, and listening to *Siouxsie and the Banshees* needed something special to mark it.

Naturally I wanted to hold this party in the cellar. I decorated the rooms appropriately, and as a centre-piece erected an altar to Atlach-Nacha, a fictional spider-god created by the peerless horror writer, Clark Ashton-Smith, in his deeply strange and disturbing 1934 story *The Seven Geases.* Atlach-Nacha was an appalling, subterranean-deity said to dwell beneath the fabled land of Hyperborea. He was constantly spinning a gargantuan web, and it was said that if this mammoth-task was ever finished, the world would come to an end. Ashton-Smith describes him thus:

> *"A darksome form, big as a crouching man, but with long spider-like members...He saw there was a kind of face on the squat ebony body, low down amidst the several jointed legs. The face peered out with a weird expression of doubt and enquiry and terror crawled through the veins of the bold huntsman as he met small crafty eyes that were circled about with hair."*

The altar consisted of an antique set of drawers, upon which sat a chalice of black ichor (in fact nothing more esoteric than black food-colouring), sat in the centre of a red pentagram. At each point of this five pointed star was a black candle. Below it was attached a script written in Arabic - the translation of which runs:

*"That is not dead which may eternal lie.
Yet in strange aeons even death may die."*

This was written by Ashton-Smith's friend and contemporary author, Howard Phillips Lovecraft, and refers to the eternal "Great Old Ones" - a group of appalling alien gods he and others created, in their bizarre writings in the early part of the 20th century.

Below the altar stood a porcelain nun with an inverted crucifix, and above it hung a large web made of muslin. The god himself, who lurked within the web, was represented by nothing more exotic than a large toy spider, a battery-powered device activated by sound. In response to a sharp or loud sound, it would scurry down a wire with its eight-inch legs flailing in a predatory dance.

The party was a roaring success. We had a fire-eater from Manchester who managed to burn through our washing line, and set our rubbish bags alight (causing much curtain-twitching from the neighbours). I masqueraded, like Erik in The Phantom of the Opera - as Edgar Allen Poe's 'Red Death'. We listened to goth bands and had black punch that I dubbed "Giant Squid Ink", and, being students, all got very drunk.

The bad thing about parties is the realisation, with cold sobriety, that one must clean up. The next morning - after fifteen bin-liners full of assorted rubbish had been collected - I came upon the altar-room. I just didn't have the heart to take down my lovely (well in my gothic eyes), creation. Then the idea struck me to leave the altar in situ, and use it as a focus for a series of thought-form experiments.

Each evening, I would descend to the cellar, with a ritual-sword and dressed in black robes. These accoutrements were not necessary, but helped to focus the mind, like an actor's dress rehearsal. I would open a circle - a practice common in magick, whereby a "magickal space" is opened for the user to work in. Then I would concentrate on visualising the spider. I would empty my mind of everything but my objective (an act that is far more difficult than you would think). I visualized a massive, slate-grey crab-spider with legs three-feet across. It had a bulbous body the size of a human-head and marked with a pattern resembling a grinning skull. Its own head was as large as an orange, and furnished with six green, bioluminescent eyes.

Its one inch fangs were constantly dripping green venom. I imagined it squatted on the altar, its head jerking from side to side as if constantly listening. Sometimes I visualized it scurrying about the room in my peripheral vision, in the odd shadows cast by the candles, accompanied by a sound akin to the rustling of dead leaves. I poured mental-energy into the thing. I also experimented with other forms of energy raising. I chanted the creature's name over and over, until the words ran into one another, and became a mantra, *"Atlach-Nacha, AtlachNacha, AtlachNachaAtlachNacha."* I tried Dervish-whirling, and staring into dancing flames. My fire-eating friend had left behind a strobe-light at the party. I found this most efficacious in the aid of entering altered mental states. After experimenting with the strobe's settings, I found one that was particularly hypnotic.

After a few weeks, a distinct change came over the cellar. It was noticeably colder. When one was in the altar room, a feeling of being watched prevailed, and I would often find the hairs on my neck standing on end upon entering. When the lights were turned off the feeling of not being alone was palpable. To get a second opinion I called a friend of mine; Steve Jones. Steve is a practising witch and psychic, and one of his talents is the apparent power to detect spirits in any given place. He has used this in the investigation of several haunted buildings. On entering the altar-room, Steve declared that he could definitely sense something, but that it seemed unfocused. I tried switching off the lights, and this had a dramatic effect. Steve instantly got the unpleasant feeling that his bare fore-arms were being wrapped in cobwebs. Steve, who is an arachnophobe found this most distressing.

A couple of weeks later, there was an interesting development. I entered the room as normal only to be greeted by an image of a gigantic white spider upon the far wall. The image seemed two-dimensional, flat like a photograph. It was about four feet across, and did not much resemble the spider I had been visualising. I had imagined a crab-spider with angular, spindly-legs, this thing more closely resembled a tarantula with thick legs. As I looked to one side, the vision followed my moving eyes and appeared on the right-hand wall. As I looked to the ceiling, it followed and appeared there. It was as if the spider was attached to my retina. The effect was akin to the one you get by staring for a long time at a white image on a black background then, if you turn away, the image is retained for a few seconds. Whether it had any objective-reality is debatable, but it meant one thing to me, I was succeeding!

A few days later I saw it again! This time it was outside the altar-room, at the bottom of the stairs leading down into the cellar. This meant the tulpa was growing stronger and more independent. However, the summer-holidays loomed, and I had to return to the Midlands. Without me on hand to feed it, I thought that the tulpa would become weakened. I was very wrong.

Upon my return, I went immediately to the cellar to find that the creepy atmosphere remained unchanged. I also found out that my "offspring" had been busy in my absence. Over the summer of 1997, Leeds, Wakefield, and the surrounding areas had suffered a plague of giant spiders. Panicked locals called the police and took captured specimens to wildlife experts, believing them to be dangerous exotics. In fact they turned out to be *Teginaria domestica* - the common house spider, but these specimens had grown to twice their natural size. If this were not enough, a worker at a Morrison's warehouse in Wakefield, was in for a nasty jack-in-the-box style surprise. Michael Haigh was unpacking a consignment of bananas from Brazil, when in his own words:

"All of a sudden I realized that a giant spider was attacking my face. It was

*so huge it blocked my vision totally making everything dark. The spider bit
me on the left cheek. It was terrifying!"*

A Brazilian huntsman spider (*Hetropoda* spp.), with a nine inch leg span, had leapt out
of the bananas and buried its fangs in Mr Haigh's cheek. He was taken to the Tropical
Medicine Department of Saint James' Hospital in Leeds, where he was kept waiting for
eight hours! The doctor's brilliant diagnosis was that if he was still alive the spider
could not have been a very poisonous one! Mr Haigh decided to sue the banana
company for damages, and mental stress. His solicitor said:

> *"Frankly, if the day has come that a man cannot go to his place of
> work without being attacked by giant spiders, then frankly something
> is drastically wrong."*

That must have been one of the strangest lines in legal history, but it didn't impress the
banana company, who insisted it was Mr Haigh's fault for doing such a dangerous job
as unpacking bananas in the first place! In a statement that reminds one of the Monty
Python sketch 'How to defend yourself against assailants armed with pieces of fruit',
they issued this statement: *"He exposed himself to sustaining an accident such as he
alludes befell him."*

It seems to be pushing coincidence too far to think that all of these odd, spider-related
stories, occurred at the same time that I was involved in spider thought-form
experiments. It was as if the experiments had swollen beyond the bounds of the cellar
and were effecting the mindset of the whole population, as well as having physical
effects on the area. The best, however, was still to come. It soon rolled round to the
next year and I threw another party. This time I had the express intent of getting the
whole crowd of my guests involved in a mass visualization of the creature. I soon had
thirty people sitting around the altar. These included several girls who had been passing
by the house. One of the other guests had asked them if they wanted to come and join
the party. I'm sure these poor unwitting girls still tell stories about what happened next.
The whole scene was reminiscent of the classic Dr Who story *Planet of the Spiders*, in
which Buddhist monks summon giant, psychic, interdimensional spiders into an
English country house, via arcane rituals. I began by describing Atlach-Nacha in detail,
and telling the crowd to visualize it on the altar. Then, they began to chant the spider's
mantra over and over. Such a multiple raising of energy would empower a tulpa
greatly.

I realized, that this being my last year at university, I would soon have to move out, and
would no longer be able to continue feeding the tulpa. I needed to find a secondary,
way of keeping belief and energy flowing into my creation. I came up with some novel
"triggers" to keep my child in the pink.

In my bedroom, there was a tiny door in one of the walls. When unscrewed, this led to
a crawl space, some six feet long, by two feet wide, by three feet tall. I knew that it

would be human nature for any inquisitive mind to do what I had done, and look behind the door. I procured a rusty old biscuit-tin into which I placed a number of pseudo-occult items. These included pieces of jewellery, inscribed in Sanskrit and Arabic, a huge dead spider that I covered in black wax, spells written on parchment that had been dyed and singed to give the illusion of age, and an occult diary. The diary was by far the most important of these.

The diary, in a way, was a fictionalised account of my experiments. It followed a young student who moved into the house and began a series of rituals. He finds a tome in an antiquarian bookshop in Hebden Bridge, and uses it to contact a spider-demon. Starting in an understated manner, with only the suggestion of something unnatural afoot, like scratching and scuttling noises, it builds slowly up into an ever-mounting crescendo of horror. The diary charts the student's descent into the black pit of madness. He glimpses vistas of hell described in stomach-churning detail, page after page, and finally kills himself on the remotest wind-blasted Yorkshire moors. But not before the spider god has clawed its way into our world.

I left this 'box of delights' in the crawl-space, and screwed the tiny door back into place. It was my ardent hope that someone of a nervous-disposition would find them. At the same time I was preparing another trigger in the cellar.

I decided to leave the altar-room intact, for my successor, in all its mind-blasting glory. I added to the already unwholesome décor with giant red pentagrams on the walls. I rigged the light-bulb in the room to explode upon being switched on. I also left a convenient torch beside the door. I intended the next tenants to open the door, and fumble for the light switch and in the brief illumination of the explosion to see the Spider-God's altar. Then upon finding the torch they would play the wan, shaky beam over my play-pen of evil.

It worked better than I could have ever hoped for. I revisited Leeds several months after my course was over. A friend of one of my old housemates, who had stayed on in Leeds, related some interesting information. Apparently the student that had been given my room that term was a born-again Christian. He had indeed opened the little doorway to the alcove. He had indeed found the box of esoteric items. And he had indeed read the madman's diary.

This gullible fellow had descended to the cellar and in the exact sequence of events I had planned for, he found the alter of Atlach-Nacha. He utterly freaked out and ended up having two priests in to exorcise the house and the altar was burnt in the back-yard!

This may all sound like a student prank in poor-taste, but it was done for a reason. The belief that this farcical affair would have generated in the minds of the residents would be huge, and tulpas thrive on belief. You are feeding the spider-god now by reading these words. It doesn't matter if you don't believe in it as long as you think about it. As for the exorcism, in many cases this makes matters worse. If the exorcist is not properly trained, and does not possess sufficient-skill, it can stir up a whole psychic hornet's nest. A poor exorcist will also confirm a person's belief that something uncanny is afoot, thus feeding the tulpa.

All of this begs some questions.

Firstly if an untrained and cluttered mind can achieve such results, what could a clear, well-trained, concentrated mind achieve? Secondly, what could minds work in a gestalt, perhaps unknowingly create on a huge scale?

The 'Spider Cult' has become part of Leeds urban legend. It is still discussed today. Not so long ago, on an online forum about ghosts in Leeds, somebody asked if they remembered the story of the 'weird cult that summoned up a giant spider '. Many people answered, including one of the girls from that second party! The phantom giant spider is now part of the zeitgeist of Leeds.

The point of this story is that if my mind, as cluttered as Albert Steptoe's rag and bone yard can call forth a tulpa then what could the collective, unconscious mind of the human race create?

THE NUMINOUS

Numinous is a term coined by German theologian, Rudolph Otto, in his 1917 book *The Idea of the Holy*. Otto defined this as the feeling one gets when encountering what he terms *"the wholly other"* - *"something which has no place in our scheme of reality but belongs to an absolutely different one"*

Otto defined the *mysterium fascinans* and *mysterium tremendum*, the beautiful and frightening aspects of the numinous. He also defined a third state, that of dread - the *mysterium horrendum*, or the "negative numinous," which can be utterly terrifying.

> *"The feeling of it may at times come sweeping like a gentle tide pervading the mind with a tranquil mood of deepest worship. It may pass over into a more set and lasting attitude of the soul, continuing, as it were, thrillingly vibrant and resonant, until at last it dies away and the soul resumes its "profane," non-religious mood of everyday experience. It has its crude, barbaric antecedents and early manifestations, and again it may be developed into something beautiful and pure and glorious. It may become the hushed, trembling, and speechless humility of the creature in the presence of— whom or what? In the presence of that which is a Mystery inexpressible and above all creatures."*

Ottos writings influenced Swiss psychoanalyst, Carl Jung, and fantasy writer C.S Lewis, who summed up the numinous in a most eloquent way in his 1940 book *The Problem of Pain.*

"Suppose you were told there was a tiger in the next room: you would know that you were in danger and would probably feel fear. But if you were told "There is a ghost in the next room," and believed it, you would feel, indeed, what is often called fear, but of a different kind. It would not be based on the knowledge of danger, for no one is primarily afraid of what a ghost may do to him, but of the mere fact that it is a ghost. It is "uncanny" rather than dangerous, and the special kind of fear it excites may be called Dread. With the Uncanny one has reached the fringes of the Numinous. Now suppose that you were told simply "There is a mighty spirit in the room," and believed it. Your feelings would then be even less like the mere fear of danger: but the disturbance would be profound. You would feel wonder and a certain shrinking—a sense of inadequacy to cope with such a visitant and of prostration before it—an emotion which might be expressed in Shakespeare's words "Under it my genius is rebuked." This feeling may be described as awe, and the object which excites it as the Numinous."

It is exactly this feeling that attaches itself to high strangeness cases, and indeed it may be one of the features by which we define them.

THE OCCURENCE OF GLOWING LIGHTS

As we have wandered through this collection of weird encounters the reader will have noticed that strange lights have accompanied many of the phenomena. Bigfoot has been seen holding glowing orbs Another time when witnesses tried to take photos of bigfoots, all they got on their pictures was balls of glowing light. Strange lights have been seen over Loch Ness, haunt of a water dragon. A glowing sphere transformed into a serpent in a stone circle. Many UFOs manifest not as solid craft, but balls of light. A ball of such light burned Joao Prestes Filho to death. Dragons in Asia were said to hold glowing balls or pearls in their claws. The 'wall doughnuts' case began with a single light and ended up as a light show. Green lights manifested in Harley House where the phantom insects were seen. Fairies have been associated with strange lights, such as willow-the -wisp and phantom lanterns.

Paul Devereux, Senior Research Scholar with the International Consciousness Research Laboratories at Princetown University, has researched weird light phenomena for many years. Devereux termed them 'earth-lights' and theorized that stresses and strains in the Earth's crust could produce them. For example, he suggests that the pressure of bodies of water on underlying geology, can provide sufficient tectonic forces, to produce luminosities in the atmosphere. The electrical and magnetic properties of earth lights suggests to Devereux that they are constructed from plasma, a state of matter in which an ionized substance becomes highly electrically conductive. The magnetic fields associated with plasma are known to cause hallucinations. Devereux thinks that the plasma itself has strange properties.

"One thing that has struck me in pouring over witness reports from

different periods of time and parts of the world is the similarity of descriptions stating that earth lights sometimes behave as if they have a rudimentary intelligence, like inquisitive animals. (Intriguingly, it was fairly recently announced that scientists in Romania had created laboratory plasmas that they observed behaved exactly like living cells, and long before them, the late David Bohm, who was recognised as laying the foundations of plasma physics, observed that once electrons were in a plasma they stopped behaving like individuals and started behaving as if they were part of a larger and interconnected whole – he remarked that he frequently had the impression that the sea of electrons was "in some sense alive.

Another type of observation noted in witnesses' reports from all times and places that has impressed me is that the lights sometimes display illogical effects, such as, and particularly, being visible from one side but not the other. This makes me suspect that earth lights may be macro-quantal events – phenomena that should exist only at the sub-atomic quantum level, but have somehow manifested on our larger macro-scale of experience. More modest macro-quantal phenomena have already been produced in the laboratory, and I think earth lights, produced in the greatest laboratory of them all (the one that belongs to Mother Nature) have remarkable lessons to teach us.

There is one further implication. If earth lights actually are geophysical-based manifestations of consciousness, then they represent an older form than biologically-based consciousness. In effect, they are ancestor lights. Perhaps it is time we got to know the ancestors a whole lot better."

He also links earth lights with Fortean phenomena.

"Earth lights are or were known to traditional and ancient peoples: they were fairies to the Irish and other Celtic peoples, though also harbingers of death known as 'corpse candles' to the Welsh; the disembodied heads of women who had died in childbirth to Malaysians, and known as pennangal; the lanterns of the chota admis, the little men, according to the people of the Darjeeling area; manifestations of Bodhisattvas to Chinese and Tibetan Buddhists, who built temples where such sightings occurred, and they were devils to people in western Africa, and so on and so forth."

Could something akin to earth lights be the raw material from which Fortean entities are formed? Maybe some kind of plasma energy is what these entities construct themselves from. Perhaps a tulpa in its rawest form *is* a ball of energy. It constructs itself a more physical form, when the mind of a witness shapes them, or indeed the collective unconscious of the human race moulds the plasma energy into a dragon, a spaceship, a fairy a ghost or a Bigfoot. Maybe this energy is latent in the earth and can be triggered by interaction with the human mind under certain conditions. If David Bohm's idea of electrons in plasma behaving like they were alive, maybe, like cells, it is the power of our fossil memories, that shape the energy into members of the Global Monster Template.

RITUAL MAGICK
It seems that, to a certain extent, their entities can be summoned and controlled by ritual magick.

Doctor John Dee, the court astrologer to Queen Elizabeth I, whom we met in the first chapter, was said to have performed magick, by the control of spirits.

William Dury, said that he had conjured up the tormenting spirits in the case of the Drummer of Tedworth, using spells from a book he got from a wizard.

According to Mike Hallowell's research, a dragon worship cult was carrying out human sacrifices on the north east coast of England, well into the 20th century.

The famous parapsychologist, Guy Lyon Playfair saw evidence of the control of spirits via magick during his time in Brazil, in the late 1960s and early 1970s.

Guy's friend Hernani Guimaraes Antrde, head of the Brazilian Institute for Psycho-biophysical Research, had been investigating a case in Guarulhos where furniture had been mysteriously slashed. The family consisted of Marcos, a builder, his wife Noemia, his parents Pedro and Judite and his children. One day Pedro saw what was causing the slashes. He saw a muscular arm, covered in reddish hair and a hand armed with 6 inch claws that were curved, black and shiny. In the house next door, Noemia was working a sewing machine. Seeing slashes appear on a mattress she looked up and saw an animal like a gorilla. A neighbour who later saw the beast's arm and claw, fainted on the spot.

Money and keys vanished from Marcos' house and stones fell on the roof, in typical poltergeist fashion. All members of the family were scratched by invisible claws and clothes and bedding were slashed to bits.

Marcos got a plumbing job in the town of Taubate. He rented a temporary house there and his family went to stay with relatives. The haunting followed him. His bibles were slashed to bits, money vanished and a fire broke out inside his toolbox.

Marcos built himself a new house and moved his wife and children in. The phantom followed, bringing fires, scratches, vandalism and spiriting money away. Prayer sessions only made matters worse.

One day, when Noemia was alone in the house, two strange women turned up asking if they could use the toilet and have a glass of water. One woman had a plastic bag full of candles and cuttings of rosemary. They said they had been sent to bless the house, but would not tell Noemia who had sent them. They seemed determined to get into the house but Noemia would not let them. Finally they left, but said they would be back.

A couple of weeks later the pair returned. This time they brought cake for the children but Noemia refused to take it. After they left she discovered two knives under her bed placed in the form of a cross. Marcos was convinced that they were under a *trabalho* or hex.

Weirdly Hernani had investigated another case, in the Villa Labanesa district of San Paulo, where children had been scratched by a hairy animal with big claws, and strange fires broke out.

In 1984, over a decade after the events started, Hernani visited Marcos and his family again. Marcos told him that after services by the Assembly of God church, the haunting stopped, but not before Noemia had a vision of a hideously deformed person she thought was Satan. Marcos thought that it was one of his two ex-girlfriends who had cursed him. One of them had a mother who he said practised 'black magick'.

In another case Guy investigated a haunting in Ipiranga, another suburb of Sao Paulo. The family consisted of a mother, her daughter Iracy, one of her sons and his wife Nora. In 1968 they found an offering of bottles, candles and cigars in their garden indicating somebody was casting a *trabalho* upon them.

Nora became violently possessed and tried to throw herself out of a second story window. She was put into a hospital but escaped and came home. She had no memory of the events.

Bangs and crashed were heard at night and strange fires broke out. The family moved twice and the haunting followed. In their third house, wardrobes, mattresses, sheets and Nora's pyjamas all caught fire.

Four passport sized photos of Iracy turned up on the kitchen floor. They showed her at different ages and were all pierced with a needle and thread. Then leather pouches began to turn up all over the place. They contained live cockroaches, disfigured photos of Nora's husband or notes telling him that Nora was seeing other men.

Investigating, Guy and his friend Suzuko Hashizume, both heard violent banging in the house and saw objects thrown from upstairs windows.

Catholic priests brought in to exorcize the house had zero effect. The family finally called in a man from *Candomble,* an African cult brought over by slaves and faithfully preserved in the city of Salvador. Called a 'father-in-sainthood', he brought with him assistants and proceeded to burn huge amounts of incense and exhort the spirit to leave.

After this, nothing more occurred at the house. The father-in-sainthood said that it had been a particularly heavy case of black magick.

Tony Healy and Paul Cropper uncovered a similar case on the other side of the world. They were investigating a poltergeist case in Humpy Doo, Northern Territory in 1998. 90 McMinns Drive was home to Andrew and Kirsty Agius and their 10 month old daughter Jasmine, Dave Clark and his partner Jill Summerville and their friend Doug Murphy. The events started with showers of gravel apparently from the long gravel drive that ran up to their house. The gravel was always dry and hot even if it had been raining outside. Soon other objects like knives, bottles, spanners and batteries began to fly about.

No less than three priests were called in. The clerics had bibles and crosses ripped from their hands and one of them had his arm twisted behind his back by unseen hands.

Tony and Paul saw objects flying around, including a knife that narrowly missed Tony.

On April 21st, Kristy was home alone with Jasmine, when she saw two aboriginal men digging under a mango tree in the garden. When challenged they ran away to a parked car and drove away. They left behind a hole 6 feet long, two feet wide and 8 inches deep.

Andrew told Tony and Paul that he thought the haunting was part of an aboriginal curse. On another occasion, Doug's friend Brett Styles, saw an object flying along above the gravel path. It was a black sphere slightly smaller than a clenched fist. As it flew past him Brett saw that it was trailing a two foot long trail of gravel behind it. The object seemed to be transporting the gravel to the house.

The creation of tulpas is very much ritual magick. The visualization and the feeding of an entity on mental energy, is ritual magick at its purest. I can attest this from my own experience. One wonders how many ghosts and monsters are tulpas that broke free of their creators or were set free on purpose?

The strange story of The Cult of the Moon Beast, does not sound quite so farfetched now. The members of the cult spoke of summoning monsters in the forms of great serpents, wolf-men, ape-like beasts, black dogs, giant birds and panthers, the roll call of the Global Monster Template. His informant, Rob, told Nick Redfern that the monster's true forms were balls of energy. This fits with the earth light theory as the raw material of monster creation.

But dealing with monsters can be a dangerous game. Aleister Crowley spent months near Loch Ness attempting a vast ritual. As we have seen, if the ritual goes wrong, dark powers can be set free. These powers drove some of Crowley's servants mad and the great magician himself had to leave Boleskine House. Doc Shiels says he experienced 'psychic backlash' due to his dealings with lake monsters. Another Doc, Ted 'Doc'

Holliday thought he was under psychic attack from lake dwelling dragons.
We have seen in the last chapter that the tokoloshe is thought to be created by and controlled by witch doctors, but it is not the only African monster said to be used by the sangomas.

A monster nearly identical to the Owlman, is said to haunt the jungles of Senegal in West Africa. The natives know it as Kikyaon and describe it as being half-owl half-man. They believe that it exists both physically and psychically. Witch-doctors use the beast to destroy their enemies. On the physical plane Kikyaon destroys men's bodies with its beak and talons, and on the spiritual plane it eats their souls. Its name means "soul cannibal."

Jon Downes, cryptozoologist and Fortean researcher came up with a novel theory. He had noticed that certain 'window areas', places where Fortean phenomena are concentrated were often once places of religious worship that had fallen out of use. His idea was, that centuries of worship had either created or fed tulpas, once the worship stopped, the source of power stopped, and the tulpas had to feed themselves, which they did by creating fear. In his 1906 book *Puck of Pook's Hill* Rudyard Kipling tells the story of two children who meet with the sprite Puck, who shows them various times in England's history. He also explains about The Old Things, diminished gods who, through lack of worship, have now become ghosts and fairies.

> *'All sorts of sacrifices,' said Puck. 'If it wasn't men, it was horses, or cattle, or pigs, or metheglin—that's a sticky, sweet sort of beer. I never liked it. They were a stiff-necked, extravagant set of idols, the Old Things. But what was the result? Men don't like being sacrificed at the best of times; they don't even like sacrificing their farm-horses. After a while, men simply left the Old Things alone, and the roofs of their temples fell in, and the Old Things had to scuttle out and pick up a living as they could. Some of them took to hanging about trees, and hiding in graves and groaning o' nights. If they groaned loud enough and long enough they might frighten a poor countryman into sacrificing a hen, or leaving a pound of butter for them. I remember one Goddess called Belisama. She became a common wet water-spirit somewhere in Lancashire. And there were hundreds of other friends of mine. First they were Gods. Then they were People of the Hills, and then they flitted to other places because they couldn't get on with the English for one reason or another."*

Examples are Stonehenge and Avebury, ancient stone circles where Fortean phenomena are reported, from weird lights to strange creatures. Mark Shackleman met a wolfman at an Indian burial mound.

On September 30th, 1965, Maureen Ford was driving home along the A85 between Dundee and Perth in Scotland. It was about 11.30 at night. Looking out of the side window she saw a strange object. It was a long grey object. The body had no visible legs, a long neck and a cow-like head. Mrs Ford thought she saw ears on the head. It was lying on the bank of the River Tay estuary.

About an hour later, another car, driven by Robert Swankie, owner of the National Bar in Arbroath. He had been down to London on business. Driving along the same road as Maureen Ford, his headlights illuminated an object in the road.

> *"The head was more than two feet long. The body-which was about 20 feet long- was humped like a giant caterpillar. It was moving very slowly, and made a noise like someone dragging a heavy object through wet grass."*

Overlooking the A85 is Kinnoull Hill. Witches were supposed to perform their rites on the 729ft peak of the hill. The witch cult was savagely suppressed during the reformation in the 16th century.

Henry Adamson (1581–1637), the Scottish historian, writes in *The Muses Threnodie* published a year after his death, of 'dragon's hole' in the side of the hill. Coincidence?

The hill has become a noted area for people throwing themselves to their deaths. In 2015 it became "highlighted as a national area of concern for completed suicides."

Carl Jung, the celebrated Swiss psychiatrist, believed that UFOs were projections of the human mind or unconscious monsters. He felt that they were modern substitutes for religious symbols.

F. W Holliday thought that there 'was a disc culture' that venerated the flying objects. He also felt there was a serpent cult that worshipped dragons. He thought that this cult, which he dubbed the Serpent People, could be traced back to ancient Babylon. Holliday writes in *"The Dragon and The Disc"*, that in Ur, the Sumerian city state, that there was a cult that worshipped a serpent dragon held in a temple. When Cyrus the Great incorporated Ur into the Persian Achaemenid realm in 539BCE, the priests escaped the Persians by fleeing north into the mountains of Asia Minor, taking their god with them. They settled in Pergamon in Lydia, western Turkey.

Romans annexed the area into their empire in 133BCE, after the death of Attalus II who left the kingdom to the Romans in his will. A plague broke out about this time and prayers to the Roman gods failed to halt it. It was decided to appeal to the dragon god at Pergamon. When the plague swiftly stopped it was attributed to the serpent that the priests of Ur still held in their temple. So popular was the god that native snakes began to be kept as pets in people's houses. They were known as *dracunculi* or little dragons.

The dragon worship cult persisted after the conversion of Rome to Christianity. Tertullian the early Christian writer from Carthage wrote...

> *"These heretics magnify the serpent to such a degree as to prefer him even to Christ himself: for he, they say, gave us the first knowledge of good and evil."*

The point is that these things seem to feed off emotion. They fed on worship once, but now it seems they feed on fear. Think about the high levels of fear engendered in the encounters we have looked at. When they are not causing blind terror, they are causing utter confusion and bafflement. They need to provoke extreme reactions in humans and it seems they somehow 'feed' on these emotions.

OTHER DIMENSIONS

There is, however, another possibility about the nature of monsters, ghosts UFOs and other strange things - a disquieting one. Namely that they are truly real, not products of our minds but fully independent and from another dimension.

The idea of other realities that co-exist with ours, is not new. In the dark ages this place was called "fairyland." In the Victorian era it was known as the "astral plane." Modern writers have many names for it, John Keel calls it the "super-spectrum", Jerome Clark the "outer-edge" and F. W. Holliday the "goblin-universe." This elsewhere is the postulated domain of just about every monster, phantom, and weird-entity ever reported.

Irish poet and author, W.B Yeats, studied esoteric and occult subjects and spoke to The Ghost Club several times about his experiences.

> *"Fairyland actually exists as an invisible world within which the visible world is immersed like an island in an unexplored ocean, and is peopled by more species of living beings than this world, because it is incomparably more vast and varied in its possibilities."*

Perhaps this hypothetical dimension is separated from our own world by speed. We know that atoms oscillate at a certain frequency. Is it possible then, that other realities are composed of atoms that oscillate at different rates, either faster or slower than the norm? Such atoms could conceivably co-exist in the same space as the atoms in our dimension. Normally the objects made from these other atoms would be invisible to the naked eye. Perhaps these atoms can occasionally speed up or slow down for a time, and hence the things they compose, become visible to us.

In Islamic lore there is a race of daemons known as djinn. This is where the western concept of the genie is derived from. Djinn were not one type of creature, but came in as many kinds as there are animals in our dimension. The Koran devotes a whole chapter to them. Djinn are made from 'smokeless fire' which could be translated in modern parlance to energy. They are said to inhabit our world, but are usually invisible to us, and we to them. Sometimes the veil slips and the djinn are seen.

A Muslim friend of mine, Mohammed Bhula, related several stories he had heard from djinn in India. One was seen sitting beneath a rose-bush. It grew from a tiny baby to an old man in seconds and then returned to its child form and repeated the process. Another witness reported that he had slipped through to the djinn's reality, and could see hundreds of robed figures apparently attending the funeral of one of their peers. Finally, one man reported encountering a djinn on a remote beach. He saw a trail of footprints apparently stopping with no-one to make them. Looking back he saw a figure walking away in the distance and knew it was a djinn - for their feet are supposed to point backwards, at least in the human looking ones. This last point may seem odd but it re-occurs all over the world in many different traditions.

Djinn come in many forms, other Muslims have told me, quite independently, that the main djinn forms are dragons and huge serpents, huge black dogs with glowing eyes, ape-like beasts, huge birds and massive black cats. Ring any bells? They are right out of the Global Monster Template.

The Hindus call the same entities rakshasas, shape shifting daemons.

Jerome Clark and D. Scott Rogo point out an interesting factor of monster sightings, in their book *Earth's Secret Inhabitants*. They note that most reported monsters resemble real animals, either living or extinct.

> *"Odd caricatures of the types of life-forms that populate the earth...these creatures represent the outcome of some evolutionary process paralleling life on this planet, but not exactly corresponding to it."*

In other words, the monsters come not from some other sphere, but a parallel reality within our own. Central to the idea of magick is the idea of other worlds that co-exist in a subtle way with ours. Magickal theory speaks of various levels of being, that are interconnected. The western mind, with its material obsessions, finds the concept hard to grasp. The physical level is the most dense and as the name suggests, it is the realm of physical matter that we can see and touch. The realm of flesh and blood. The etheric level is the level of life force and linked with breath and the subtle body. The part of the body that lies between the physical body and the spirit. The astral level is the realm of dreams and imagination which are, to magicians, realities in their own right. The mental level is concerned with abstract consciousness, the laws of logic and mathematics. The spiritual level is the one from which all other levels emanate and return. It corresponds with the spiritual core of the self. All patterns of being are said to run down from the spiritual to the other levels.

The etheric level is close to physical matter. It exists in time and space and can effect the physical. It is call *ch'i* in China and *ki* in Japan and those who can channel it can perform amazing feats. The energies of the etheric level can be shaped by imagination and will. These energies can become both tangible and intangible. Etheric energy could be another material from which tulpas are created, along with earth light energies mentioned earlier. Equally, it could be the substance that visitors from another dimension build their bodies, quite independently of us.

So, here is my stab at explaining these cases of high strangeness.

With monsters we have the most complicated case as there are, quite obviously real, flesh and blood mystery animals. But the stranger kinds, the ones that seem to fly in the face of biology and ecology seem to be paranormal in nature.

Ghosts in the main, I feel, have little to do with the spirits of dead human beings. There is a grey area where they seem to cross over with monsters. They fit better into the idea of thought forms or interdimensional interlopers.

Equally, UFO's make zero sense as visitors from space. They do not act in a logical or scientific fashion. They have more in common with fairies. Their surreal antics also have a dream like quality, that may be linked to the subconscious.

The miscellaneous entities are like a mixture of all of the above. They are also so strange they seem like dreams or surreal shows put on simply to baffle the witness.

I think these things need us in some way. As noted above, I think they feed on our worship or fear or confusion and they take this from us these days by manifesting in frightening ways.

These manifestations form themselves from the earth energies that form earth lights, that in themselves act as if alive or from etheric energy. They can be manifested by our collective unconscious or by ritual magic. In the latter case, some of them can be controlled, perhaps in return for the fear food they gather, though raising some of the more powerful forms, like dragons, can be dangerous. Alternatively, they could be completely independent of us and are the inhabitants of another level of reality, such as the etheric realm. They could still be manifested by ritual magick and they may still feed on fear and awe. Such beings are recorded from all cultures since records began and they are still being seen today.

The strange and uncanny intrude upon everyday life much more than most of us would like to admit. I think most people have at least one weird occurrence in their lives but most of them do not report or talk about their 'woo moments'. If you haven't encountered the weird yet, your experience may be just round the next dark bend in the road, the next shadowy woodland grove or even your own bedroom. Its just a matter of time.

Oh, and I never did find out what the deal with 1973 was.

BIBLIOGRAPHY

CHAPTER ONE-MONSTERS

BOOKS
Alexander, Marc, The Devil Hunter, Sphere, 1978
Bord, Janet and Board, Colin, Alien Animals, Panther, 1985
Bord, Janet and Board, Colin, The Bigfoot Casebook, Stackpole Books, 1982
Brooksmith, Peter, Creatures from Elsewhere, Orbis, 1984
Clark, Jerome and Coleman, Loren, Creatures of the Outer Edge, Warner Books, 1978
Coleman, Loren, Mysterious America, Faber & Faber, 1984
Coleman, Loren, Curious Encounters, Faber & Faber, 1989
Crowley, Alistair, The Confessions of Alistair Crowley: Routledge & Kegan Paul. Corrected edition, 1979
Cutchin, Joshua and Renner, Timothy, Where the Footprints End Volume I Folklore
Cutchin, Joshua and Renner, Timothy, Where the Footprints End Volume II Evidence
Dinsdale, Tim, The Leviathans, Futura Publications, 1976.
Downes, Jonathan, The Owlman and Others, CFZ Press, 1988
Downes, Jonathan, Monster Hunter, CFZ Press, 2004
Edgar, Zelia, Just Another Tinfoil Hat Presents, Beyond The Fray, 2022
Freeman, Richard, Dragons More Than a Myth? CFZ Press, 2005
Freeman, Richard, Explore Dragons, Heart of Albion, 2006
Freeman, Richard, CFZ Expedition Report, Gambia, CFZ Press, 2006
Gerhard, Ken, Encounters with Flying Humanoids, Llewellyn Publications, 2013
Godfrey, Linda S, The Beast of Bray Road, Trails Books, 2003
Godfrey, Linda S, Hunting the American Werewolf, Trails Books, 2005
Godfrey, Linda S, Unexplained Research Publishing Company, 2010
Godfrey, Linda S, American Monsters, Tracher, 2014
Godfrey, Linda S, Monsters Among Us, Penguin Publishing Group 2016
Healy, Tony and Paul, Cropper, Out of the Shadows, The Mystery Animals of Australia, Ironbark, 1994
Healy, Tony and Paul, Cropper, The Yowie, Anomalist Books, 2006
Healy, Tony and Paul, Cropper, Australian Poltergeist, Paul Cropper, 2014
Holiday, F W, The Dragon and the Disc, WW Norton & Company, 1973
Holiday, F W and Wilson, Colin, The Goblin Universe, Llewellyn Worldwide Ltd, 1986
Hunt, Gerry, Bizarre America Berkley Books, 1998

Huvelemans, Bernard, On The Track of Unknown Animals, Rupert-Hart-Davis, 1958
Huvelemans, Bernard, In The Wake of the Sea Serpents, Rupert-Hart-Davis, 1968
Keel, John, The Mothman Prophecies, Saturday Review Press, 1975
Keel, John, Strange Creatures from Time and Space, Sphere, 1976
McEwan, J, Mystery Animals of Britain and Ireland, Robert Hale, 1986
Randles, Jenny, Mind Monsters, The Aquarian Press, 1990
Redfern, Nick, Man Monkey: In Search of Britain's Bigfoot, CFZ Press, 2007
Redfern Nick, Wildman, CFZ Press, 2012
Rogo, D Scott and Clark, Jerome, Earth's Secret Inhabitants, Tempo Books, 1979
Sheils Tony, Monstrum,, A Wizard's Tale, Fortean Tomes, 1990
Shuker, Karl, The Unexplained, Carlton Books, 2003
Slate, Alan B and Berry, Alan, Bigfoot, Bantam, 1976
Thomas, Lars, Weird Waters, CFZ Press, 2011
Thomas, Lars Curious Countries, CFZ Press, 2018

ONLINE RESOURCES
Bedtime Stories
https://www.youtube.com/watch?v=vLV-gGn5Y5g
https://www.youtube.com/watch?v=z6kBEGXmU0c

Beyond Creepy
https://www.youtube.com/watch?v=zCMupT70yNk&t=199s

The Cryptonaut Podcast
https://www.youtube.com/watch?v=YfCCXI9k358
https://www.youtube.com/watch?v=wuG5h5kRV-s&t=4s
https://www.youtube.com/watch?v=o7LJkcEc0Wo
https://www.youtube.com/watch?v=3PxAw97p1KE

Just Another Tinfoil Hat
https://www.youtube.com/watch?v=0MqsaPCKumc
https://www.youtube.com/watch?v=bo0dQFMf5Os
https://www.youtube.com/watch?v=mRfe7agoQzw
https://www.youtube.com/watch?v=wTZXvykZHlM

Loch Ness Mystery Blogspot.
https://lochnessmystery.blogspot.com/2014/11/a-story-from-loch-morar.html
https://lochnessmystery.blogspot.com/2013/02/nessie-on-land-fordyce-case_14.html
https://lochnessmystery.blogspot.com/2016/04/nessie-on-land-macgruer-cameron-case.html

Mysterious Universe
https://mysteriousuniverse.org/2021/04/cryptozoology-ufology-ghost-hunting-all-parts-of-one-much-bigger-mystery/
https://mysteriousuniverse.org/2019/02/a-strange-history-of-real-dragons/?fbclid=IwAR1IJPDBY9cVn2zFN-

mf_OwFfvWE9t7XheM1iRxrYMKc1ErOByMrGOm1n_8
https://mysteriousuniverse.org/2021/04/cryptozoology-ufology-ghost-hunting-all-parts-of-one-much-bigger-mystery/

Phantoms and Monsters
https://www.phantomsandmonsters.com/2016/07/dragon-flying-over-south-jersey.html
https://www.phantomsandmonsters.com/2016/07/another-dragon-over-new-jersey.html

Real Monsters
https://www.youtube.com/watch?v=6q9eqLVZDJo&t=832s

Strange Company
https://strangeco.blogspot.com/2018/07/hexhams-hexed-heads.html?m=1&fbclid=IwAR3AflRTbtbYGX8Ez-itsdj43vLnJCeEUETC0qFezgzdVxjabTRVbVtra9s

PERSONAL COMMENTS
Bellemain, Dr Eva
Hallowell, Mike

CHAPTER TWO: GHOSTS

BOOKS
Arnold, Neil, Kent Urban Legends: The Phantom Hitch-Hiker and Other Stories, History Press, 2013.
Bord, Janet and Board, Colin, Alien Animals, Panther, 1985
Chambers, Robert, The Book of Days, Lippincott & Co, 1863
Clark, Jerome and Coleman, Loren, Creatures of the Outer Edge, Warner Books, 1978
Davis, Richard, I've Seen A Ghost, Hutchinson, 1979
Finkle, Irving, The First Ghosts, Hodder & Stoughton , 2021
Freeman, Richard, The Great Yokai Encyclopaedia, CFZ Press, 2010
Furman, Robin and Martingale, Moira, Ghostbusters UK, Hale, 1991
Glanvill, J. Saducismus Triumphatus, Roger Tuckyr, 1681
Healy, Tony and Paul, Cropper, Australian Poltergeist, Paul Cropper, 2014
Hewett, Sarah, Nummits & Crummits, Burleigh, 1900
Hole, Christina, English Folklore, B.T Batsford, 1945
Jones, Rev Edmund, A Relation of the Apparitions and Spirits in the County of Monmouth and the Principality of Wales, E Lewis, 1813.
Lee, F.G, More Glimpses of the World Unseen, Chatto & Windus, 1878
Mac Manus, Dermot, The Middle Kingdom: The Faerie World of Ireland, Max Parish, 1959

McEwan, J, Mystery Animals of Britain and Ireland, Robert Hale, 1986

Munn, Debra, Ghosts on the Range: Eerie True Tales of Wyoming, Purett Publishing, 1991

Norman, Mark, Black Dog Folklore, Troy Books, 2016

O'Donell, Eliot, Some Haunted Houses of England and Wales, Eveleigh Nash, 1908.

O'Donell, Eliot, Trees of Ghostly Dread, Rider & Company, 1959.

Owen, Dr A.R.G and Sim, Victor, Science and the SpookGarrett Publications, 1971.

Passen, Pierre Van, Days of Our Years, 1939, William Heinmann.

Pertwee, Jon, Moon Boots and Dinner Suits, David & Charles, 1985.

Playfair, Guy Lyon, The Flying Cow. Research into the Paranormal Phenomena In the World's Most Psychic Country, Souvenir Press, 1975

Price, Harry, Poltergeist Over England, London Country Life Ltd, 1945

Rosenthal, Eric, They Walk by Night, George Allen & Unwin, 1949.

Ross, Caitrien, Haunted Japan, Tuttle Publishing, 2020

Sieveking, Paul and Ogilvie, Jen, It Happened To Me; Real Life Tales of The Paranormal Volume 1, Dennis Publishing 2008.

Sutton, David, It Happened To Me; Real Life Tales of The Paranormal Volume 6, Dennis Publishing, 2013

Tucker, S.D, Paranormal Merseyside, Amberley, 2013

Williams, M. Supernatural Dartmoor, Bossiney Books, 2003

PERIODICLES
Gandy, Rob, Highway Horrors, Ghost Club Journal Issue Two 2020-2021

Gandy, Rob, The Ruskington Horror, Fortean Times issue 401 and 402

ONLINE RESOURCES
Bedtime Stories
https://www.youtube.com/watch?v=sQDUvHnMEHM&t=255s
https://www.youtube.com/watch?v=an4nQFn1AOo
https://www.youtube.com/watch?v=MkRY9uKbLd4&t=6s
https://www.youtube.com/watch?v=84J7YYbT9vI&t=5s

Cryptopia
https://www.cryptopia.us/site/2015/03/sky-spitter-new-york-usa/
https://www.cryptopia.us/site/2015/02/kinderhook-blob-new-york-usa/
https://www.cryptopia.us/site/2015/02/sugar-bundle-cuba/

Curious Fortean
https://thecuriousforteanweb.wordpress.com/2016/11/15/portishead-poltergiest-diary/comment-page-1/?fbclid=IwAR2pk_oW7AqIuRexZ1xUX_hojK4PY27B-6N4wNsFV89YmnLbHkXdt-ZvjCc#comment-1296

Cutting the Caboose
https://www.youtube.com/watch?v=eKItw94IM5M&t=1178s

Dayne's Discoveries

https://skolmen.wordpress.com/2020/01/16/the-ghost-of-uniondale/

Fortean Entity Catalogue
https://forteanentitycatalogue.tumblr.com/post/173062230285/boneless-encounter-1950s-september-longdendale

Forteana Forums
https://forums.forteana.org/index.php?threads/black-dogs.16163/page-5
https://forums.forteana.org/index.php?threads/the-buzzing-misty-entity-at-helens-bay-beach.68113/

Geocaching
https://www.geocaching.com/geocache/GC68HCE_the-demonic-cat-at-killakee-house?guid=3ca11f2d-b7ad-41ec-afee-f12ee719f345

Ghosts, Ghouls & God
https://ghostsghoulsandgod.co.uk/2020/04/the-epworth-poltergeist-1-the-wesley-home/

Historic Mysteries
https://www.historicmysteries.com/belmez-faces/

James M Deem
https://jamesmdeem.com/stories.ghost.abbeyhouse.html

Legendary Dartmoor
https://www.legendarydartmoor.co.uk/hairy_hands.htm

Mysteries & Monsters
https://mysteriesandmonsters.com/2020/10/19/mysteries-and-monsters-episode-91-british-ghosts-with-ruth-roper-wylde/

Official Shadow People Archive.
https://shadowpeople.org/main/archive.html

The Paranormal Database
https://www.paranormaldatabase.com/reports/roaddata.php?pageNum_paradata=1&totalRows_paradata=1078
https://www.paranormaldatabase.com/reports/roaddata.php?pageNum_paradata=8&totalRows_paradata=1078

Paranormal World Wiki
https://paranormal-world.fandom.com/wiki/In_Cold_Blood:_The_Qiu_Mansion_Haunting
https://paranormal-world.fandom.com/wiki/Infinity_in_Pennsylvania:_A_Serpent_Tale

South Dublin Libraries Local Studies
https://localstudies.wordpress.com/2013/10/23/haunted-happenings-at-killakee/

Uncanny
https://www.bbc.co.uk/programmes/m0010wp9?
fbclid=IwAR2QWf_Y8JcqaDPpAeCWZ9mvEUfzem6GzaD8SVVFqSbppLxGT8Yj
YdAdEAI
https://www.bbc.co.uk/sounds/play/m0011rlm
https://www.bbc.co.uk/sounds/play/m0013zd6

PERSONAL COMMENTS
Furman, Robin
Gedes Ward, Neil
Pertwee, Jon, Kolchak
Wylde, Ruth Roper

UFOS

BOOKS
Davis, Isabel, and Bloacher,Ted, Close Encounter at Kelly and Others of 1955, Center for UFO Studies, 1978.
Druffle, Ann, How to Defend Yourself Against Alien Abduction, Three Rivers Press, 1998
Druffle, Ann. Encounter on Daple Gray Lane, in UFO Abductions edited by D. Scott Rogo, Signet, 1980
Edgar, Zelia, Just Another Tinfoil Hat Presents, Beyond The Fray, 2022
Hickson, Charles and Mendez, Wiliam, UFO Contact at Pensagoula, Wendel C. Stevens Publishing, 1983.
Hopkins, Budd, Missing Time: A Documented Study of UFO Abductions. Richard Marek, 1981
Huyghe, Patrick, The Field Guide to Extra-terrestrials, New English Library, 1996
Lorenzen, Coral E, Flying Saucers: The Startling Evidence of the Invasion from Outer Space, Signet, 1966.
Lorenzen, Coral E, UFO Occupants in the United States in The Humanoids edited by Charles Bowen , Futura, 1974.
Lorenzen, Coral E, UFO Abduction in Brazil, in UFO Abductions edited by Scott D. Rogo, signet, 1980.
Lorenzen, Coral E and Lorenzen, Jim, Flying Saucer Occupants, Signet, 1967.
Randels, Jenny, The Pennine UFO Mystery, Grenada, 1983.
Steiger, Brad, Strangers from The Skies, Award, 1966
Vallee, Jacques, Anatomy of a Phenomenon, Ace Books, 1965.
Vallee, Jacques, UFO Chronicles of the Soviet Union: A Cosmic Samizdat, Bellantine, 1992
Webb, David, 1973, Year of The Humanoids, Richard H. Hall, 1976.

PERIODICLES
Boccone, Luciano, Italian Night-Watchman Kidnapped by UFO in Flying Saucer Review, Volume 26, Issue 1, 1980.
Bowen, Charles, Grand UFO Spectaculars, in The Unexplained, volume 2, 1980.
Chalker, Bill, An Extraordinary Encounter Dandenog Foothills in International UFO Reporter, volume 19 , number 5, 1994
Collins, Andrew, The Aveley Abductions Parts 1 & 2 in Flying Saucer Review volume 23 issue 6, 1978.
Collins, Andrew, The Aveley Abductions Part 3 in Flying Saucer Review volume 23 issue 7, 1978.
Creighton, Gordon, Healing from UFOs, in Flying Saucer Review volume 15 number 5. 1969
Fredrickson, Sven-Olaf, A Humaniod Seen at Imjarvo, in Flying Saucer Review volume 16 number 5, 1970.
Jordan, Peter, UFO Assault in Scotland in Fate, June 1983.
Morris, Eileen, The Winged Beings of Bluestone Walk, in Flying Saucer Review, volume 25 number 6.
Ruekert, Carla L, Kentucky Close Encounter, in Flying Saucer Review, volume 23, issue 3, 1977
Smith, Gary, Unspeakable Secret in The Washington Post, January 3rd, 1988.
Takanashi, Jun-Ichi, Humanoid Encounter on Japanese Mountaintop in MUFON UFO Journal issue 132, 1978.

ONLINE RESOURCES
Bedtime Stories
https://www.youtube.com/watch?v=EZzcBbuML2M

Beyond Creepy
https://www.youtube.com/watch?v=wTwI_Bt_v1M

Cryptid Wiki
https://cryptidz.fandom.com/wiki/Tuscumbia_Space_Penguins

Cryptopia
https://www.cryptopia.us/site/2015/02/alien-octopoids-spain/
https://www.cryptopia.us/site/2016/05/jelly-man-spain/
https://www.cryptopia.us/site/2015/02/garson-invaders-canada/
https://www.cryptopia.us/site/2011/10/cycloptic-aliens-of-harrah-washington-usa/
https://www.cryptopia.us/site/2015/02/gargantuan-gliders-nevada-usa/

Far Out
https://faroutmagazine.co.uk/the-bizarre-true-story-behind-creature-feature-the-blob/

Freak Lore
https://medium.com/freaklore/spanish-farmer-sees-alien-octopoids-while-feeding-his-cattle-874fbcc80d77

Mysterious Universe
https://mysteriousuniverse.org/2012/06/death-from-above-part-one-the-horrible-melting-man/

Obscuran Legends Wiki
https://obscurban-legend.fandom.com/wiki/Pendeli_Egg
https://obscurban-legend.fandom.com/wiki/Old_Saybrook_Blockheads

Portal UFO
https://ufo.com.br/noticias/relembrando-o-caso-joao-prestes-filho-65-anos-depois.html

Scott Net
https://www.sott.net/article/189082-England-Lair-of-the-Beasts-The-Great-Worm-of-Avebury

Shukernature
http://karlshuker.blogspot.com/2012/10/my-top-ten-strangest-aliens-close.html
https://karlshuker.blogspot.com/2011/12/sky-beasts-not-space-craft-unmasking.html

Theresa's Haunted History of the Tri-State
http://theresashauntedhistoryofthetri-state.blogspot.com/2021/01/the-vegetable-man.html

UFO Casebook
https://www.ufocasebook.com/dasilva.html

UFO Insight
https://www.ufoinsight.com/aliens/reptilians/reptilian-abduction-superstition-mountains

PERSONAL COMMENTS
Jones, Steve

CHAPTER FOUR: OTHER ENTITIES

BOOKS
Beardsworth, Timothy, Sense of Presence: Phenomenology of Certain Kinds of Visionary and Ecstatic Experience. Religious Experience Research Centre, 1977.
Bord, Janet, Fairies: Real Encounters with, Little People. Michael O'Mara Books, 1997
Cavendish, Richard, The Powers of Evil, Routledge & Kegan Paul, 1975
Cliffe, Steve, Shadows: A Northern Investigation of the Unknown, Sigma, 1993
Dash, Mike, Spring-heeled Jack: To Victorian Bugaboo from Suburban Ghost, in Fortean Studies, vol. 3, Moore, Steve [ed.], John Brown Publishing, 1996.

Edgar, Zelia, Just Another Tinfoil Hat Presents, Beyond The Fray, 2022

Farson, Daniel, Vampires, Zombies and Monster Men, Jupiter Books, 1976.

Fodor, Nandor, Between Two Worlds, Parker Publishing Company, 1964.

Forman, Joan, The Haunted South, Robert Hale, 1978.

Freeman, Richard, The Great Yokai Encyclopaedia, CFZ Press, 2010.

Gerhard, Ken, Encounters with Flying Humanoids. Llwellyn, 2013

Guirdham, Arthur, Obsession: A Foot in Both Worlds: A Doctor's Autobiography of Psychic Experience. C W Daniel Co Ltd, 1974.

Greer, John Michael, Monsters; An Investigation into Magical Beings, Llewellyn Publications, 2001

Haining, Peter, The Legend and Bizarre Crimes of Spring Heeled Jack, Frederick Muller, 1977.

Hallowell, Michael. J, Invizikids: The Curious Enigma of 'Imaginary' Childhood Friends, Heart of Albion 2007.

Hurwood, Bernard. J, Ghosts, Ghouls & Other Horrors, Scholastic, 1977.

Josiffe, Christopher, Gef! The Strange Tale on an Extra -Special Talking Mongoose, Strange Attractor, 2017.

King, F, Astral Projection, ritual Magic and Alchemy, Spearmam, 1971

Landes, Ruth, The Ojibwa Woman, University of Columbia Press, 1938.

Lewis, Chad and Lee Nelson, Kevin, Wendigo Lore: Monsters, Myths and Madness, On The Road Publications, 2020.

Maple, Eric, The Realm of Ghosts, Robert Hale, 1964.

Martin, Minnie, Basutoland: Its Legends and Customs. Nichols & Company, 1903

Matthews. John. The Mystery of Spring Heeled Jack: From Victorian Legend to Steampunk Hero, Destiny Books, 2016.

MacManus, Dermot, Between Two Worlds: True Ghost Stories of the British Isles, Colin Smythe, 1977.

McNally, Raymond. T A Clutch of Vampires, New York Graphic Society, 1974

Nunnelly, Barton, The Inhumaniods, Triangulum Publishing, 2017.

O'Donell, Eliot, Byways of Ghostland, Rider, 1911.

Randles, Jenny, Supernatural Isle of Man, Robert Hale, 2003.

Redfern, Nick, Man Monkey: In Search of Britain's Bigfoot, CFZ Press, 2007.

Stiger, Brad, Bizarre Cats, Pan, 1993.

Sutton, David, It Happened To Me; Real Life Tales of The Paranormal Volume 3, Dennis Publishing, 2010.

Sutton, David, It Happened To Me; Real Life Tales of The Paranormal Volume 4, Dennis Publishing, 2011.

Tiecher, Morton I, Windigo Psychosis: A Study of the Relationship Between Belief and Behaviour among the Indians of North Est Canada, University of Washington Press, 1961

Tucker, S.D, Paranormal Merseyside, Amberley, 2013

Tucker S,D, Terror of the Tokoloshe, CFZ Press, 2015.

Underwood, Peter, Deeper into the Occult, Harrap, 1975.

Warren, Joshua. P, and Saarkoppel, Andrea, It Was Dark And Creepy Night, The Career Press, 2014

Watson. W.T, Mysteries in the Mist, Beyond the Fray, 2022.

PERIODICALS
Creighton, Gordon, A Weird Case from the Past, in Flying Saucer Review, volume 16 issue 4, 1970
Deegan, Gordon, Fairy Bush survives the motorway planners, in The Irish Times, Sat May 29 1999.
Staff Writer, Prophet Mboro rescues woman from sex-hungry tokoloshes, in The Zimbabwe Mail, 23rd December 2019

ONLINE RESOURCES
Bedtime Stories
https://www.youtube.com/watch?v=6riTvZlneMo&t=929s

Beyond Creepy
https://www.youtube.com/watch?v=csvDD6O_ffI&t=582s

The Cryptonaut Podcast
https://www.youtube.com/watch?v=dWTIzwYEwMo&t=15s

Cryptopia
https://www.cryptopia.us/site/2015/01/octo-squatch-spain/
https://www.cryptopia.us/site/2018/11/sam-the-sandown-ghost-clown-england/

Evil Science and Magic Buddies
https://www.youtube.com/watch?v=iSgCVaJpA_o

The Fairy Faith
https://www.youtube.com/watch?v=tSvqpVH3U-I

Gef:The Eigth Wonder of the World
http://gefmongoose.blogspot.com/p/nandor-fodor-investigates.html

Paranormal World Wiki
https://paranormal-world.fandom.com/wiki/A_Brief_History_of_Killer_Smurfs

Unexplained Mysteries.com
https://www.unexplained-mysteries.com/forum/topic/52847-evil-muppets/

Your Ghost Stories.com
https://www.yourghoststories.com/real-ghost-story.php?story=16991

PERSONAL COMMENTS
Gedes Ward, Neil
Hare, Jon
Laws, Peter

CHAPTER FIVE: WHAT THE HELL IS GOING ON?

BOOKS
Bentley, Gerald Eades and Bentley Jr., G.William Blake. Oxford University Press 1996,
Brown, Nathan Robert. The Complete Idiots Guide to the Paranormal, Penguin, 2010
Dance, S.Peter, Animal Fakes and Frauds, M.E. Sharp, 1976.
David, Neel, Alexandra, Mystiques et Magiciens du Tibet, Édition Plon, 1929.
Finucane, R.C, Appearances of the Dead: A Cultural History of Ghosts, Prometheus Books, 1984.
Fuller, John G, The Day of St. Anthony's Fire, Macmillan, 1968.
Gould, R.T, Oddities; A Book of Unexplained Facts, Philip Allan & Co. 1928.
Guiley, Rosemary Ellen. The Encyclopaedia of Ghosts and Spirits. Facts on File, 1992.
Higgs, John, William Blake Vs The World, Weidenfeld & Nicolson, 2021.
Holiday, F W, The Dragon and the Disc, WW Norton & Company, 1973.
Holiday, F W and Wilson, Colin, The Goblin Universe, Llewellyn Worldwide Ltd, 1986.
Kipling, Rudyard, Puck of Pook's Hill, Macmillan, 1906.
Lewis, C. S, The Problem of Pain, The Centenary Press, 1940
Otto, Rudolph, The Idea of the Holy (English translation), Oxford University Press, 1936
Playfair, Guy Lyon, The Flying Cow. Research into the Paranormal Phenomena In the World's Most Psychic Country, Souvenir Press, 1975
Rogo, D Scott and Clark, Jerome, Earth's Secret Inhabitants, Tempo Books, 1979.
Underwood, Peter, Deeper into the Occult, Harrap, 1975.

PERIODICALS
Kirby, Jan, Clearwater Can Relax, Monster is Unmasked, St Petersburg Times, April 11th, 1988.
Roberts, Andy and Clarke, Dr David, Santilli's Alien Autopsy Film, Fortean Times, May 2006.
Sheldrake, Rupert, A New Science of Life, J.P. Tarcher, 1981.

ONLINE RESOURCES
BBC News
https://www.bbc.co.uk/news/uk-england-41110193
http://news.bbc.co.uk/1/hi/world/south_asia/2205194.stm

Cryptid Archives
https://cryptidarchives.fandom.com/wiki/Frank_Searle

Dread Central
https://www.dreadcentral.com/editorials/358897/the-philip-experiment-we-want-ghosts-to-be-real/

Ghost Village.
http://www.ghostvillage.com/resources/2004/resources_10312004.shtml

The Guardian
https://www.theguardian.com/science/2003/oct/16/science.farout

The Hellblazer Index
http://www.qusoor.com/hellblazer/Sting.htm

How Stuff Works
https://history.howstuffworks.com/historical-events/10-strangest-mass-hysterias.htm

Liveabout
https://www.liveabout.com/how-to-create-a-ghost-2594058

Live Science
https://www.livescience.com/55597-joan-of-arc-voices-epilepsy.html
https://www.livescience.com/12868-top-10-spooky-sleep-disorders.html

Mental Floss
https://www.mentalfloss.com/article/641697/objects-mistaken-ufos

Mysterious Universe
https://mysteriousuniverse.org/2020/07/the-strange-saga-of-a-seance-and-a-classic-neanderthal

National Alliance on Mental Illness
https://www.nami.org/Blogs/NAMI-Blog/May-2020/What-Is-It-Like-to-Hallucinate

National Library of Medicine
https://www.ncbi.nlm.nih.gov/pmc/articles/PMC3105559/

NHS
https://www.nhs.uk/conditions/charles-bonnet-syndrome/

NUVO
https://nuvomagazine.com/culture/canadian-urban-legends-newfoundlanders-can-tell-you-all-about-the-old-hag

Paul Devereux
https://pauldevereux.co.uk/earth-lights.html

Ranker
https://www.ranker.com/list/ergot-medieval-outbreak/genevieve-carlton

Rationalists International (archived)

https://web.archive.org/web/20150924084816/http://www.rationalistinternational.net/archive/en/rationalist_2001/72.htm

Reuters
https://www.reuters.com/article/usa-bigfoot-death-idINDEE87S00720120829

Saltwire
https://www.saltwire.com/nova-scotia/lifestyles/amazing-horror-content-st-johns-filmmakers-inundated-with-old-hag-experiences-306393/

Scientific American
https://www.scientificamerican.com/article/sleep-paralysis-and-the-monsters-inside-your-mind/

Sleep Foundation
https://www.sleepfoundation.org/how-sleep-works/hypnagogic-hallucinations

Tabula Rasa
http://www.tabula-rasa.info/AusComics/Hellblazers.html

Vulture
https://www.vulture.com/2014/10/secret-history-of-john-constantine.html

The Walrus
https://thewalrus.ca/nightmare-on-george-street/

PERSONAL COMMENTS
Bhula,Mohammed
Hallowell, Mike
Tonks, Jackie

STILL ON THE TRACK OF UNKNOWN ANIMALS

T he Centre for Fortean Zoology, or CFZ, is a non profit-making organisation founded in 1992 with the aim of being a clearing house for information, and coordinating research into mystery animals around the world.

We also study out of place animals, rare and aberrant animal behaviour, and Zooform Phenomena; little-understood "things" that appear to be animals, but which are in fact nothing of the sort, and not even alive (at least in the way we understand the term).

Not only are we the biggest organisation of our type in the world, but - or so we like to think - we are the best. We are certainly the only truly global cryptozoological research organisation, and we carry out our investigations using a strictly scientific set of guidelines. We are expanding all the time and looking to recruit new members to help us in our research into mysterious animals and strange creatures across the globe.

Why should you join us? Because, if you are genuinely interested in trying to solve the last great mysteries of Mother Nature, there is nobody better than us with whom to do it.

We publish a journal *Animals & Men*. Each issue contains nearly 100 pages packed with news, articles, letters, research papers, field reports, and even a gossip column! The magazine is Royal Octavo in format with a full colour cover. You also have access to one of the world's largest collections of resource material dealing with cryptozoology and allied disciplines, and people from the CFZ membership regularly take part in fieldwork and expeditions around the world.

The CFZ is managed by a board of trustees, with a non-profit making trust registered with HM Government Stamp Office. The board of trustees is supported by a Permanent Directorate of full and part-time staff, and advised by a Consultancy Board of specialists - many of whom are world-renowned experts in their particular field.   We have regional representatives across the UK, the USA, and many other parts of the world, and are affiliated with other organisations whose aims and protocols mirror our own.

You'll find that the people at the CFZ are friendly and approachable. We have a thriving forum on the website which is the hub of an ever-growing electronic community. You will soon find your feet. Many members of the CFZ Permanent Directorate started off as ordinary members, and now work full-time chasing monsters around the world.

Write to us, e-mail us, or telephone us. The list of future projects on the website is not exhaustive. If you have a good idea for an investigation, please tell us. We may well be able to help.

We are always looking for volunteers to join us. If you see a project that interests you, do not hesitate to get in touch with us. Under certain circumstances we can help provide funding for your trip. If you look on the future projects section of the website, you can see some of the projects that we have pencilled in for the next few years.

In 2003 and 2004 we sent three-man expeditions to Sumatra looking for Orang-Pendek - a semi-legendary bipedal ape. The same three went to Mongolia in 2005. All three members started off merely subscribers to the CFZ magazine. Next time it could be you!

We have no magic sources of income. All our funds come from donations, membership fees, and sales of our publications and merchandise. We are always looking for corporate sponsorship, and other sources of revenue. If you have any ideas for fund-raising please let us know. However, unlike other cryptozoological organisations in the past, we do not live in an

intellectual ivory tower. We are not afraid to get our hands dirty, and furthermore we are not one of those organisations where the membership have to raise money so that a privileged few can go on expensive foreign trips. Our research teams, both in the UK and abroad, consist of a mixture of experienced and inexperienced personnel. We are truly a community, and work on the premise that the benefits of CFZ membership are open to all.

Reports of our investigations are published on our website as soon as they are available. Preliminary reports are posted within days of the project finishing.

Each year we publish a 200 page yearbook containing research papers and expedition reports too long to be printed in the journal. We freely circulate our information to anybody who asks for it.

We have a thriving YouTube channel, CFZtv, which has well over two hundred self-made documentaries, lecture appearances, and episodes of our monthly webTV show. We have a daily online magazine, which has over a million hits each year.

From 2000—2016 we held our annual convention - the Weird Weekend. It went on hiatus because of the illness of several of the major personnel and the eventual death of one of them. But we plan to bring it back soon. It is three days of lectures, workshops, and excursions. But most importantly it is a chance for members of the CFZ to meet each other, and to talk with the members of the permanent directorate in a relaxed and informal setting and preferably with a pint of beer in one hand. Since 2006 - the Weird Weekend has been bigger and better and held in the idyllic rural location of Woolsery in North Devon.

Since relocating to North Devon in 2005 we have become ever more closely involved with other community organisations, and we hope that this trend will continue. We have also worked closely with Police Forces across the UK as consultants for animal mutilation cases, and we intend to forge closer links with the coastguard and other community services. We want to work closely with those who regularly travel into the Bristol Channel, so that if the recent trend of exotic animal visitors to our coastal waters continues, we can be out there as soon as possible.

Apart from having been the only Fortean Zoological organisation in the world to have consistently published material on all aspects of the subject for over a decade, we have achieved the following concrete results:

- Disproved the myth relating to the headless so-called sea-serpent carcass of Durgan beach in Cornwall 1975
- Disproved the story of the 1988 puma skull of Lustleigh Cleave
- Carried out the only in-depth research ever into the mythos of the Cornish Owlman.
- Made the first records of a tropical species of lamprey
- Made the first records of a luminous cave gnat larva in Thailand

- Discovered a possible new species of British mammal - the beech marten
- In 1994-6 carried out the first archival Fortean zoological survey of Hong Kong
- In the year 2000, CFZ theories were confirmed when a new species of lizard was added to the British List
- Identified the monster of Martin Mere in Lancashire as a giant wels catfish
- Expanded the known range of Armitage's skink in the Gambia by 80%
- Obtained photographic evidence of the remains of Europe's largest known pike
- Carried out the first ever in-depth study of the ninki-nanka
- Carried out the first attempt to breed Puerto Rican cave snails in captivity
- Were the first European explorers to visit the `lost valley` in Sumatra
- Published the first ever evidence for a new tribe of pygmies in Guyana
- Published the first evidence for a new species of caiman in Guyana
- Filmed unknown creatures on a monster-haunted lake in Ireland for the first time

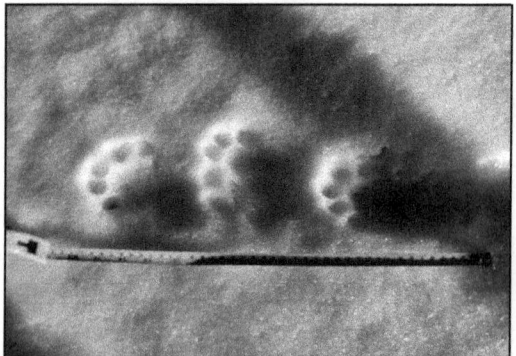

- Had a sighting of orang pendek in Sumatra in 2009
- Found leopard hair, subsequently identified by DNA analysis, from rural North Devon in 2010
- Brought back hairs which appear to be from an unknown primate in Sumatra
- Published some of the best evidence ever for the almasty in southern Russia

CFZ Expeditions and Investigations include:
- 1998 Puerto Rico, Florida, Mexico (Chupacabras)
- 1999 Nevada (Bigfoot)
- 2000 Thailand (Naga)
- 2002 Martin Mere (Giant catfish)
- 2002 Cleveland (Wallaby mutilation)
- 2003 Bolam Lake (BHM Reports)
- 2003 Sumatra (Orang Pendek)

- 2003 Texas (Bigfoot; giant snapping turtles)
- 2004 Sumatra (Orang Pendek; cigau, a sabre-toothed cat)
- 2004 Illinois (Black panthers; cicada swarm)
- 2004 Texas (Mystery blue dog)
- Loch Morar (Monster)
- 2004 Puerto Rico (Chupacabras; carnivorous cave snails)
- 2005 Belize (Affiliate expedition for hairy dwarfs)
- 2005 Loch Ness (Monster)

- 2006 Gambia (Gambo - Gambian sea monster , Ninki Nanka and Armitage's skink
- 2006 Llangorse Lake (Giant pike, giant eels)
- 2006 Windermere (Giant eels)
- 2007 Coniston Water (Giant eels)
- 2007 Guyana (Giant anaconda, didi, water tiger)
- 2008 Russia (Almasty)
- 2009 Sumatra (Orang pendek)
- 2009 Republic of Ireland (Lake Monster)
- 2010 Texas (Blue Dogs)
- 2010 India (Mande Burung)
- 2011 Sumatra (Orang-pendek)
- 2012 Sumatra (Orang Pendek)
- 2014 Tasmania (Thylacine)
- 2015 Tasmania (Thylacine)
- 2016 Tasmania (Thylacine)
- 2017 Tasmania (Thylacine)
- 2018 Tajikistan (Gul)
- 2020 Forest of Dean (Lynx)

For details of current membership fees, current expeditions and investigations, and voluntary posts within the CFZ that need your help, please do not hesitate to contact us.

The Centre for Fortean Zoology,
Myrtle Cottage,
Woolfardisworthy,
Bideford, North Devon
EX39 5QR

Telephone: 01237 431413
Fax: +44 (0)7006-074-925
email: cfzjon@gmail.com

Websites:

www.cfz.org.uk
www.weirdweekend.org

THE WORLD'S WEIRDEST PUBLISHING COMPANY

ANIMALS & MEN
ISSUES 16-20
THE JOURNAL OF THE CENTRE FOR FORTEAN ZOOLOGY

NEW HORIZONS

Edited by Jon Downes

BIG CATS
LOOSE IN BRITAIN

PREDATOR DEATHMATCH

NICK MOLLOY
WITH ILLUSTRATIONS BY ANTHONY WALLIS

TER!
THE ... ZOO... M PHENOMENA

Edited by
Jonathan Downes and Richard Freeman

FOREWORD BY Dr. KARL SHUKER

A DAINTREE DIARY
Tales from Travels ... Daintree
tropical North ...

CARL PORTMAN

THE COLLECTED POEMS
Dr Karl P. N. Shuker

STRANGELYSTRANGE
...ly normal

an anthology of writings by
ANDY ROBERTS

HOW TO START A PUBLISHING EMPIRE

Unlike most mainstream publishers, we have a non-commercial remit, and our mission statement claims that "we publish books because they deserve to be published, not because we think that we can make money out of them". Our motto is the Latin Tag *Pro bona causa facimus* (we do it for good reason), a slogan taken from a children's book *The Case of the Silver Egg* by the late Desmond Skirrow.

WIKIPEDIA: "The first book published was in 1988. *Take this Brother may it Serve you Well* was a guide to Beatles bootlegs by Jonathan Downes. It sold quite well, but was hampered by very poor production values, being photocopied, and held together by a plastic clip binder.

In 1988 A5 clip binders were hard to get hold of, so the publishers took A4 binders and cut them in half with a hacksaw. It now reaches surprisingly high prices second hand.

The production quality improved slightly over the years, and after 1999 all the books produced were ringbound with laminated colour covers. In 2004, however, they signed an agreement with Lightning Source, and all books are now produced perfect bound, with full colour covers."

Until 2010 all our books, the majority of which are/were on the subject of mystery animals and allied disciplines, were published by `CFZ Press`, the publishing arm of the Centre for Fortean Zoology (CFZ), and we urged our readers and followers to draw a discreet veil over the books that we published that were completely off topic to the CFZ.

However, in 2010 we decided that enough was enough and launched a second imprint, `Fortean Words` which aims to cover a wide range of non animal-related esoteric subjects. Other imprints will be launched as and when we feel like it, however the basic ethos of the company remains the same: Our job is to publish books and magazines that we feel are worth publishing, whether or not they are going to sell. Money is, after all - as my dear old Mama once told me - a rather vulgar subject, and she would be rolling in her grave if she thought that her eldest son was somehow in `trade`.

Luckily, so far our tastes have turned out not to be that rarified after all, and we have sold far more books than anyone ever thought that we would, so there is a moral in there somewhere...

Jon Downes,
Woolsery, North Devon
July 2010

CFZ PRESS

CFZ Press is our flagship imprint, featuring a wide range of intelligently written and lavishly illustrated books on cryptozoology and the quirkier aspects of Natural History.

CFZ Classics is a new venture for us. There are many seminal works that are either unavailable today, or not available with the production values which we would like to see. So, following the old adage that if you want to get something done do it yourself, this is exactly what we have done.

Desiderius Erasmus Roterodamus (b. October 18th 1466, d. July 2nd 1536) said: "When I have a little money, I buy books; and if I have any left, I buy food and clothes," and we are much the same. Only, we are in the lucky position of being able to share our books with the wider world. CFZ Classics is a conduit through which we cannot just re-issue titles which we feel still have much to offer the cryptozoological and Fortean research communities of the 21st Century, but we are adding footnotes, supplementary essays, and other material where we deem it appropriate.

http://www.cfzpublishing.co.uk/

Fortean Words is a new venture for us. The F in CFZ stands for "Fortean", after the pioneering researcher into anomalous phenomena, Charles Fort. Our Fortean Words imprint covers a whole spectrum of arcane subjects from UFOs and the paranormal to folklore and urban legends. Our authors include such Fortean luminaries as Nick Redfern, Andy Roberts, and Paul Screeton. . New authors tackling new subjects will always be encouraged, and we hope that our books will continue to be as ground-breaking and popular as ever.

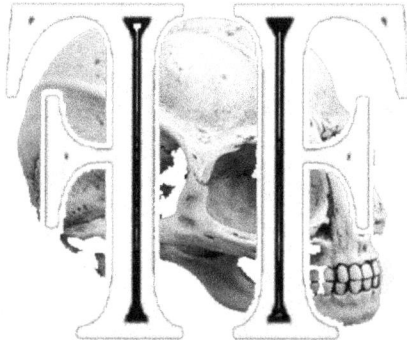

Just before Christmas 2011, we launched our third imprint, this time dedicated to - let's see if you guessed it from the title - fictional books with a Fortean or cryptozoological theme. We have published a few fictional books in the past, but now think that because of our rising reputation as publishers of quality Forteana, that a dedicated fiction imprint was the order of the day.

http://www.cfzpublishing.co.uk/

www.ingramcontent.com/pod-product-compliance
Lightning Source LLC
Chambersburg PA
CBHW060958280326
41935CB00009B/752